Supercritical Fluid Technology
in Oil and Lipid Chemistry

Supercritical Fluid Technology in Oil and Lipid Chemistry

Editors

Jerry W. King
Gary R. List
National Center for
Agricultural Utilization Research
Peoria, Illinois

AOCS
PRESS

Champaign, Illinois

AOCS Mission Statement

To be a forum for the exchange of ideas, information, and experience among those with a professional interest in the science and technology of fats, oils, and related substances in ways that promote personal excellence and provide high standards of quality.

Library of Congress Cataloging-in-Publication Data

Supercritical fluid technology in oil and lipid chemistry/editors,
   Jerry W. King, Gary List.
      p. cm.
   Includes bibliographical references and index.
   ISBN 0-935315-71-3 (alk. paper)
    1. Oils and fats. 2. Lipids. 3. Supercritical fluid extraction.
  I. King, Jerry W. II. List, Gary R.
TP670.S83 1996
664′.3—dc20   96-15367
                                                CIP

Printed in the United States of America with vegetable oil-based inks.

00 99 98 97 96 5 4 3 2 1

Preface

Supercritical fluid extraction (SFE) represents a rapidly growing and useful technology to analytical chemists, food processors, and chemical engineers. Since the early 1980s the use of SFE has been extensively applied to oilseeds and lipids. This volume covers all aspects of SFE of lipids. The book begins with the thermodynamics of supercritical fluids followed by a discussion of the solubility of lipids in supercritical fluids. Other chapters discuss the SFE of oilseeds and lipids from natural products. The supercritical fractionation of lipids is discussed as well as oilseed solubility and extraction modeling. The design and economics of oilseed SFE extraction processes are discussed. Other chapters deal with SFE and the extraction of fish oil, cocoa and cocoa products, eggs, meat, animal fats, and the extraction of algae. Supercritical fluid chromatography (SFC) represents an allied technique for the lipid chemist. Several chapters on determination of trace components in fats and oils and the use of SFE for oleochemicals are included as well as an overview of the applications of SFC to lipid analysis. A chapter on the use of supercritical fluids for enzymatic synthesis is included and the effects of SFC on enzymes, proteins, and amino acids are discussed.

J.W. King
G.R. List

Contents

Chapter 1

Thermodynamics of Supercritical Fluids with Respect to Lipid-Containing Systems

Carl T. Lira

Department of Chemical Engineering, Michigan State University, East Lansing, MI 48824-1226

Supercritical fluid extraction relies on control of solubility via manipulation of temperature and pressure. Solubility behavior in supercritical fluids follows patterns that depend on similarities and differences in the thermodynamic and structural properties of the solute and the solvent.

Mixtures processed from natural materials contain multiple components and frequently are not well identified or characterized. Solubility and extractability of these mixtures are currently difficult to predict quantitatively; however, significant knowledge regarding solubility trends can be obtained by studying simpler binary and ternary systems. Solubility represents a saturation condition; therefore, solubility is represented as a boundary on a phase diagram. Systematic study of binary and ternary systems shows that the phase boundaries of ternary systems are intermediate to the constituent binary systems, and many of the same trends continue in multicomponent systems, although fundamental study of this phenomenon is the topic of ongoing research (1).

Throughout this chapter the term *supercritical* will be used to refer to extractions above the critical point of a solvent, such as CO_2. In the strictest sense, when an extraction from oilseeds occurs above the critical point of the solvent, a mixture of gas and oil usually exists as two distinct, partially miscible phases below the critical point *of the mixture*. Stahl et al. (2) have suggested the use of the term *dense gas* as more acceptable; however, use of the term *supercritical* has predominated in the literature and causes little confusion if the temperature and pressure are explicitly stated. Below the critical point of the solvent, extractions are termed *near-critical* or *liquid* extractions, based again on the properties of the pure solvent rather than the properties of the mixture.

In the early part of the 20th century, experiments in phase behavior were primarily driven by a need to develop applications in distillation and crystallization. Therefore, most publications addressed only vapor-liquid or solid-liquid equilibria. The one exception occurred in the rather specialized characterization of cryogenic systems, which were studied to learn how to avoid solid formation in fluid-handling equipment. The majority of vapor-liquid studies involved components of interest to the developing petroleum industry. However, in the course of these experiments, the extraction of fatty oils with CO_2 was studied first. In 1931 Auerbach patented a process for fractionation of fatty oils, terpenes, tar oil, resin oils, ester oils, and the like, with CO_2 (3). Shortly thereafter, Wilson et al. (4) developed a process for near-critical propane deasphalting of petroleum, which led to the techniques used today. These two processes differ significantly in the manner in which the separation is performed due to differences in the phase behavior. The Solexol process for propane fractionation of fatty oils (5–7) was developed from earlier propane deasphalting studies. These processes may be best understood by studying fundamental binary phase behavior.

Classes of Binary Phase Behavior

Several types of phase behavior may occur in binary systems. The types are usually summarized by the projection of their phase boundaries onto two-dimensional pressure-temperature diagrams. Figure 1.1 summarizes the major types of phase behavior. Subtle variations of these major types arise from a shifting of the placement, curvature, or both, of the phase boundary lines (8). Type VI phase behavior, which is not shown in Figure 1.1, is restricted to aqueous systems (9). Figure 1.1 illustrates only fluid phase behavior. The superposition of solidus lines on these diagrams will be discussed later.

Since 1970, there have been several reviews and classifications of high-pressure phase behavior (1, 8, 9, 10–18). Classifications are based on experimental studies and may require revision if additional phase transitions are found to occur (16,19). Unfortunately, a universal labeling of the phase behavior types does not exist, and the references must be consulted carefully when comparing classifications. For this discussion, the convention of van Konynenberg and Scott (8) will be followed for the phase behavior of Types I through V.

The Gibbs phase rule is helpful in reading the P-T projections. The Gibbs phase rule for a nonreactive system is written

$$F = C - P + 2 \qquad [1.1]$$

where F is the number of intensive degrees of freedom, C is the number of components, and P is the number of coexisting phases. In a C-component system, if P phases exist, then F is the number of intensive properties that may be varied (over a finite range) with assurance that P phases will continue to exist. *Intensive* properties—including pressure, temperature, molar volume, and molar composition—characterize a specific phase and are *independent of mass*. Overall system volume or overall mole fraction are *extensive* variables, which do not contain enough information to specify the state of the mixture. In a pure system, if two phases coexist, then one degree of freedom is available; therefore a two-phase coexistence appears as a line on a P-T projection. At the triple point, solid, liquid, and gas coexist, and no degrees of freedom are available; therefore the condition is a point on the P-T projection. At critical points all intensive properties of two phases become identical, so the number of degrees of freedom is reduced. This also appears as a point for pure systems. To avoid misuse of the phase rule, only intensive variables are used for the degrees of freedom. Also, the intensive variables must be varied over a finite range; for example, two fluid phases are impossible above the critical temperature of a pure substance. More detailed discussions of the correct use of the degrees of freedom are available (20,21).

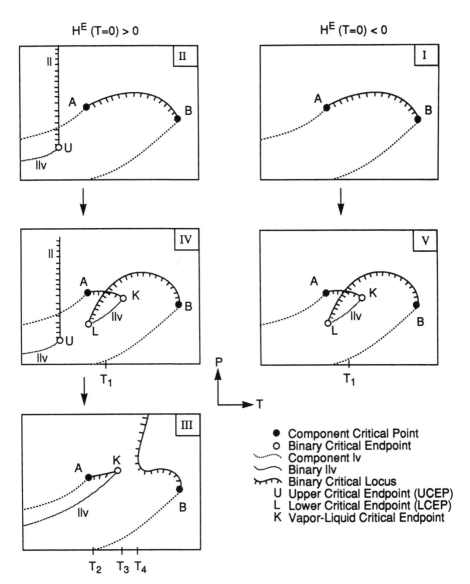

Fig. 1.1. Theoretical progressions of binary phase behavior with increasing molecular asymmetry according to van Konynenberg and Scott (8).

In a binary system, liquid-vapor phase behavior may occur with two degrees of freedom. Liquid-vapor (*lv*) critical behavior occurs with one degree of freedom. Therefore, liquid-vapor behavior appears within a region on the *P-T* projection, and critical behavior occurs along a line known as the *critical locus*. In Fig. 1.1, pure-component *lv* lines are indicated by the dotted lines. Pure-component critical points are indicated by solid circles. Invariant critical points in the binary are indicated by open circles, and critical lines are indicated by solid lines. Critical lines are hashed to indicate the side on which two phases coexist. Since the critical locus is a projection on the *P-T* diagram, the diagram and phase rule indicate that two phases coexist on one side of the curve *over a finite composition range.* The diagram does not imply that two phases will coexist at *all* compositions below the critical locus. Also, in some areas, two critical curves are superimposed over the same temperature-pressure ranges, as in Types IV and V; however, the overlaps are due to projection onto the two-dimensional diagram—the two overlapping regions of critical behavior occur at different composition ranges, as shown in the three-dimensional plots of Fig. 1.2, where *x* is the concentration of component B. The critical locus represents conditions where two phases become identical, but this does not necessarily imply that only one phase exists above the critical locus, because (1) the two phases that become identical at the critical locus may coexist at temperatures or pressures slightly above the critical locus as nonidentical phases (22), and (2) if three phases exist below the critical line, two phases will coexist above the critical line. Also, the absence of a critical line on a region of a *P-T* diagram does not imply that only one phase is present; it simply means that no two phases become identical within the range of the diagram.

Figure 1.1 summarizes theoretical progressions of *P-T* projections of phase behaviors that occur as the molecular asymmetry of the mixtures increases. Figure 1.2 shows additional detail for some of the types, and Fig. 1.3 shows some isothermal sections. Molecular asymmetry may arise from size differences (molecular weight) for functionally similar molecules or from polarity or functional differences for molecules of similar molecular weight. As the molecular asymmetry of the system increases, the critical points of the species generally move farther from each other on the *P-T* traces. With increasing disparity of critical points, all phase behavior begins to occupy increasingly larger areas of *P-T* space. For all the phase diagrams shown here the lighter component, A, the *supercritical solvent,* which appears in the back of Fig. 1.2, has the lower critical temperature and appears at the left in Fig. 1.1.

The theoretical progressions of phase behavior shown in Fig. 1.1 depend on whether the lowest-temperature heat of mixing is endothermic or exothermic (8). The lowest-temperature heat of mixing is calculated as

$$\lim_{T \to 0} H^E = \left(\frac{a_1}{b_1^2} - \frac{2a_{12}}{b_1 b_2} + \frac{a_2}{b_2^2} \right) \frac{x_1 x_2 b_1 b_2}{x_1 b_1 + x_2 b_2} \qquad [1.2]$$

where the *a*s and *b*s are the van der Waals parameters. Endothermic systems may be expected to theoretically follow the progression on the left and exothermic systems may be expected to theoretically follow the progression on the right. Although the five *P-T* projections in Fig. 1.1 appear quite different, they have many similarities in behavior. Note that the isothermal sections near the critical point of B are similar for all systems in Fig. 1.2. Also note that isothermal sections in Fig. 1.2 for Types II and

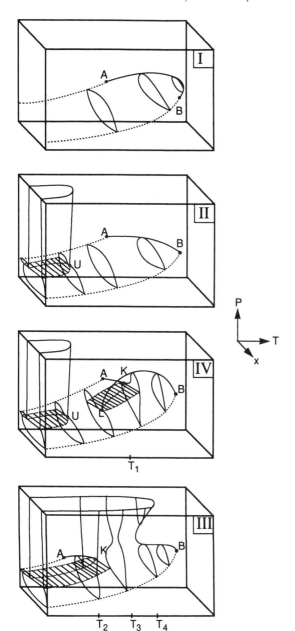

Fig. 1.2. Progression of phase behavior illustrated on three-dimensional diagrams. Symbols are the same as in Fig. 1.1.

IV at temperatures below the upper critical endpoint (UCEP) temperature may look similar to isothermal sections for Type III below the critical temperature of A. Also note the similarities shown in Fig. 1.3 for Type IV at T_1 and Type III at T_2. The only qualitative difference is that in Type III the liquid-liquid region extends to extremely

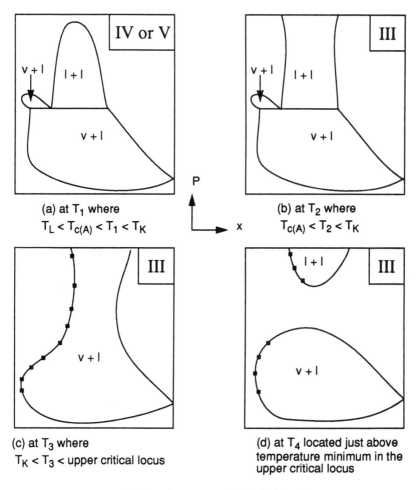

Fig. 1.3. Illustration of selected isothermal *P-x* sections for Types IV, V, and III. Squares are representative of the solubility of component B in the A-rich phase. Other symbols are the same as in Fig. 1.1.

high pressures. Figure 1.4 illustrates the progression of critical endpoints with increasing molecular asymmetry for low-temperature endothermic systems.

When a gas and a liquid phase coexist near atmospheric pressure, the less dense phase is naturally the gas phase. For Type I this nomenclature is adequate, but for Types II through V additional nomenclature is needed because an additional fluid phase forms along the three-phase line. By convention, the third fluid phase of intermediate concentration in A is called a *liquid* phase, and the three-phase line is the *llv* line. In Types III, IV, and V this additional *liquid* phase may exist at temperatures

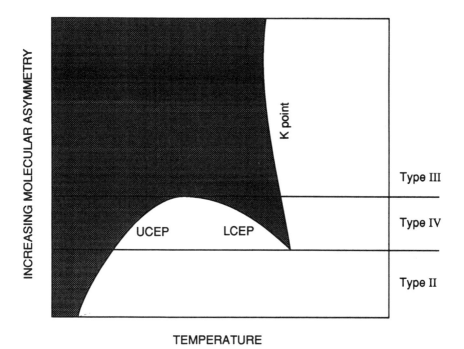

Fig. 1.4. Illustration of trends in critical points as the molecular asymmetry of systems increases for systems that are endothermic as defined by Eq. [1.2].

above the critical temperature of A and becomes identical to the *vapor* phase at *K*. At high pressures, the nomenclature for gas and liquid phases is strictly a labeling convention, rather than a suggestion of what would be considered a "normal" liquid or gas at atmospheric pressure.

Types I and II occur when the molecules are fairly similar in structure or critical properties. Most nonpolar systems are expected to have endothermic heats of mixing and might be expected to be Type II; however, *they are classified according to the experimentally observed phase behavior.* Type II phase behavior includes a liquid-liquid immiscibility below the critical temperature of both components. The liquid-liquid behavior is often relatively insensitive to pressure when the liquids are incompressible far below the critical temperature. In the liquid-liquid region, as the temperature is increased, the liquids become increasingly miscible in each other until they become identical at the UCEP. The UCEP occurs at extremely low temperatures for small endothermic heats of mixing and is frequently not found experimentally for endothermic systems, because of lack of experimental data at low temperatures or because the liquids freeze before they become immiscible. Chemistry and chemical engineering undergraduate textbooks present Type I and II phase behaviors, and the *P-x* and *T-x* projections for these types are common in the study of distillation and flash calculations (22). Azeotrope behavior is a subset of Types I and II and is not shown explicitly here. While a significant number of liquid-liquid experiments have been performed to locate liquid-liquid upper critical endpoints at atmospheric pressure (23), phase behavior characterization near the critical point of the lighter component is necessary to permit classification as Type II or IV.

For the vapor-liquid-liquid (*llv*) line of Type IV or V, at temperatures and pressures just below L, vapor and liquid phases coexist. At the temperature of the LCEP, the liquid phase is at incipient instability and will split into two phases at higher temperatures along the *llv* line. Between the LCEP and the vapor-liquid critical endpoint, K, an isothermal P-x projection at T_1 exists, as shown in Fig. 1.3a. The phase behavior at pressures immediately above the *llv* line appears as both vapor-liquid and liquid-liquid regions. As the temperature is increased, the vapor-liquid envelope above the three-phase line shrinks in size on the P-x projection, while the liquid-liquid envelope enlarges. This trend continues until, at the K point, the vapor and intermediate liquid phases become identical.

The phase behaviors of Types IV and V differ only in the presence of the liquid-liquid region of Type IV. Because freezing may occur (which is not shown) at temperatures above the UCEP and near the LCEP, the lower temperature liquid-liquid region of Type IV may be experimentally inaccessible. Freezing may occur at temperatures above or below the LCEP.

Schneider (13,15) suggests that Types II, IV, and III follow a trend that can be understood if the phase behaviors are considered. When a homologous series of systems is studied, with only the heavier compound varying, the first system to exhibit Type III behavior has a strong maximum and minimum in the critical line extending from the heavier component critical point. If this strong minimum exists for less dissimilar systems, it is expected to intersect the *llv* line in two places, giving Type IV behavior. These trends also appear consistent with the calculations of Chai (24) using the Peng–Robinson equation.

Another trend is obvious that also supports this theory. As the dissimilarity of components increases, the upper critical endpoint temperature at U increases, tending to decrease the difference in the temperatures of U and L of Type IV systems until they merge, resulting in a Type III diagram (Fig. 1.4). Hence, it is possible for the phase behavior to go directly from Type II to Type III (8).

Experimental Studies of Homologous Series

Ethane/n-Alkane Systems. Studies of the ethane family are summarized by Peters et al. (16) and Miller and Luks (11). Type I behavior exists up through *n*-heptadecane. UCEP's have not been reported, possibly because the components freeze before they become immiscible. Beginning with *n*-octadecane a liquid-liquid-vapor region develops near the ethane critical point, characteristic of Type V. Once again, a UCEP is not found experimentally. This phase behavior continues through *n*-tricosane (C_{23}). With *n*-tetracosane (C_{24}) and *n*-pentacosane (C_{25}) a modification of Type V or III occurs where the $s_B l_2 v$ line interferes with the fluid behavior, as shown in Fig. 1.5, and is labeled Type III' (this author's designation). The three-phase $s_B l_2 v$ line extends from the triple point of the heavier component and intersects the llv line at a quadruple point, Q, where four phases coexist. Beginning with n-hexacosane (C_{26}), the $s_B l_2 v$ line moves to temperatures above the K point and Type VII (this author's designation) phase behavior begins (16), which is shown in Fig. 1.5. In Types III' and VII phase behavior, the line labeled $s_B l_2 v$ denotes the conditions where the solid melts over some composition ranges (12). Schneider (14) reports that squalene, a branched C_{30} hydrocarbon, exhibits Type IV, which does not fit the pattern for n-alkanes having the same carbon number.

CO₂/n-Alkane Systems. Studies of the CO_2 family are summarized by Schneider (13,14), by Miller and Luks (11), and by Enick et al. (19). Pure CO_2 freezes at a relatively high reduced temperature ($T_r = 0.712$), whereas ethane freezes at a comparatively low reduced temperature ($T_r = 0.295$). Therefore, liquid-liquid immiscibility may initially seem less likely to be observed in CO_2 systems, because of interference from freezing; however, the opposite is found. Although the liquid-liquid region is not seen for the lightest alkanes, n-heptane clearly shows Type II behavior (11). Type II behavior continues through n-dodecane. Enick et al. (19) show that n-tridecane is Type IV and that, beginning with n-tetradecane, Type III' begins. Schneider (14) reports Type III' for n-hexadecane. Type III' is also exhibited at n-heneicosane (25), and Type VII appears with n-docosane (11,26). Schneider (14) reports that squalene, a branched C_{30}, is Type III and, as with ethane, varies from the pattern exhibited by n-alkanes.

The homologous series of n-alkanes with CO_2 exhibits liquid-liquid behavior at much higher reduced temperatures compared to those exhibited with ethane. (For comparative purposes, the reduced temperature is calculated based on the critical temperature of ethane or CO_2, for the respective systems.) This may be understood by considering the heat of mixing, using the propane system as an example. Using Eq. (1.1) to approximate the low-temperature heat of mixing and the normal van der Waals geometric mean combining rule for the cross parameter a_{12}, both the ethane/propane and CO_2/propane systems are seen to be endothermic; however, the heat of mixing in the CO_2/propane system is approximately 100 times the heat of mixing for the ethane/propane system. Therefore, the excess Gibbs energy of mixing will also be considerably larger, and liquid-liquid behavior would be expected to occur to higher temperatures. In general, the n-alkane series with CO_2 shows greater asymmetry than the same ethane series, as might be expected. Therefore the solubilities are lower at a given temperature and pressure. The progression of behavior from Type II to VII also occurs for lower carbon numbers, even though the critical temperatures of ethane and CO_2 are approximately the same.

CO₂/Alcohol. Schneider (14) has reported Type II systems for CO_2 with 2-hexanol and 2-octanol and Type III systems for CO_2 with 2,5-hexanediol and 1-dodecanol (13). As expected, the addition of a hydroxyl group to component B increases the

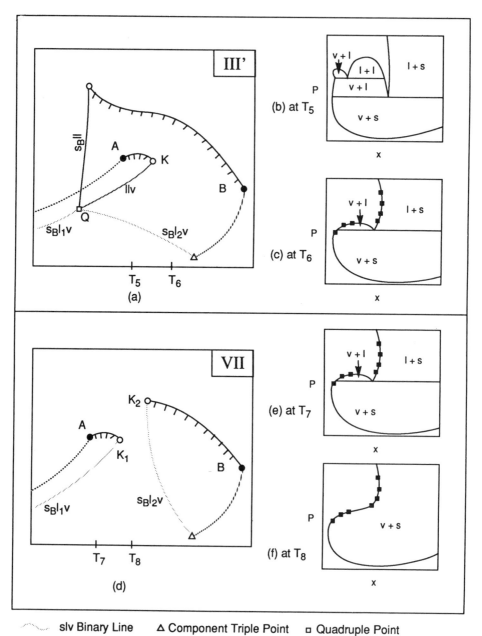

slv Binary Line   △ Component Triple Point   ▫ Quadruple Point

Fig. 1.5. Pressure-temperature projections of phase behavior when the triple point of B is in the vicinity of the critical point of A. Symbols are the same as in Fig. 1.1 except as noted.

asymmetry of the system for solutes of the same carbon number by raising the critical point of B.

CO_2/Aromatics and Ethane/Aromatics. Both CO_2 and ethane have been studied with *n*-alkylbenzenes through C_{21} and C_{20}, respectively (11). The overall trends in behavior are very similar to the comparisons made in the *n*-alkane series. Polycyclic aromatics have also been studied and typically exhibit Type VII behavior (12).

N_2O/n-Alkanes. Nitrous oxide has been studied with the *n*-alkanes, and the phase behavior is very similar to the ethane series (10) for carbon numbers where the phase behavior changes, and in the critical endpoint temperatures.

CO_2/Triglycerides/Fatty Acids. Many common triglycerides and fatty acids are solids at 30°C, so they will exhibit Type III' or Type VII behavior in binaries with CO_2. Compounds that are normally liquids will probably be Type IV, V or III. Relatively few studies of model systems are available, and in all studies only solubilities are reported (27–33). Bamberger et al. (33) did determine that lauric and myristic acids, trilaurin, and trimyristin melted over the average pressure of their experiments, but the experiments were insufficient to characterize the phase behavior types as Type III' or VII. Tripalmitin is reported to remain a solid, although the triple point is only 5°C above that of trimyristin. Also, most of the mixtures involving tripalmitin were reported to remain solid, which would not occur if the solids formed an eutectic mixture (12).

Other publications have not provided any characterization of the melting behavior of solids. Unfortunately, disagreements of up to an order of magnitude exist in a few of the solubility measurements. The disagreements have been attributed in part to the purity of the materials (29,31). Nilsson et al. (31) report solubility data for mono- and diglycerides. One interesting conclusion from the study of Czubryt et al. (32) is that stearic acid appears to be dimerized in the CO_2 phase. Phase behavior of oil mixtures will be discussed in the section regarding solubility behavior.

Propane/Triglycerides/Fatty Acids. Hixson et al. published most of the available information on these systems in the 1940s (34–38), and more complete information has recently been published on the propane/tripalmitin system (39). Propane refining of oils was practiced industrially as the Solexol process, and descriptions are available (5–7). As in the case of stearic acid in CO_2 previously discussed, the acids appear to dimerize in the fluid phase. This hypothesis is supported by the correlation of phase behavior with the effective molecular weight, which is double the molecular weight of an acid and equal to the molecular weight of an ester or triglyceride. Binary mixtures of propane with lighter-molecular-weight acids and esters is Type I (the UCEP was not located), whereas the triglycerides and heavier acids and esters exhibit Type IV or V behavior. For example, for lauric acid (effective MW = 400.6) or myristic acid (effective MW = 456.7) an LCEP has not been located at propane concentrations

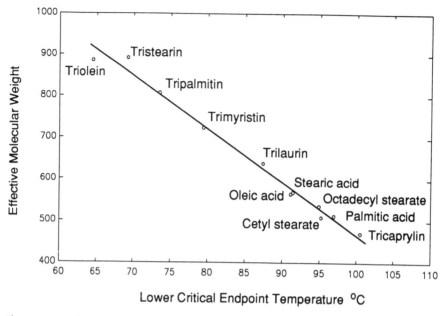

Fig. 1.6. Trend in the lower critical endpoint (LCEP) for components in binary mixtures with propane (Type IV or V systems).

between 25 and 95 wt% (37). Tricaprylin (MW = 470.7) and higher-molecular-weight compounds exhibit Type IV or V diagrams. The correlation of the LCEP with temperature is shown in Fig. 1.6 (compare with Fig. 1.4). The correlation is based on only saturated and monounsaturated compounds.

The Solexol process is based on refining of the oils using the llv region in Type IV systems between the LCEP and the vapor-liquid critical endpoint, K. As might be expected, the compounds with the lowest effective molecular weight have the greatest solubility in the propane-rich liquid, and for a given chain length, solubility increases with decreasing *effective* molecular weight: that is, triglyceride—fatty acid—ester. Hixson's studies concentrated on the measurement of *ell* solubilities on the three-phase *llv* line in binaries and on extension of the measurements to model ternary systems. Hixson mentions the incorporation of a staged countercurrent flow operation, the importance of density differences for effective countercurrent flow, and internal column recycle using temperature gradients, which have been recently discussed in the literature pertaining to CO_2 extractions (2,40–42). These engineering concepts remain important in industrial applications of supercritical technology. In the Solexol process, virtually all the feed oil leaves the top of the first tower dissolved in propane, and the stream components are fractionated according to molecular weight. The majority of color bodies leave the bottom of the first column. The conjugated unsaturated components are generally less soluble than the saturated or monounsaturated components. One of the distinct capabilities of the Solexol process was the concentration of vitamin A, but as alternative methods for obtaining vitamin A became available, this option was abandoned.

Solubility Behavior

As the number of components increases, so does the number of degrees of freedom, which results in complicated three-dimensional plots that are inadequate for complete representation of *T-P-x* data. Isothermal or isobaric behavior is frequently represented in ternary systems as "stacked" ternary diagrams (35,43,44). Another additional complication is that three-phase *llv* or *slv* behavior will occur over regions, rather than along lines, in the *P-T* projection (1,12).

Solubility behavior in mixtures frequently follows the trends in binary systems. Mixtures of naturally occurring oils with CO_2 appear to follow an analogy to Type III behavior. As noted by Stahl (45,46), isothermal increases in pressure result in a solubility maximum for vegetable oils. It may be expected that at some temperature, if the analogy of Type III is followed, the oil would become completely miscible above the critical locus (Fig. 1.3*d*). The upper region of Fig. 1.3*d*, labeled "*l* + *l*," exists only for the Type III systems that exhibit a temperature minimum in the critical line extending from the critical point of B; unfortunately detailed behavior of triglycerides is not known.

Solubility behavior of solids may also exhibit maxima with pressure, as shown in Fig. 1.5*f*. In fact, if the only phase sampled by experiments in a vessel where the phases are not observed is the phase rich in component A, "solubility" measurements from Figs. 1.3*c*, 1.5*c*, 1.5*e*, and 1.5*f* will all appear similar, as shown by the small squares on the phase boundaries. Hence, classification of the phase behavior is not possible.

In general, at temperatures above the critical point of A, the solubility of higher-molecular-weight compounds first decreases with increasing pressure, then increases dramatically, and then may decrease at even higher pressures. The exception to this behavior is shown in Figs. 1.3*a*, 1.3*b*, and 1.5*b*, below the vapor-liquid *K* point. Typically the *K* point is within about 5°C of the critical point of A. In the case of propane refining below the *K* point, the concentration of solute in the propane-rich liquid phase decreases as the *K* point is approached, while the solubility in the gas phase increases. However, the propane-rich gas phase is not used in the process.

Either temperature or pressure changes may be used to effect fractionation, separation, or both. The location of the solubility minimum is important if the fluid is to be recycled, because total depressurization is energetically wasteful and also results in more extract in the recycle stream. Also, the solubility is frequently related to the density of the fluid (47). Temperature affects the vapor pressure of the solute as well as the density of the solvent, so both factors are important. By understanding the structure of the phase diagrams, one may develop some guidelines.

The effect of solute structure has been studied by Hyatt (48) and by Dandge et al. (49) and summarized by Rizvi et al. (50). A large number of systems have also been studied by Francis (51), and although limited to 25°C, the studies provide guidelines as to whether an *llv* region is likely to exist above the critical point of CO_2.

One final point of great importance regarding solubility has been summarized by List et al. (52). The "measured" solubility may be a function of the ratio of total solute available for dissolution in the pressure vessel to the CO_2 required to fill the vessel. The maximum effluent concentration from an extractor is fixed at this value, regardless of the actual solubility limit. Knowledge of this limit is particularly important with seeds having low oil content.

Adsorption

Extractions from an oilseed or other natural substance may also provide an apparent solubility significantly less than the solubility of the pure solute, because of its adsorption on the sample matrix. In this case, the solid matrix may be functioning as a solid adsorbent material. Adsorption is a competitive partitioning process that forms the basis of a separation process such as supercritical fluid chromatography. Adsorption phenomena with supercritical fluids have been studied in the context of using the adsorbent as a separating agent (53–55) or in the context of using the supercritical fluid to regenerate the adsorbent (18,56). Analytical chromatography is usually performed in the dilute region, where adsorption is proportional to bulk concentration; however, additional studies of adsorption in more concentrated bulk solutions need to be conducted.

A separation process in the presence of an adsorbent is expected to be strongly influenced by the proximity of the experiment to the fluid's critical point. As discussed by King (53) and Parcher (57), adsorption of a pure fluid exhibits a strong maximum near the fluid critical point. More studies are needed to understand how this behavior affects adsorption in mixtures. King reports an unusually strong pressure dependence for adsorption in mixtures at supercritical conditions (54). Current adsorption models (58,59) are difficult to apply for routine calculations at high pressure, although some progress has been made (60).

Cosolvents

Solubilities in supercritical may be increased with the addition of a *cosolvent* (or *entrainer*) (61–63). It has been suggested that use of an entrainer may be preferred, because of the energy savings (64,65). Kim and Johnston (66) have used spectroscopic data to show that the local composition surrounding the solute is enriched in the cosolvent relative to the bulk solution. When cosolvents are added to supercritical fluids, care must be used to verify that additional phases are not formed, because the cosolvent increases the degrees of freedom in the system. In addition, when a cosolvent is used in a batch extraction of liquids or oilseeds, it may be expected that some of the cosolvent will dissolve in the liquid phase or adsorb on the solids. If the quantity is significant, the effluent stream from the extractor will be of a different entrainer concentration than the inlet; hence, the ratio of entrainer to supercritical fluid should be monitored. In addition, the entrainer may be soluble in products that are extracted, so subsequent fractionation may be required.

Solubilities of less volatile components are enhanced when they are extracted from mixtures with more soluble species. Bamberger et al. (33) showed that the solubilities of trimyristin and tripalmitin are enhanced when extracted with trilaurin. Nilsson et al. (31) noted a similar effect with mono-, di-, and trioleins. The same behavior is exhibited in studies with mixtures of polycyclic aromatics (67). The most soluble species typically does not exhibit significant solubility enhancement; however, it acts as a solubility enhancer for the less soluble species.

Molecular Structure

The classification scheme of van Konynenburg and Scott, summarized in the figures in this chapter, does not strictly classify solubility behavior. Solubilities increase and decrease with temperature/pressure in different regions of the diagrams. Solubility increases are determined by the thermodynamic partial molar properties. Petsche and Debenedetti (68) have proposed classification of mixtures as repulsive, weakly attractive, and attractive, based on the following equation:

$$C_{12}^{\infty} = 1 - \frac{1 + \rho(b_1 - b_2)}{(1 - \rho b_2)^2} + \frac{2\rho a_{12}}{kT} \qquad [1.3]$$

where the as and bs are the van der Waals parameters, ρ is the number density, and subscript 1 denotes the solute. When this relationship, called the direct correlation function is greater than 1, the infinite dilution partial molar volume and partial molar enthalpy will be negative. The solubility will increase with pressure (at constant temperature) and decrease with temperature (at constant pressure). Note that the direct correlation function is a function of density and is expected to change with pressure and temperature. At constant temperature, the solubility may decrease at low pressure, increase at intermediate densities, and then begin to decrease at higher densities. A given binary mixture may be weakly attractive, attractive, and repulsive over various pressure and temperature regions.

Quantitative predictions of these transitions require the equation parameters, and if these parameters were known, the solubility could be predicted directly with the equation of state. Therefore, the correlation functions are primarily an interpretive tool. Also, the van der Waals equation is known to provide poor quantitative predictions of solubility behavior. Petsche and Debenedetti also consider the effect of chain length on the classification of mixtures using a lattice gas model and find that the effect is important.

If a mixture is attractive, then solvent molecules tend to aggregate in the immediate vicinity of the solute molecules. This is a requirement to have negative partial molar volumes, which result in an increase in the solubility of the solute. Molecular clustering of this kind has also been studied by molecular simulation (69–71) and fluorescence spectroscopy (72–74).

Conclusions

Phase behavior in supercritical solutions may be characterized by a natural progression of phase behaviors that occurs as the molecular asymmetry of the system is increased. Depending upon the type of phase behavior, solubilities may increase or decrease with increases in temperature or pressure. Phase behaviors in multicomponent systems follow trends analogous to binary systems, although multiphase behavior may occur over larger regions of *P-T* space. Both experimental and theoretical studies are important in developing our understanding of the extraction processes that occur with supercritical fluids. More studies of CO_2/triglyceride/fatty acid systems will be helpful in classification of their phase behavior.

List of Symbols

$1 - 2 = 3$	Phases 1,2,3 coexist, and phases 2 and 3 are identical at a critical state
$1 = 2 - 3$	Phases 1,2,3 coexist, and phases 1 and 2 are identical at a critical state
A	Component with lower critical temperature
B	Component with higher critical temperature
K	A *K* point, a critical endpoint where $l - l = v$
l	Liquid phase
L,LCEP	A lower critical endpoint, where $l = l - v$
Q	A *Q* point, or quadruple point, which has four phases in equilibria; herein, the four phases are *sllv*
s	Solid phase
U,UCEP	An upper critical endpoint, where $l - l = v$
v	Vapor phase

References

1. Luks, K.D., *Fluid Phase Equil. 29:*209 (1986).
2. Stahl, E., K.-W. Quirin, and D. Gerard, *Dense Gases for Extraction and Refining,* Springer-Verlag, New York, 1988.
3. Auerbach, E.B., U.S. Patent 1,805,751 (1931).
4. Wilson, R.E., P.C. Keith, and R.E. Haylett, *Ind. Eng. Chem. 28:*1065 (1936).
5. Passino, H.J., *Ind. Eng. Chem. 41:*280 (1949).
6. Dickinson, N.L., and J M. Meyers, *J. Am. Oil Chem. Soc. 29:*235 (1952).
7. Moore, E.B., *J. Am. Oil Chem. Soc. 27:*75 (1950).
8. van Konynenberg, P.H., and R.L. Scott, *Phil. Trans. Roy. Soc. London, Ser. A 298:1442:*495 (1980).

9. Streett, W.B., in *Chemical Engineering at Supercritical Fluid Conditions,* edited by M.E. Paulaitis, J.M.L. Penninger, R.D. Gray, and P. Davidson, Ann Arbor Science, Ann Arbor, MI, 1983, pp. 3–30.

10. Jangkamolkulchai, A., D.H. Lam, and K.D. Luks, *Fluid Phase Equil. 50:*175 (1989).

11. Miller, M.M., and K.D. Luks, *Fluid Phase Equil. 44:*295 (1989).

12. White, G.L., and C.T. Lira, *Fluid Phase Equil. 78:*269 (1992). See also White, G.L., and C.T. Lira, in *Supercritical Fluid Science and Technology,* edited by K.P. Johnston and J.M.L. Penninger, American Chemical Society, Washington, D.C., 1989, pp. 111–120.

13. Schneider, G.M., *Pure and Applied Chem. 63:*1313 (1991). Also published as Schneider, G.M., *J. Chem. Therm. 23:*301 (1991).

14. Schneider, G.M., Angew. *Chem. Intl. Ed. Engl. 17:*716 (1978).

15. Schneider, G.M., *Adv. Chem. Phys. 1:*1 (1970).

16. Peters, C.J., R.N. Lichtenthaler, and J. de Swaan Arons, *Fluid Phase Equil. 29:*495 (1986).

17. Estrera, S.S., M.M. Arbuckle, and K.D. Luks, *Fluid Phase Equil. 35:*291 (1987).

18. McHugh, M.A., and V.J. Krukonis, *Supercritical Fluid Extraction: Principles and Practice,* Butterworths, Stoneham, MA, 2nd edn., 1994.

19. Enick, R., G.D. Holder, and B.I. Morsi, *Fluid Phase Equil. 22:*209 (1985).

20. Modell, M., and R. Reid, *Thermodynamics and Its Applications,* 2nd edn., Prentice-Hall, Englewood Cliffs, N.J., 1983, pp. 259–264.

21. Denbigh, K., *The Principles of Chemical Equilibria,* 4th edn., Cambridge University Press, New York, 1981, pp. 188–190.

22. Smith, J.M., and H.C. Van Ness, *Introduction to Chemical Engineering Thermodynamics,* 4th edn., McGraw-Hill, New York, 1987, pp. 363–373.

23. Hildebrand, J.H., J.M. Prausnitz, and R.L. Scott, *Regular and Related Solutions,* Van Nostrand, New York, 1970.

24. Chai, C.-P., Phase Equilibrium Behavior for Carbon-Dioxide and Heavy Hydrocarbons, Ph.D. Thesis, University of Delaware, 1981.

25. Fall, D.J., J.L. Fall, and K.D. Luks, *J. Chem. Eng. Data 30:*82 (1985).

26. Fall D.J., and K.D. Luks, *J. Chem. Eng. Data 29:*413 (1984).

27. King, M.B., T.R. Bott, M.J. Barr, and R.S. Mahmud, *Sep. Sci. Techn. 22:*1103 (1987).

28. Brunetti, L., A. Daghetta, E. Fedeli, I. Kikic, and L. Zanderighi, *J. Am. Oil Chem. Soc. 66:*209 (1989).

29. Goncalves, M., A.M.P. Vasconcelos, E.J.S. Gomes de Azevedo, H.J. Chaves das Neves, and M. Nunes da Ponte, *J. Am. Oil Chem. Soc. 68:*474 (1991).

30. Chrastil, J., *J. Phys. Chem. 86:*3016 (1982).

31. Nilsson, W.B., E.J. Gauglitz, and J.K. Hudson, *J. Am. Oil Chem. Soc. 68:*87 (1991).

32. Czubryt, J.J., M.N. Myers, and J.C. Giddings, *J. Phys. Chem. 74:*426 (1970).

33. Bamberger, T., J.C. Erickson, C.L. Cooney, and S.K. Kumar, *J. Chem. Eng. Data 33:*23 (1988).

34. Hixson, A.W., and A.N. Hixson, *Trans. Amer. Inst. Chem. Eng. 37:*927 (1941).

35. Hixson, A.W., and J.B. Bockelmann, *Trans. Amer. Inst. Chem. Eng. 38:*891 (1942).

36. Drew, D.A., and A.N. Hixson, *Trans. Amer. Inst. Chem. Eng. 40:*675 (1944).

37. Bogash, R., and A.N. Hixson, *Chem. Eng. Prog. 45:*597 (1949).

38. Hixson, A.N., and R. Miller, U.S. Patent 2,219,652 (1940); U.S. Patent 2,344,089 (1944); and U.S. Patent 2,388,412 (1945).

39. Coorens, H.G.A., C.J. Peters, and Arons J. de Swaan, *Fluid Phase Equil. 40:*135 (1988).

40. Eisenbach, W., *Ber. Bunsenges. Phys. Chem. 88:*882 (1984).
41. Nilsson, W.B., E.J. Gauglitz Jr., J.K. Hudson, V.F. Stout, and J. Spinelli, *J. Am. Oil Chem. Soc. 65:*109 (1988).
42. Nilsson, W.B., E.J. Gauglitz, and J.K. Hudson, *J. Am. Oil Chem. Soc. 66:*1596 (1989).
43. Spee, M., and G.M. Schneider, *Fluid Phase Equil. 65:*263 (1991).
44. Paulaitis, M.E., R.G. Kander, and J.R. DiAndreth, *Ber. Bunsenges. Phys. Chem. 88:*869 (1984).
45. Stahl, E., K.W. Quirin, A. Glatz, D. Gerard, and G. Rau, *Ber. Bunsenges. Phys. Chem. 88:*900 (1984).
46. Stahl, E., K.W. Quirin, and D. Gerard, *Fette Seifen Anstrichm. 85:*458 (1983).
47. Lira, C.T., in *Supercritical Fluid Extraction and Chromatography: Techniques and Applications,* edited by B.A. Charpentier and M.R. Sevenants, American Chemical Society, Washington, DC, 1988, pp. 1–25.
48. Hyatt, J.A., *J. Org. Chem. 49:*5097 (1984).
49. Dandge, D.K., J.P. Heller, and K.V. Wilson, *Ind. Eng. Chem. Proc. Des. Dev. 24:*162 (1985).
50. Rizvi, S.S.H., J.A. Daniels, A.L. Benado, and J.A. Zollweg, *Food Techn. 40(7):*57 (1986).
51. Francis, A.W., *J. Phys. Chem. 58:*1099 (1954).
52. List, G.R., J.P. Friedrich, and J.W. King, *Oil Mill Gazetteer (12):*28 (1989).
53. King, J.W., in *Supercritical Fluids: Chemical and Engineering Principles and Applications,* edited by T.G. Squires and M.E. Paulaitis, American Chemical Society, Washington, DC, 1987, pp. 150–171.
54. King, J.W., R.L. Eissler, and J.P. Friedrich, in *Supercritical Fluid Extraction and Chromatography: Techniques and Applications,* edited by B.A. Charpentier and M.R. Sevenants, American Chemical Society, Washington, DC, 1988, pp. 63–88.
55. Zosel, K., U.S. Patent 4,247,570 (1981).
56. deFillipi, R.P., and R.J. Robey, EPA Report 600/2-83-038 (1983).
57. Strubinger, J.R., H. Song, and J.F. Parcher, *Proceedings of the 2nd International Symposium on Supercritical Fluids,* edited by M.A. McHugh, Boston, MA, May 20–22, pp. 213–215.
58. Teletzke, G.F., L.E. Scriven, and H.T. Davis, *J. Chem. Phys. 77:*5794 (1982).
59. Kung, W.C., L.E. Scriven, and H.T. Davis, *Chem. Phys. 149:*141 (1990).
60. Subramanian, R., H. Pyada, and C.T. Lira, 3830 (1995).
61. Dobbs, J.M., J.M. Wong, R.J. Laheire, and K.P. Johnston, *Ind. Eng. Chem. Res. 26:*56 (1987).
62. Schmitt, W.J., The Solubility of Monofunctional Organic Compounds in Chemically Diverse Supercritical Fluids, Ph.D. Thesis, Massachusetts Institute of Technology, 1984.
63. Walsh, J.M., G.D. Ikonomou, and M.D. Donohue, *Fluid Phase Equil. 33:*295 (1987).
64. Peter, S., Ber. Bunsenges. *Phys. Chem. 88:*875 (1984).
65. Brunner, G., and S. Peter, *Sep. Sci. Technol. 17:*199 (1982).
66. Kim, S., and K.P. Johnston, *AiChE J. 33:*1603 (1987).
67. Kurnik, R.T., and R.C. Reid, *Fluid Phase Equil. 8:*93 (1982).
68. Petsche, I.B., and P.G. Debenedetti, *J. Phys. Chem. 95:*386 (1991).
69. Lee, L.L., P.G. Debenedetti, and H.D. Cochran, in *Supercritical Fluid Technology: Reviews in Modern Theory and Applications,* edited by T.J. Bruno and J.F. Ely, CRC Press, Boca Raton, FL, 1991, pp. 193–225.

70. Knutson, B.L., D.L. Tomasko, C.A. Eckert, P.G. Debenedetti, and A. Chialvo, "Local Density Enhancement in Dilute Supercritical Mixtures: Comparison Between Theory and Experiment," paper 204e presented at the 1991 Annual AIChE Meeting, November 17–22, Los Angeles, CA.

71. Chialvo, A.A., and P.G. Debenedetti, *Ind. Eng. Chem. Res. 31:*1391 (1992).

72. Eckart, M.P., J.F. Brennecke, and C.A. Eckert, in *Supercritical Fluid Technology: Reviews in Modern Theory and Applications,* edited by T.J. Bruno and J.F. Ely, CRC Press, Boca Raton, FL, 1991, pp. 163–192.

73. Brennecke, J.F., D.L. Tomasko, J. Peshkin, and C.A. Eckert, *Ind. Eng. Chem. Res. 29:*1682 (1990).

74. Brennecke, J.F., D.L. Tomasko, and C.A. Eckert, J. Phys. Chem. 94:7692 (1990).

Chapter 2

Solubility Measurements of Lipid Constituents in Supercritical Fluids[1]

R.J. Maxwell

Eastern Regional Research Center, Agricultural Research Service, U.S. Department of Agriculture, 600 East Mermaid Lane, Philadelphia, PA 19118

Solubility measurements of solutes in supercritical fluids have been of interest to investigators for over a century. The first report of solubility phenomena above the critical point was by Hannay and Hogarth in 1879 (1), who measured the solubility of several inorganic salts in supercritical ethanol. Two decades later Villard (2) published the first description of the solubility of a fatty acid (stearic acid) in supercritical fluid carbon dioxide, ethylene, and nitrous oxide. An extensive study by Frances in 1954 (3) reported the qualitative solubilities of a series of carboxylic acids from acetic through oleic, under near-critical conditions (25°C, 65.5 bar).

The study of lipid solubilities in supercritical fluids has expanded considerably over the past ten years. Recent publications have described the solubilities of a diverse array of lipid constituents such as hydrocarbons, fatty acids and esters, fatty alcohols, glycerides, and sterols. Because of their very limited solubility in the commonly used supercritical solvents such as CO_2 and N_2O, few measurements have been reported for phospholipids.

Many of these measurements have been made to meet the need for fundamental data for process design purposes as well as analytical applications. When a process is designed for the isolation of lipids from matrices such as oilseeds or tissue, it is critical to understand the requisite solute-solvent phase equilibria. These phase equilibria are important in determining the optimal operating conditions, the solvent-to-feed ratio, and the selectivity of the extracted solutes in engineering-scale supercritical fluid extraction (SFE). Such experimental data may also be used to develop solubility correlation models.

Many types of instruments have been designed to measure supercritical phase behavior. These many types of apparatus are needed because solids and liquids have different properties in the critical state, so not all compounds can be studied using an apparatus of common design. For instance, supercritical fluids are generally not regarded as being soluble in a solid solute, so a single phase can be assumed to exist above the solute. However, in the case of liquids the supercritical fluid may be soluble in the liquid phase, as well as the liquid solute being soluble in the supercritical fluid phase. In that case, both phases have to be sampled to determine the liquid–supercritical fluid

---

[1]Mention of brand or firm names does not constitute an endorsement by the U.S. Department of Agriculture over others of a similar nature not mentioned.

equilibria. Instrumentation has been designed to allow both phases to be sampled simultaneously.

Regardless of the physical state of the solute being investigated, the experimental systems used to determine solubilities may be divided into three categories: dynamic (flow-through), static, and recirculating. The three techniques for measuring solubilities all use the same principle: attainment of equilibrium between the two phases. However, they differ in operation. In a *flow-through* or *dynamic* system the supercritical fluid is passed through the sample at selected flow rates, and instantaneous equilibrium between the fluid and the sample is usually assumed. Static determinations are performed in a vessel containing a fixed amount of solute, and the pressure or temperature is adjusted until all, or a portion, of the sample dissolves in the supercritical phase. *Recirculating* devices operate with a fixed fluid volume, which is continuously recirculated throughout the system until equilibrium is achieved. Many variations on these experimental designs have been reported, and although these techniques differ in operation, good agreement is reported in experimental results for lipid-SF systems from one experimental system to another.

This chapter reports on solubility determinations carried out on lipid constituents, and obtained using dynamic, static, and recirculating techniques. An overview of the various types of instruments used in these investigations is also provided. Tables are included that list typical solubility data for lipid materials. These tables list only solubility data on individual lipid constituents and do not contain solubility data for complex mixtures such as oils. Discussions of oil solubilities in supercritical fluids are covered in detail by others in this volume.

Dynamic Techniques for Determining Solubilities

Since no commercial supercritical fluid (SF) instrument has been introduced specifically for determining equilibrium solubilities, most investigators have assembled their own instruments for this purpose. By far the majority of the solubility measurements reported for solid lipids have been carried out in dynamic (flow-through) instruments. Although the designs of these instruments may differ significantly they are all based on a common principle.

In measuring solubilities in flow-through instruments it is assumed that a steady-state solute-solvent equilibrium is achieved as the supercritical fluid passes over the solute (4). A simplified schematic diagram of such a dynamic flow apparatus is shown in Fig. 2.1. This diagram is a composite of many experimental designs reported by various investigators (5,6). In using this apparatus, carbon dioxide is delivered from the gas cylinder into a compressor or pump, where it is compressed and heated to give the desired operating pressure (>73 bar for supercritical operation). In this case, the compressor is used to deliver CO_2 (solvent) continuously throughout the entire experiment. The fluid passes from the compressor to a section of coiled tubing mounted in a constant-temperature bath or oven, which is used to ensure that the fluid is at the required temperature, before it enters the extraction cell, which contains the lipid solute. The CO_2 solvent becomes imbued with the solute as it passes through the extraction cell; then it reaches the heated micrometering valve, where the SF phase is expanded to atmospheric pressure. The micrometering valve serves to control the flow rate of the expanded gas as well as the pressure on the extraction cell. The solute precipitates from the expanded gas and is collected in a flask or some other type of receiver. Finally, the expanded gas passes through a flow meter and flow totalizer, where the flow rate and the total gas volume (as a function of time) are recorded.

If the solute is a solid, such as stearic acid or tristearin, it is generally mixed with glass beads, glass wool, or some other inert support (5,7) in order to disperse the material in the extraction cell. Glass or polypropylene wool plugs or sintered metal filters are typically placed at the exit port of the cell to prevent loss of the sample from the cell. Solubilities of lipids in the liquid state have also been determined with the type of device shown in Figure 2.1. Nilsson et al. (8) among others have measured the solubilities of methyl oleate, oleic acid, and triolein using such a flow-through device. They suspended the liquid solutes on glass wool while filling the remainder of the cell volume with a stainless steel packing material (8).

Before beginning solubility measurements with flow-through devices, one should measure the amount of solute extracted at several fluid flow rates. These experiments are required to ensure that the fluid has reached equilibrium with the test solute and that the determined solubilities are not flow-dependent. If the solute and solvent are at equilibrium, the differences between the extracted amounts of solute measured at the various flow rates should be less than 2% (8).

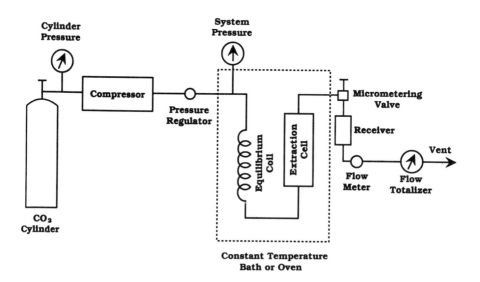

Fig. 2.1. Schematic diagram of dynamic (flow-through) apparatus for solubility determinations in supercritical fluids.

McHugh and Krukonis (4) have compiled a list of advantages and disadvantages associated with the use of such flow-through devices. Advantages include the following:

- The apparatus can be assembled with off-the-shelf equipment.
- Sampling procedures can be straightforward.
- Solubility data are accumulated rapidly.
- Equilibrium, stripping, or fractionating data can be obtained.

The following disadvantages have been encountered:

- Liquid or solid solute may obstruct the metering valve.
- Entrainment may occur with liquid solutes.
- Undetected phase changes may occur in the equilibrium cell.
- High pressures can cause the density of the SF phase to become greater than the density of the solute-rich liquid phase, which can result in extrusion of the liquid solute from the cell, particularly if the cell is held in a vertical configuration.
- Because only the SF phase is sampled, the solubility of the SF in the liquid phase cannot be measured.

Investigators have made many improvements to the designs of flow-through instruments that overcome some of the disadvantages described (4). An apparatus that incorporates many of these improvements has been reported by Dobbs et al. (9) and is shown in Fig. 2.2. This apparatus utilizes a second equilibrium cell (saturator) in line with the first cell. In this design, both cells are charged with solute. The additional cell ensures that the solute-laden SF phase exiting this cell arrangement is saturated. Dobbs and coworkers also included a second compressor in their system for the addition of cosolvent to the mobile phase. The concentration of cosolvent in the SF phase is measured by means of a vibrating-tube densitometer (ρI) placed in line before the saturators. The micrometering valve shown in Fig. 2.1 has been replaced in this design by an air-actuated switching valve (Fig. 2.2) contained within a constant-temperature bath. As the SF phase exits the saturators, it passes through the sample loop of the sampling valve, whose use was first described by McHugh and Paulaitis (10), before being vented from the system. The system remains at constant temperature and pressure while the contents of the sample loop are switched to the receiver, thus obviating the problem of solute loss, which is frequently encountered with the apparatus shown in Fig. 2.1. The solute is then swept from the sample loop into receiver A by the addition of a few mL of solvent through port B. If a calibrated sample loop is used in the switching valve, the amount of solute in the loop may be calculated from the volume of the loop. Wong and Johnson used this apparatus to measure solubilities of sterols, with and without cosolvents, up to pressures of 350 bar. Using this apparatus, they found that the addition of small amounts of dipolar aprotic and protic cosolvents to SC-CO_2 increased the solubility of certain sterols by one or two orders of magnitude (11).

Dynamic experimental systems have been used to measure the solubilities of members of several lipid classes. Those lipids that have been reported in the literature

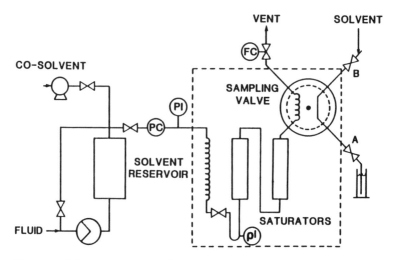

Fig. 2.2. Schematic diagram of dynamic apparatus with cosolvent addition and microsampling valve. (Reprinted with permission from *Journal of Chemical Engineering Data 31*:303, ©1986 American Chemical Society.)

and whose solubilities have been determined using dynamic techniques are listed in Table 2.1 (5,7,8,11–26), grouped under their respective class subheadings. For each compound in the table, the ranges of temperature and pressure used in the solubility measurements are given. It will be necessary for the reader to consult the references to obtain the solubility data for the individual entries, because it is beyond the scope of this work to include those values in the table.

Examples of both liquid and solid lipids are included in the table. In several examples in the table the investigators have compared their solubility results with those of others using different experimental techniques (13,14). It should also be noted that solubility determinations on the same lipid (β-carotene, 1-octadecanol, cholesterol, oleic acid, tristearin, etc.) have been made by several investigators. Such data allows a critical comparison to be made among the various instrumental techniques employed.

Static and Recirculating Techniques for Determining Solubilities

Solubility measurements obtained from static or recirculating systems generally involve more elaborate instrumentation than those designed for dynamic studies. This point is illustrated by the apparatus shown in Fig. 2.3, which was assembled by McHugh et al. (26) to measure the solubility of a solid hydrocarbon, octacosane, in SF media. The main component of this apparatus is a variable-volume equilibrium view cell, which allows for the visual determination of phases when equilibrium between phases has been attained. The cell can withstand pressures up to 690 bar. Solubilities

TABLE 2.1 Lipid Solubilities Determined in Carbon Dioxide Using Dynamic (Flow-Through) SF Extractors

Solute	Temperature (°C)	Pressure range (bar)	SF Cosolvent[a]	References
Hydrocarbons				
Carotene, lutein	40	100–700	—	15
β-carotene	40	240	—	17
Squalene, C_{18}–C_{24} alkanes	37–47	110–280	—	16
Octacosane	35–52	80–325	—	26
Alcohols, sterols				
1-octadecanol	40	274–1925	—	18
1-octadecanol	29–70	110–364	—	16
1-hexadecanol, 1-octadecanol	45–65	140–467	—	5,9
1-hexadecanol	35	89–218	—	7
α-tocopherol	25–40	100–180	—	19
Cholesterol, progesterone, testosterone	35–55	86–253	N_2O	20
cholesterol, ergosterol, stigmasterol	35–60	101–354	A,E,M	12
35 steroids	40	80–200	—	21
Acids, esters				
Oleic acid	35–45	96–186	—	13
Stearic acid	40	274–1925	—	18
Stearic acid	37–47	114–364	—	16
Palmitic, stearic acids	45–65	142–575	—	5,9
Lauric, myristic, palmitic acids	50	70–249	—	22
Palmitic acid	25–40	80–187	—	19
Myristic, palmitic	35	81–228	—	7
Oleic acid, methyl oleate ethyl palmitate, ethyl oleate, ethyl eicosapentaenoate and ethyl docosahexaenoate	50–60 25–55	110–206 69–172	— —	8 23
Methyl palmitate, methyl stearate, methyl arachidate and methyl behenate	50	78–147	EA	24
Glycerides				
Triolein, tristearin	30–70	98–270	A,D,E, EA,EE	24,25
Trilaurin, trimyristin, tripalmitin	50	91–304	—	22
Tripalmitin	25–40	80–166	—	19
Mono–, di, and triolein	50–60	150–309	—	8
Triolein, tristearin	50–70	98–270	EA	24,25
Trilaurin, trimyristin, tripalmitin, tristearin,	35–55	83–370	—	14

[a]A = acetone, D = dichloromethane, E = ethanol, EA = ethyl acetone, EE = ethyl ether.

are determined in the following manner: First, the solute is charged in the equilibrium cell, and the system is filled with gaseous CO_2. Then a piston is used to increase the pressure in the cell isothermally until all of the solute has solubilized. At this point a clear, single fluid phase is present in the view cell. The mixture is then decompressed until solid precipitates and two phases are visible in cell. Solute solubility occurs in the interval between the two-phase and the previous single-phase state. If a liquid solute is being studied, the vapor-liquid phase transition is measured in a similar manner (26).

McHugh and coworkers also studied the solubility of octacosane using a dynamic system similar to the apparatus illustrated in Fig. 2.2 (26). They compared their results using these two techniques in a graph, shown in Fig. 2.4, in which the mole fractions of octacosane in the SF obtained by both techniques are plotted as a function of pressure. The mole fractions obtained by both techniques have an experimental error of less than 2%, indicating that excellent agreement between dynamic and static techniques is possible even though the experimental systems differ considerably in design and operation.

McHugh and Krukonis (4) note the advantages of the variable-volume view-cell apparatus as follows:

- Equilibrium phases are visually determined.
- Phase transitions and inversions are easily detected.
- Solubilities in binary systems are obtained without sampling.
- Minimum amounts of solute and CO_2 are used.
- The pressure of the mixture can be adjusted continuously to a fixed composition and temperature.

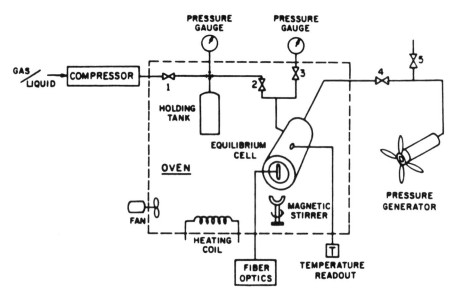

Fig. 2.3.   Schematic diagram of an apparatus used to obtain static solubility measurements. (Reprinted with permission from *Industrial and Engineering Chemistry Fundamentals* 23:493 ©1984 American Chemical Society.)

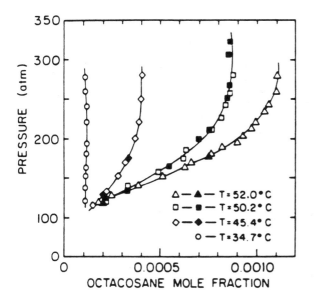

Fig. 2.4. Solubility of octacosane in supercritical CO_2 at 34.7, 45.4, 50.2, and 52.0°C. The open symbols represent data obtained with the flow apparatus, and the solid symbols represent data obtained with the view cell apparatus. (Reprinted with permission from *Industrial and Engineering Chemistry Fundamentals 23:*493, ©1984 American Chemical Society.)

The apparatus in Fig. 2.3 is capable of measuring a single phase transition, the solubility of a solute in the supercritical fluid, or the vapor phase composition only. Experimental systems of a different design are needed when solubility data are required on oils, since at high pressures a large amount of CO_2 may dissolve in the liquid phase, causing changes in its physical properties (27). A recirculation-type apparatus design by Zou et al. (28), shown in Fig. 2.5, permits measurement of the mole fraction of solute and solvent in both the vapor and liquid phases. This recirculating apparatus, like the apparatus shown in Fig. 2.3, contains a view cell (EC). The liquid solute is charged into the cell (EC), and the system is filled with supercritical CO_2 through a high-pressure pump (HPP). Pressure in the system decreases as CO_2 is added until the liquid phase is saturated. Carbon dioxide is then added until the desired pressure is reached, and then the magnetic pump (MRP) is activated to circulate each phase. The vapor phase is recirculated to the view cell, where it bubbles up through the liquid. When equilibrium has been reached, representative samples of the vapor and liquid phases are taken from the recirculating system into the respective sampling loops (VSL and LSL), where they are subsequently expanded into collection vessels for further analysis.

Zou et al. used this system to study binary liquid supercritical–fluid equilibria in CO_2 for oleic and linoleic acids, methyl oleate, and methyl linoleate at 40 and 60°C. A similar apparatus was used by Cheng et al. (29) to study the solubility of methyl oleate in supercritical ethane and carbon dioxide. They noted that the molar volume of the vapor phase becomes smaller than that of the liquid at higher pressures. This result is due to the very large differences in molecular weights of the solute and solvent.

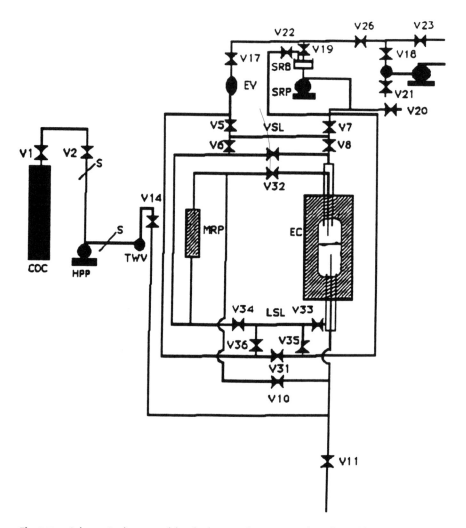

Fig. 2.5. Schematic diagram of the dual-recirculation vapor-liquid equilibrium apparatus. Components are labeled as follows: equilibrium cell (EC); expansion volume (EV); vapor sampling loop (VSL); liquid sampling loop (LSL); recirculation beaker for solvent (SRB); solvent recirculating pump (SRP); magnetic recirculating pump (MRP); high-pressure pump (HPP); three-way valve (TWV); carbon dioxide cylinder (CDC). (Reprinted with permission from *J. Supercrit. Fluids 3*:23, © 1990, PRA Press, Polymer Research Associates, Inc.)

None of the solubility determinations described thus far have been carried out using commercial instrumentation designed specifically for that purpose. Most commercial analytical supercritical fluid extractors have flow-through designs that use orifice restrictors rather than metering valves for gas expansion. This makes them less suitable than laboratory-built devices for the study of phase equilibria. One commercial instrument, the SPA (Sample Preparation Apparatus, LCD Analytical, Riviera Beach, FL), has been used for solubility measurements of lipids and other compound classes. The design of the SPA incorporates a recirculating-loop configuration that can be used for the measurement of vapor-phase equilibria in a manner similar to several laboratory-built recirculating systems (30–35). In addition, the SPA recirculation loop contains two air-actuated valves, which are used in a fashion similar to those in the apparatus shown in Fig. 2.2 to transfer an aliquot of the recirculation loop into the mobile-phase stream of a high-performance liquid chromatograph (HPLC). This apparatus was used first by Schäfer and Baumann to measure the equilibrium solubilities of a series of pesticides in SF (35) in an on-line system, where an aliquot of the SF phase was injected directly into the mobile phase of an HPLC. The concentration of the pesticide in the aliquot then was determined from the response of the system in-line UV detector.

In the present author's laboratory the SPA has been used by Cygnarowicz et al. to determine the solubility of β-carotene (36) in supercritical CO_2. In this application the solute was collected off-line. The graph in Fig. 2.6 illustrates the results obtained by Cygnarowicz et al. using the SPA for the measurement of the solubility of β-carotene in carbon dioxide at three temperatures. The experimental data in Fig. 2.6 have been correlated to a modified Peng-Robinson equation of state (36).

Further experiments in our laboratory indicated the need to redesign the SPA recirculating loop system to improve the efficiency of the apparatus for the measurement of other lipid classes (37). A schematic of the modified SPA extractor is shown in Fig. 2.7. In the modified design, fluid flows from the recirculation pump through injection valve 1, which incorporates a sample loop, to the extraction cell containing the solute, and then through a UV-vis flow cell, through a loop control valve, and finally back to the recirculation pump to begin the cycle again. In the modified SPA shown in Fig. 2.7 an aliquot of the sample loop can be injected directly into injection valve 2 and then through a mobile-phase stream to a volumetric flask receiver, thus changing the system from on-line to off-line solute recovery. These modifications, reported elsewhere (37), resulted in improved accuracy and reproducibility for solubility measurements over those obtained using the SPA's original flow design.

Further improvements have increased the maximum operating pressure of the apparatus from 345 bar to 517 bar (R.J. Maxwell and J.W. Hampson, unpublished results). The modified SPA extractor has been used to measure solubilities of several lipid classes including sterols, fatty acids, and triglycerides (J.W. Hampson and R.J. Maxwell, unpublished results). Unfortunately, the SPA is no longer commercially available.

A listing of the equilibrium solubilities of lipids determined using static or recirculating experimental systems is given in Table 2.2. In addition to the column headings used in Table 2.1, Table 2.2 includes an additional heading describing the type of apparatus used to obtain the data. It is apparent from Table 2.2 that many of the lipids investigated using flow-through instruments (Table 2.1) have also been studied with

R.J. Maxwell

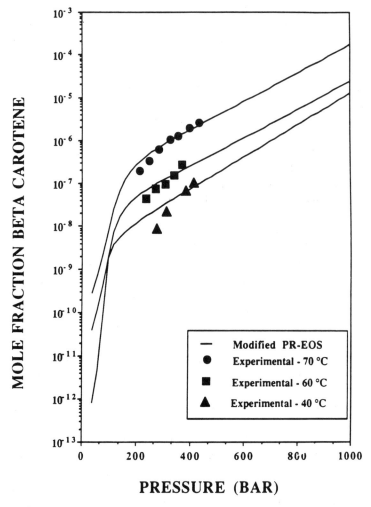

Fig. 2.6. Mole fractions of β-carotene in supercritical CO_2 as a function of pressure at 40, 60, and 70°C: experimental data and theoretical estimates. (Reprinted with permission from *Fluid Phase Equil. 59*:57, ©1990 Elsevier Science Publishers.)

static or recirculating systems. Most of the reported lipid solubility measurements have used carbon dioxide as the SF solvent. However, a few studies have reported lipid solubilities in other solvents, such as ethane or propane (29,40,41). Only one lipid, β-carotene, has been studied in a recirculating apparatus using cosolvents (36). In general, liquid lipids, such as methyl esters, have been investigated using static or recirculating rather than flow-through systems. This is not surprising, since both

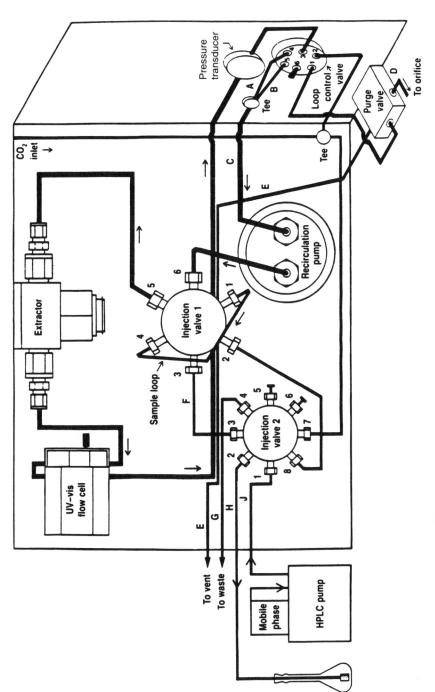

Fig. 2.7. Schematic diagram of the modified SPA extractor showing the direction of flow through the recirculating loop. (Reprinted with permission from *LC · GC 9*:788, ©1991 Advanstar Communications, Inc.)

TABLE 2.2 Lipid Solubilities Determined Using Static or Recirculating SF Extractors

Solute	Tempera-ture, °C	Pressure range, bar	SF Solvent	Cosolvent[a]	Static (S) or recircu-lating (R)	References
Hydrocarbons						
β-carotene	40–70	200–439	CO_2	E,D,M	R	36
Octacosane	35–52	80–325	CO_2	—	S	26
$C_{12,14,16,18}$ alkanes	25,32	172	CO_2	—	S	38
Sterols						
Cholesterol	40,60	138–345	CO_2	—	R	*
α–tocopherol		150–350	CO_2	—	R	32
α-tocopherol, cholesterol	40–80	101–253	CO_2	—	S	39
Acids, Esters						
Palmitic, stearic acids	40–55	414	CO_2	—	R	*
Decanoic acid	25–32	172	CO_2	—	S	38
Myristic, oleic, stearic acids	90–104	21	Propane	—	S	40
Myristic, oleic, palmitic, stearic acids	35–60	116–300	CO_2	—	R	33,34
Oleic acid	50–90	150–300	CO_2	—	R	27,30
Behenic, oleic, stearic acids, behenyl behenate, palmityl behenate	40–60	81–253	CO_2	—	S	39
Oleic and linoleic acids, methyl oleate, methyl linoleate	40–60	41–284	CO_2	—	R	28
	20–197		CO_2	—	R	29
Methyl oleate	50–80	23–122	Ethane	—	R	
Glycerides						
Tripalmitin, tristearin	40–60	130–310	CO_2	—	R	*
Tributyrin, trilinolein, triolein, tripalmitin	40–80	81–253	CO_2	—	S	39
Triolein	25–75	200	CO_2	—	R	31
Triolein, tristearin	40–60	200–300	CO_2	—	R	33
Triolein	50,70	150–300	CO_2	—	R	27
Monoolein	35,60	104–190	CO_2	—	R	30
Tripalmitin	22–157	3–140	Propane	—	S	41
Triolein, tristearin	85–95	16–21	Propane	—	S	42

[a]E = ethanol, D = dichloromethane, M = methanol.
*Hampson and Maxwell, unpublished results.

vapor- and liquid-phase solubilities can be studied using static systems (28,29). Where comparisons are available for lipids studied using both systems, agreement between the results, in general, has been good (26).

In recent years, investigators have initiated an impressive number of studies in which the solubilities of lipids in various supercritical fluids have been measured. They have used a wide variety of experimental systems to compile their results. Solubility measurements have now been made on selected solutes from every lipid class. Examples of comparative studies using different techniques are limited, and in many instances investigators have not compared their results with those in the literature. In the interest of improving the accuracy of lipid solubility data in SF solvents, future studies should include comparisons with the literature data.

References

1. Hanny, J.B., and J. Hogarth, *Proc. R. Soc. London 29:*324 (1879).
2. Villard, P., *J. Phys. 5:*455 (1896).
3. Francis, A.W., *J. Phys. Chem. 58:*1099 (1954).
4. McHugh, M.A., and V.J. Krukonis, *Supercritical Fluid Extraction: Principles and Practice,* Butterworth-Heinemann, Boston, MA, 2nd edn., 1994, p. 85.
5. Kramer, A., and G. Thodos, *J. Chem. Eng. Data 33:*230 (1988).
6. Van Leer, R.A., and M.E. Paulaitis, *J. Chem. Eng. Data 25:*257 (1980).
7. Iwai, Y., T. Fukuda, Y. Koga, and Y. Arai, *J. Chem. Eng. Data 36:*430 (1991).
8. Nilsson, W.B., E.J. Gauglitz, and J.K. Hudson, *J. Am. Oil Chem. Soc. 68:*87 (1991).
9. Dobbs, J.M., J.M. Wong, and K.P. Johnson, *J. Chem. Eng. Data 31:*303 (1986).
10. McHugh, M.A., and M.E. Paulaitis, *J. Chem. Eng. Data 25:*326 (1980).
11. Wong, J.M., and K.P. Johnson, *Biotech. Prog. 2:*29 (1986).
12. Kramer, A., and G. Thodos, *J. Chem. Eng. Data 34:*184 (1989).
13. Foster, N.R., S.L.J. Yun, and S.S.T. Ting, *J. Supercrit. Fluids 4:*127 (1991).
14. Pearce, D.L., and P.J. Jordan, in *Proceedings of the 2nd International Conference on Supercritical Fluids,* May 20–22, Boston, MA, edited by M.A. McHugh, Johns Hopkins University, Baltimore, MD, 1991, Vol. 2, p. 478.
15. Favati, F., J.W. King, J.P. Friedrich, and K. Eskins, *J. Food Sci. 53:*1532 (1988).
16. Schmitt, W.J., and R.C. Reid, *Chem. Eng. Comm. 64:*155 (1988).
17. Lorenzo, T.V., S.J. Schwartz, and P.K. Kilpatrick, *Proceedings of the 2nd International Conference on Supercritical Fluids,* May 20–22, Boston, MA, edited by M.A. McHugh, Johns Hopkins University, Baltimore, MD, 1991, Vol. 2, p. 297.
18. Czubryt, J.J., M.N. Myers, and J.C. Giddings, *J. Phys. Chem. 74:*4260 (1970).
19. Ohgaki, K., I. Tsukahara, K. Semba, and T. Katayama, *Intl. Chem. Eng. 29:*302 (1989).
20. Lee, C.H., E. Kosal, and G.D. Holder, *Proceedings of the 2nd International Conference on Supercritical Fluids,* May 20–22, Boston, MA, edited by M.A. McHugh, Johns Hopkins University, Baltimore, MD, 1991, Vol. 2, p. 308.
21. Stahl, E., and A. Glatz, *Fette Seifen Anstrichm. 86:*346 (1984).
22. Bamberger, T., J.C. Erickson, and C.L. Cooney, *J. Chem. Eng. Data 33:*327 (1988).
23. Liang, J.H., and A.I. Yeh, *J. Am. Oil Chem. Soc. 68:*687 (1991).
24. Ikushima, Y., K. Hatakeda, S. Ito, N. Saito, T. Asano, and T. Goto, *Ind. Eng. Chem. Res. 27:*818 (1988).

25. Ikushima, Y., N. Saito, K. Hatakeda, S. Ito, T. Asano, and T. Goto, *Chem. Ltr.* (Japan) 1789 (1985).
26. McHugh, M.A., A.J. Seckner, and T.J. Yogan, *Ind. Eng. Chem. Fundam. 23:*493 (1984).
27. Bharath, R., T. Adschiri, H. Inomata, K. Arai, and S. Saito, *Proceedings of the 2nd International Conference on Supercritical Fluids,* May 20–22, Boston, MA, edited by M.A. McHugh, Johns Hopkins University, Baltimore, MD, 1991, p. 288.
28. Zou, M., Z.R. Yu, P. Kashulines, S.S.H. Rizvi, and J.A. Zollweg, *J. Supercrit. Fluids 3:*23 (1990).
29. Cheng, H., J.A. Zollweg, and W.B. Streett, in *Supercritical Fluid Science and Technology,* edited by K.P. Johnston and J.M.L. Penninger, American Chemical Society, Washington, DC, 1989, p. 86.
30. King, M.B., D.A. Alderson, F.H. Fallah, D.M. Kassim, K.M. Kassim, J.R. Sheldon, and R.S. Mahmud, in *Chemical Engineering at Supercritical Fluid Conditions,* edited by M.E. Paulaitis, J.M.L. Penninger, R.D. Gray, Jr., and P. Davidson, Ann Arbor Science, Ann Arbor, MI, 1985, p. 31.
31. King, M.B., T.R. Bott, M.J. Barr, and R.S. Mahmud, *Sep. Sci. Tech. 22:*1103 (1987).
32. Zehnder, B., and Ch. Trepp, *Proceedings of the 2nd International Conference on Supercritical Fluids,* May 20–22, Boston, MA, edited by M.A. McHugh, Johns Hopkins University, Baltimore, MD, 1991, p. 329.
33. Brunetti, L., A. Daghetta, E. Fedeli, I. Kikic, and L. Zanderighi, *J. Am. Oil Chem. Soc. 66:*209 (1989).
34. Inomata, H., T. Kondo, S. Hirohama, and K. Arai, *Fluid Phase Equil. 46:*41 (1989).
35. Schäfer, K., and W. Baumann, *Fresenius Z. Anal. Chem. 332:*122 (1988).
36. Cygnarowicz, M.L., R.J. Maxwell, and W.D. Seider, *Fluid Phase Equil. 59:*57 (1990).
37. Maxwell, R.J., J.W. Hampson, and M. Cygnarowicz-Provost, *LC GC 9:*788 (1991).
38. Dandge, D.K., J.P. Heller, and K.V. Wilson, *Ind. Eng. Chem. Prod. Res. Dev. 24:*162 (1985).
39. Chrastil, J., *J. Phys. Chem. 86:*3016 (1982).
40. Hixson, A.W., and A.N. Hixson, *Trans. Amer. Inst. Chem. Eng. 37:*927 (1941).
41. Coorens, H.G.A., C.J. Peters, and J. De Swaan Arons, *Fluid Phase Equil. 40:*135 (1988).
42. Hixson, A.W., and J.B. Bockelmann, *Trans. Amer. Inst. Chem. Eng. 38:*891 (1942).

Chapter 3

Supercritical Fluid Extraction (SFE) of Oilseeds/Lipids in Natural Products

R. Eggers

Technische Universität Hamburg—Hamburg Arbeitsbereich Verfahrenstechnik II
Postfach 90 14 03, D-2100, Hamburg, 90 Germany

In the past four decades many potential applications of SFE for recovery and refining of oils and fats from natural substances have been investigated. These applications have resulted in patents on the extraction, refining, and fractionation of oil from natural substances by means of supercritical gases (1–5). Applications of SFE to different types of oilseeds are numerous, and these are listed in Table 3.1 (6–49).

World production of vegetable and animal fats has increased during recent decades (Table 3.2), while at the same time the price of the classic extracting agent, hexane, increased. Concern over residues in oil-derived products has also catalyzed a search for alternative processing methods such as SFE. In comparison with established methods, SFE has some important advantages, particularly its ability to achieve products that are completely free from processing residues. A disadvantage, however, is that the equilibrium solubility is low compared to that with hexane, especially in the case of CO_2, which is the preferred extracting agent because it is benign and nonflammable (Fig. 3.1). In addition, conventional solid/liquid extraction uses continuously working percolation extractors, whereas SFE of oilseeds must be performed as a high-pressure process semicontinuously; therefore, the processing of soybeans or rapeseed must employ a great number of large pressure vessels involving a series of chargings and dischargings.

Phase Equilibria and Mass Transfer in the SFE of Oilseeds

The solubility curve illustrated in Fig. 3.1 depicts the solubility of soybean oil in CO_2. The solubility increases with fluid density up to approximately 300 bar, i.e., as pressure increases or temperature decreases. Above approximately 300 bar, the solubility further increases with pressure and also directly with rising temperature, up to a maximum at approximately 1000 bar. This fundamental behavior is observed for all vegetable and animal fats in densified CO_2, with only slight solubility differences among the various oilseeds, in the lower pressure range (Fig. 3.2) (7,18,32,50,51). The reason for this observed behavior is that the fatty acid composition of each oil is slightly different, resulting in different molecular weights for the respective oils.

TABLE 3.1 Survey of Publications Concerning SFE of Oilseeds

Material	Characteristics	Year of publication	Ref.
Soy, rape, sunflower	CO_2; 20 bar; 20–50°C	1979	6
Soy, rape, sunflower	Liquid CO_2; crushing of materials; fractionated separation	1980	7
Soy	CO_2; 360 bar; 50°C; quality aspects	1982	8
Soy	CO_2; 570 bar; 50°C; analysis of extract composition	1982	9
Soy, corn germ, cottonseed	CO_2; 785 bar; 70°C; influence of humidity; particle diameter; solvent ratio; geometry of extractor	1984	10
Soy, cottonseed	CO_2; 570 bar; 50°C; influence of humidity, particle diameter; analysis of fatty acid composition	1984	11
Soy, corn germ, cottonseed	CO_2; >550 bar; >60°C; solubility	1984	12
Soy	CO_2; 350 bar; 40°C; behavior of proteins	1984	13
Soy	CO_2; 570 bar; 50°C; oil quality; comparisons to conventional oils; oxidative stability	1985	14
Soy, rape	CO_2; gas mixtures; entrainers; 300 bar; 60°C	1986	15
Soy, corn germ, cottonseed	CO_2; 860 bar; 90°C; oxidative stability; accompanying substances of oil in the extract	1989	16
Soy, rape, sunflower	CO_2; 500 bar; 60°C; quality of oil; comparison to hexane extraction	1989	17
Soy, rape	CO_2; propane; gas mixtures; influence of pressure and temperature	1989	18
Soy	CO_2; ethylene; 200 bar; 50°C	1989	19
Rape	CO_2; 750 bar; 80°C; mechanical conditioning; humidity; energy study	1984	20
Rape	CO_2; 350 bar; 40°C; modeling of SFE	1984	21
Rape	CO_2; 70 bar; 25°C; modeling of mass transfer	1984	22
Rape	CO_2; propane; modeling of mass transfer	1984	23
Rape	CO_2; Propane; 750 bar; 80°C; mechanical conditioning; energy study	1985	24
Rape	CO_2; 290 bar; 40°C; solvent ratio; geometry of extractor; modeling of mass transfer; energetic study	1984; 1985	25,26
Rape	CO_2; 360 bar; 55°C; modeling of mass transfer	1986	27
Rape	CO_2; 200 bar; 75°C; modeling of mass transfer	1987	28
Rape	CO_2; 360 bar; 90°C; mechanical conditioning	1988	29

Corn germ	CO_2; 570 bar; 50°C; quality of meal	1984	30
Corn germ	CO_2; 860 bar; 90°C; quality of oil	1984	31
Corn germ, lupine seed	CO_2; 450 bar; 60°C; fractionated separation	1984	32
Corn germ	CO_2; 450 bar; 60°C; quality of oil	1987	33
Corn germ	CO_2; 500 bar; 50°C; fractionated separation; analysis of accompanying substances	1991	34
Cottonseed	CO_2; 1070 bar; 80°C	1983	35
Cottonseed	CO_2; 1070 bar; 80°C; quality of oil	1984	36
Cottonseed	CO_2; 700 bar; 135°C; separation of gossypol by fractionation and refining	1986	37
Lupine seed	CO_2; 300 bar; 40°C; analysis of accompanying substances	1981	38
Lupine seed	CO_2; 500 bar; 70°C; separation of alkaloids	1986	39
Oenothera seed	CO_2; 700 bar; 80°C; fractionated separation countercurrent refining	1986	37
Oenothera seed	CO_2; 800 bar; 40°C	1987	33
Oenothera seed	CO_2; 500 bar; 50°C; analysis and quality of oil	1988	39
Oenothera seed	CO_2; 200–700 bar; 40–60°C; kinetics	1991	40
Jojoba	CO_2; 660 bar; 80°C	1984	32
Jojoba	CO_2; 660 bar; 80°C; quality of waxes	1987	33
Wheat germ, wheat bran	CO_2; 570 bar; 50°C; quality of oil	1984	10
Wheat germ	CO_2; 300 bar; 40°C; tocopherol content	1985	41
Wheat germ	CO_2; 250 bar; 40°C; coupling of SFE-SFC	1989	42
Wheat bran	CO_2; 350 bar; 40°C; quality of bran	1986	43
Flax	CO_2; 600 bar; 60°C; quality of meal	1987	33
Copra	CO_2; 900 bar; 60°C	1983	44
Castor	CO_2; 600 bar; 80°C; countercurrent refining	1986	37
Peanuts	CO_2; 710 bar; 70°C; quality of meal	1984	11
Olive	CO_2; 110–200 bar; 30–40°C; deacidification	1991	45
Oat	CO_2; antioxidative properties of SFE oat oil	1990	46
Sage	CO_2; antioxidative activity of sage extract	1991	47
Algae	CO_2; 170–310 bar; 40°C; quality of SFE oils	1989	48
Eucalyptus	Ethylene; 55–65 bar; 15°C; countercurrent refining	1990	49

TABLE 3.2 Major Oils and Fats, World Production by Type (Thousands of Metric Tons)

	1990	1985	1980	1975	1970	1965	1960
Soybean oil	15,400	13,939	13,358	8,025	6,381	3,999	3,355
Cotton oil	3,800	3,811	3,030	2,929	2,511	2,733	2,324
Groundnut oil	3,600	3,144	2,801	2,672	2,730	2,806	2,282
Sunflower oil	7,740	6,549	5,044	3,900	3,491	3,077	1,788
Rapeseed oil	7,183	5,993	3,484	2,392	1,778	1,500	1,164
Sesame oil	590	589	502	502	499	500	419
Corn oil	1,230	1,036	765	580	475	436	370
Olive oil[a]	1,780	1,724	1,586	1,562	1,388	1,015	1,295
Coconut oil	3,350	2,630	2,722	2,593	2,019	2,032	1,949
Palm kernel oil	1,525	935	633	474	393	412	432
Palm oil	11,214	6,890	4,621	2,858	1,796	1,446	1,306
Butter (fat content)	6,580	6,314	5,706	5,346	4,793	4,796	4,215
Lard	5,696	5,278	5,025	4,268	3,901	3,613	3,142
Fish oil	1,540	1,392	1,211	1,059	1,078	783	470
Linseed oil	670	681	702	611	926	1,001	886
Castor oil	390	402	363	363	358	330	199
Tallow	6,800	6,558	6,345	5,270	4,907	4,238	3,436
Total	79,088	67,865	57,898	45,404	39,424	34,717	29,032
% increase from 5 years before	17%	17%	27%	15%	14%	20%	

[a]Season beginning November of preceding years.

Results obtained from modeling equilibrium solubilities by del Valle (50) are illustrated in Fig. 3.2 and fit the experimental data quite well. However, these equations do not describe the maxima that appear at very high pressures:

$$\ln c = 40.361 - \frac{18,708}{T} + \frac{2,186,840}{T^2} + 10.724 \ln \rho \qquad [3.1]$$

where

c = concentration of oil in CO_2 (g/L)
T = temperature (K)
ρ = density of CO_2 (g/L)

In comparison with another relation derived by Chrastil (52), the above equation is more suited to correlating the experimental data of different oilseeds under the following conditions:

- Soybean oil [data from Stahl et al. (7)]:
  293 K < T < 313 K
  150 bar < P < 340 bar

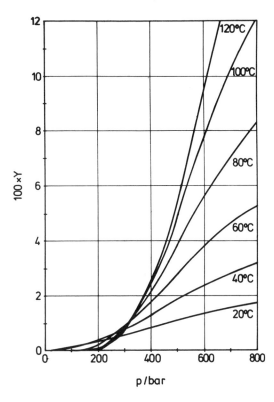

Fig. 3.1. Solubility of soybean oil in CO_2.

- Sunflower oil [data from Stahl et al. (7)]:
  $T = 313$ K
  230 bar $< P <$ 670 bar
- Soy oil [data of Friedrich and List (8)]:
  313 K $< T <$ 343 K
  200 bar $< P <$ 680 bar
- Cottonseed oil [data from Friedrich et al. (10,35)]:
  313 K $< P <$ 670 bar
- Corn germ oil [data from Friedrich and Pryde (10)]:
  $T = 353$ K
  $P = 264$ bar

Equation (3.1) describes the solubility for many oilseeds in SC-CO_2 adequately for general use in process engineering.

The kinetics of mass transfer during the SFE of vegetable oils from fixed beds are shown in Fig. 3.3. The initial portion of the extraction curve is determined by equilib-

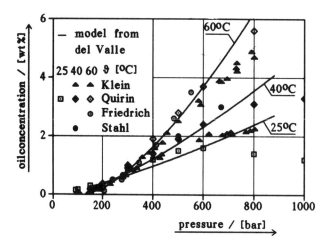

Fig. 3.2. Solubility of various seed oils in CO_2 compared with model.

rium constraints; i.e., the mass transfer resistance is in the fluid phase and results in a linear relationship between the mass of product extracted and the quantity of fluid used. The characteristics of this extraction curve depend on the primary preparation of the oilseeds (mechanical and thermal conditioning) (53), which accelerates the release of the oil bound in the plant cells (Fig. 3.4). In the second section of the curve (Fig. 3.2), mass transfer is determined by the diffusional resistance in the solid matrix.

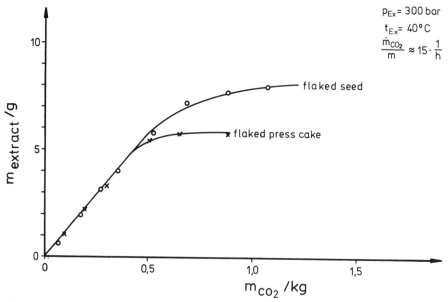

Fig. 3.3. High-pressure extraction kinetics of rapeseed.

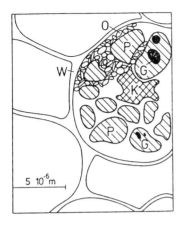

G – globoid
K – nucleus
O – oil droplet
P – protein
W – cell wall

Fig. 3.4. Microsection of rapeseed.

Modeling of the mass transfer during the SFE of oilseeds is similar to that used to describe deoiling of a fixed bed by conventional extraction practices using hexane. This is a difficult problem for several reasons. The material has irregular structure and complicated geometry. In addition, the structure of the natural substance to be extracted changes during the extraction process. This requires measurement of the porosity of the compacted solids, which can not be done during the extraction because of the high process pressure. Also, it is questionable whether the models used for conventional solid-liquid extraction (54,55) can be applied to the SFE of oilseeds; for example, the Schmidt number, which is important in describing the initial portion of the extraction curve, is around 10 (56) for densified gases, thus lower than for liquid solvents.

A one-dimensional mass transfer model has been developed for the SFE of rapeseed oil (57). The integral mass balances for the extractor are as follows:

Fluid: $\varepsilon\rho\,\dfrac{\partial y}{\partial t} + \rho u\,\dfrac{\partial y}{\partial h} = Ak(y^* - y)$ [3.2]

Solid: $(1 - \varepsilon)\rho_s\,\dfrac{\partial x}{\partial t} = -Ak(y^* - y)$ [3.3]

where:

t = time (s)
Ak = volumetric mass transfer coefficient (kg/m^{3}s)
ε = void fraction of the bed of the seed
y^* = solubility of oil in CO_2 (kg oil/kg CO_2)
y = concentration of oil in CO_2 (kg oil/kg CO_2)
x = concentration of oil in seed (kg oil/kg oil free seed)

u = superficial solvent velocity (mm/s)
ρ_s = density of seed (kg/m³)
h = axial distance along the bed of the seed (mm)
A = specific surface area of seeds (m²/m³)
ρ = density of solvent (kg/m³)

The product of the specific surface area and mass transfer coefficient, Ak, can be obtained by iterative regression to measured values and represents a value for the whole fixed bed. The time and position dependencies of the oil content in the gas phase and of the remaining oil in the solid can also be computed from the model.

Mass transfer in the linear portion of the extraction curve can also be described by application of Sherwood's empirical relations (23) coupled with the evaluation of the kinetic equilibrium model:

$$\frac{y_h}{y^*} = 1 - \exp(-kA\tau) \qquad [3.4]$$

where

y_h = solubility of oil at the outlet of the extractor
k = modified mass transfer coefficient
τ = residence time of the solvent in the fixed bed

Using Eq. (3.4), it is very easy to determine the product kA for the initial linear portion of the extraction curve. By measuring c during the extraction of rapeseed oil, King (22,28,58) was able to explain the second, asymptotic section of the extraction curve. The nonequilibrium kinetics not only depend on the diffusion resistance in the solid but also on the changing value of y^* during the course of the extraction. This change in time in the SFE of seed oil appears because the triglycerides with lower molecular weights dissolve preferentially in the SF.

One can consider the mass transfer during SFE of oilseeds as analogous to the drying of porous material. Initially the mass transfer kinetics remain constant. As the mass transfer decreases, an increasing resistance to extraction appears, because the free oil on the surface is incorporated back into the solid. In ref. (25) a mass transfer model of the drying technique is successfully adapted to SFE of rapeseed oil. In order to model the mass transfer during the SFE of oilseeds reliably, it is necessary to describe the structure of the solid after cracking of the cells. For corn germ, Fig. 3.5 shows clearly how the porosity changes during the extraction process.

Experimental Results in SFE of Oilseeds

Parameters that influence the SFE of oilseeds include both specific features of the material being extracted and the actual processing parameters (see Table 3.3).

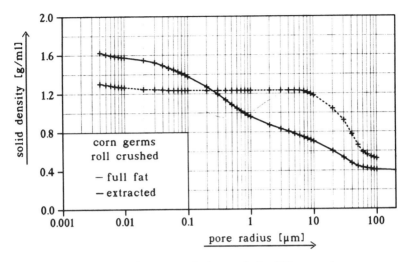

Fig. 3.5. Porosimetry of corn germ before and after SFE processing.

TABLE 3.3 Processing Parameters for the SFE of Oilseeds

Process Parameters	Specific Features of the Material
Extraction pressure	Bulk density
Extraction temperature	Oil content
Separation pressure	Specific surface
Specific solvent flow	Pore diameter
Superficial solvent velocity	Porosity
Vessel geometry	Particle geometry
Residence time	Moisture
Residual oil content	Void faction
Moisturizing	

    The oil contents of some oilseeds are summarized in Fig. 3.6. In order to improve the mass transfer, it is generally desirable to crack the oil-containing cell structure of the seed material before beginning the extraction process. Figures 3.7 and 3.8 show how different mechanical conditioning processes influence the SFE of rapeseed and corn germ, respectively (24,34). Although their particle geometries are totally different (spherical geometry vs. flat flakes), the mass transfer is favored by flaking the seed before the extraction process. For oil-rich materials, such as rapeseed, corn germ, or sunflowerseed, a prior deoiling in cage screw presses results in a good cracking of the seed material because of the high shearing and frictional forces. Flaking the press cake from the screw press improves the mass transfer of oil in subsequent SFE operations. The thickness of the flakes must be less than 0.3 mm in order to get optimum extraction results (11).

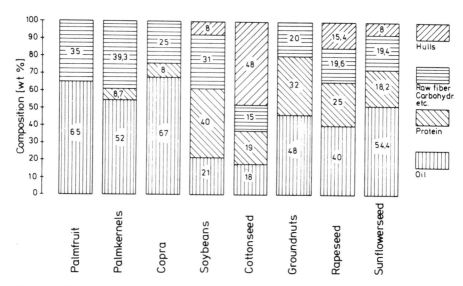

Fig. 3.6. Composition of oilseeds.

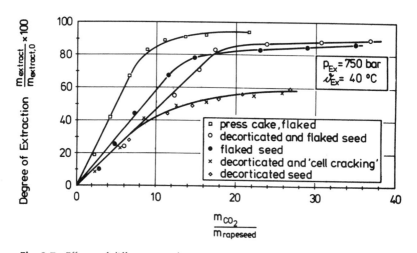

Fig. 3.7. Effects of different conditioning processes on SFE of rapeseeds.

A high moisture content in the oilseeds before starting SFE is a disadvantage. The influence of moisture on oil mass transfer is negligible in the range between 3 and 12% by weight (11,20). For extraction pressures above 200 bar the solubility of triglycerides in CO_2 is much higher than that of water (see Fig. 3.9) (59). Only at the

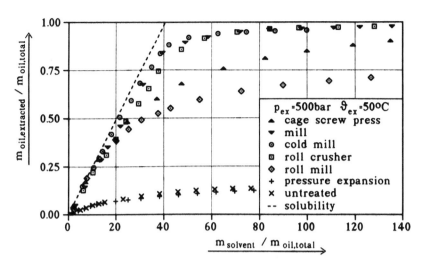

Fig. 3.8. Effects of different conditioning processes on SFE of corn germ.

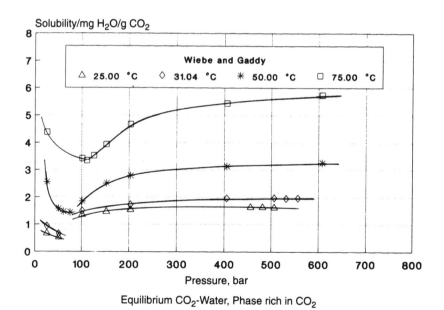

Fig. 3.9. Solubility of water in CO_2.

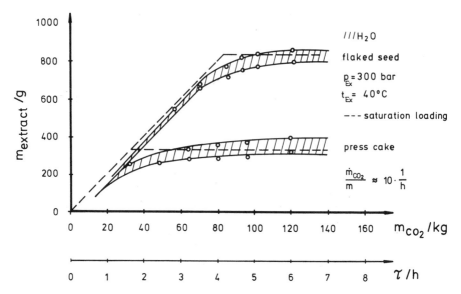

Fig. 3.10. High-pressure extraction of rapeseed oil showing the water content of   the extract.

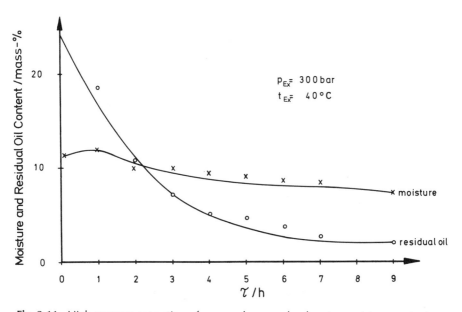

Fig. 3.11. High-pressure extraction of rapeseed press cake showing moisture content.

end of the extraction will the water content of the extract increase considerably (Fig. 3.10), but the moisture content in the meal will be reduced substantially (Fig. 3.11). In contrast to the extraction of caffeine from raw coffee, where water swells the beans and the enhanced solubility of caffeine in water influences the mass transfer during SFE, additional moisture in oilseeds leads to a prolongation of the extraction.

The effect of the seed properties, such as bulk density, size of particles, particle geometry, and porosity, on mass transfer during SFE of oilseeds must also be considered. If the bulk density is too high—for instance, due to compression of the charged material or to small particle size—inhomogeneous extraction will occur because of channeling in the fixed bed. Conversely, if the bulk density is too low, larger pressure vessels will be required.

Because of their small specific surface area, large particles lead to a distinct, diffusion-dominated extraction and long processing times. Therefore, geometries that deviate from the form of a sphere are favored (e.g., flaked material). For the diffusion process in the solid, the microstructure—porosity and diameter of pores—can be of great importance. As can be seen in Fig. 3.12, the internal porosity of oilseeds is nearly exclusively determined by macropores with diameters from 1 µm up to 20 µm. The porosity was measured by means of mercury pressure porosimetry on pelletized soy powder. Depending on the degree of pressurization, the pure densities of the solid are between 1.2 g/cm^3 and 1.6 g/cm^3 and the porosities are between 0.4 and 0.6.

The influences of the process parameters listed in Table 3.3 have been extensively investigated. They may be summarized as follows:

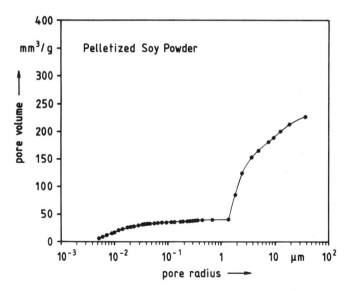

Fig. 3.12. Porosity and pore size of pelletized soy powder.

1. Increasing pressure favors extraction.

2. Up to an interim range between 300 bar and 350 bar, extraction is favored at lower temperatures, whereas at higher pressures (>350 bar), the extraction increases considerably with temperature. The reason for this crossover effect is that the influence of the vapor pressure of the oil components becomes more significant. Also, at higher temperatures the mass transfer of the oil is favored because of the viscosity characteristics of the oil and the solvent.

3. Separation of the oil in the separator should be carried out under supercritical conditions to save energy (60). The solubility of triglycerides in CO_2 below 160 bar—especially in case of increased separation temperatures—is so low that the oil can be easily separated. Another attendant advantage is that the need for a condenser to liquefy the CO_2 ahead of the pump is eliminated. There is no phase change in the CO_2, so there is no risk of cavitation in the plunger pump. However, the higher compressibility of the supercritical CO_2 compared to liquid CO_2 does reduce the pump efficiency (61).

4. The specific mass flow must be optimized per unit of weight of the oilseed to be extracted. For as long as possible, the loading of the fluid should nearly correspond to that predicted by phase equilibrium. However, the residence time of the solvent in the extraction vessel should not be too long, resulting in a long extraction time. For SFE of oilseed with CO_2, specific mass flows between 10 kg CO_2/kg • h and 50 kg CO_2/kg • h, at superficial velocities of the solvent between 1 mm/sec and 5 mm/sec, have been utilized.

5. The pressure vessel geometry of the extractor is predominantly determined by the charging volume of the solids to be extracted. The use of smaller vessel diameters is favored because homogenous extractions are more easily performed. Normal values for SFE of oilseeds are $4 < H/D < 6$.

The SFE of corn germ with CO_2 can serve as an example to illustrate the influence of the process parameters described. The results described in the following paragraph were obtained in the SFE plant shown in Figs. 3.13 and 3.14, featuring a three-stage separator.

The influence of extraction pressure is illustrated in Fig. 3.15. The dashed lines are the theoretical extraction curves based on phase equilibrium considerations. Higher extraction pressures increase the oil loading of the SC-CO_2, thereby reducing the time of extraction. At constant mass flow, the residence time in the extractor becomes longer with increasing pressure. For constant mass flow, the influence of extraction temperature is seen in Fig. 3.16. At an extraction pressure of 500 bar the slope of the kinetic curve in the linear portion increases as predicted from solubility considerations (dashed line).

Although the results shown in Figs. 3.15 and 3.16 were obtained on corn germ comminuted with a roll crusher before extraction, the influence of the solvent flow

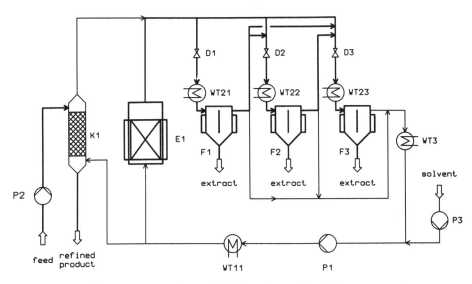

Fig. 3.13. High-pressure extraction and refining plant with multistage separation.

Fig. 3.14. SFE pilot plant with three-stage separation.

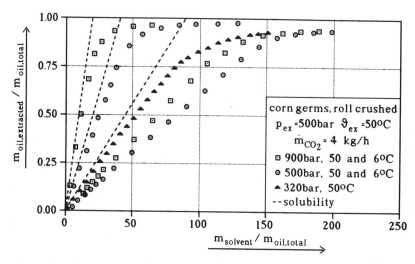

Fig. 3.15. The influence of extraction pressure on the SFE of corn germ oil.

can be shown on milled corn germ as shown in Fig. 3.17. Increasing the solvent flow reduces the residence time but increases the solvent requirement (kg solvent/kg solid). These contradictory effects show why the solvent flow must be optimized for obtaining an efficient extraction (Fig. 3.18) (62).

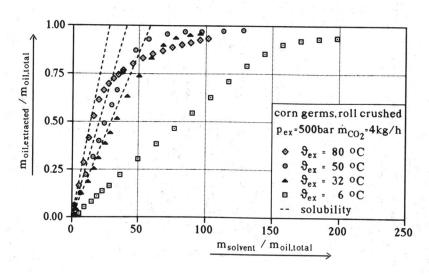

Fig. 3.16. The influence of extraction temperature on the SFE of corn germ oil.

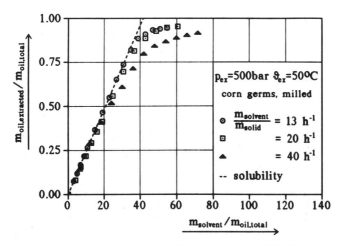

Fig. 3.17. The influence of specific mass flow on SFE of corn germ oil.

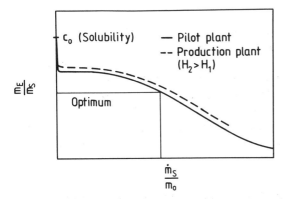

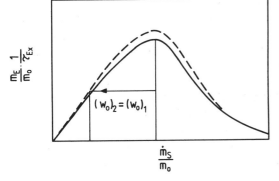

Fig. 3.18. Optimization of solvent ratio.

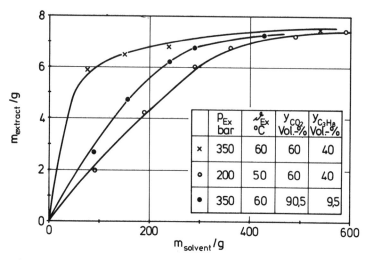

Fig. 3.19. High-pressure extraction of rapeseed press cake with a CO_2/propane gas mixture.

To avoid the use of high CO_2 pressures, gas mixtures with CO_2 as the other binary component have been investigated. Propane, nitrous oxide, and freons have been used as cofluids (18,24,63). A mixture of CO_2 and propane allows complete miscibility with the triglycerides at pressures below 300 bar and at temperatures up to 70°C (64). Figure 3.19 shows the influence of different mixtures of CO_2 and propane at different pressure on the oil extraction from rapeseed press cake (24). The use of hydrocarbons as extraction gases, however, impairs the selectivity of the triglycerides. Pigments and other lipophilic components are also extracted, necessitating further refining of the extracted oil.

Evaluation of SFE-Extracted Oils: Refining and Fractionation

SFE using CO_2 yields oils having superior properties. Oils extracted by conventional methodologies require further refining to remove phospholipids, free fatty acids, and oxidation products (see Table 3.4). SC-CO_2 can also be used to refine crude oils extracted by conventional methods. This refining with compressed CO_2 can be carried out continuously in countercurrent columns.

The somewhat lower oil yield obtained by CO_2 extraction is due to the considerably lower solubility of the phospholipids and glycolipids in CO_2. Therefore, it is possible to eliminate the degumming step in the post-SFE refining operation.

Frequently there is less fatty acid in the CO_2 extract, although this varies somewhat for the different oilseeds. For example, in the case of cottonseed, more fatty acid is found in CO_2-extracted oil than in hexane-derived oil. The resultant fatty acid con-

TABLE 3.4 Comparison of Oil Refining by Classical Methods and SFE Processing

Classical refining of vegetable oil	Refining of vegetable oil by SFE	
Raw oil (hexane-extracted; press oil)	Raw oil (hexane-extracted; press oil)	SFE-produced oil
• Separation of lecithin	• SFE in countercurrent columns (multistage)	• SF-deacidification in countercurrent columns
• Degumming	• Bleaching	• Bleaching (slight)
• Deacidification		
• Bleaching		
• Deodorizing		
• Refined oil		

tent depends strongly on the conditions under which extraction is performed as well as the storage of the material prior to SFE. For tocopherols, the CO_2 extract of cottonseed contains less tocopherol than the hexane extract, whereas the reverse is true of soybeans. The fatty acid contents in CO_2 and hexane extracts are the same.

Sensory evaluation of the extracted oils indicates that the color of the CO_2 extract is lighter, and that odor and taste are milder. Refined oils, by comparison, do not show any difference between CO_2 and hexane extracts. These features of oils obtained by SFE permit the omission of several refining steps after the extraction with CO_2; doing so reduces the consumption of alkali and minimizes loss of neutral lipids.

One disadvantage in the resulting oil compared to hexane-extracted oil is its reduced oxidation stability compared to the phospholipid content (13). Cottonseeds contain gossypol, a toxic sesquiterpene that must not be in the oil. In CO_2-extracted oil there is only a tenth of the gossypol that is in hexane-extracted oil. The toxic alkaloids of the sand lupine can better be dissolved in the CO_2-oil mixture (36).

The present author's own experiments on SFE of corn germ oil and oil from rapeseed press cake have confirmed the good quality of CO_2-extracted oil. Table 3.5 presents the data from the analysis of corn germ oil that was extracted from roll-crushed corn germ at 500 bar and 50 °C and collected in the first separator of a two-stage separation stage under supercritical conditions. The coextracted moisture is carried to the second separator in dissolved condition and separates under subcritical conditions (34). By that method an oil is obtained that is nearly free from water, and its features compare well with those of a refined edible oil; see Table 3.5.

The content of tocopherol in the extracted oil depends directly on the corn germ used in the SFE process. The tocopherols and sterines will be extracted completely along with the main fraction of oil. Water content and color number are similar to those of a finished oil. Refining of SFE-produced oils consists mostly of deacidification. The tocopherols and sterines remain in the refined oil.

TABLE 3.5 Comparison of SFE and Refined Corn Germ Oil

Comparison of quality	Raw oil produced by SFE with 2-stage separation	Refined oil from solvent extraction
Free fatty acid (wt%)	0.4–0.6	< 0.05–0.1
Content of water (wt%)	0.05–0.1	< 0.05
Color number, Lovibond (red/yellow)	8/10–12	1–2/11–12
Sterines (mg/kg)	9–12	9–12
Tocopherols (mg/kg)	900	900
Sensory	Clear light yellow	Colorless

The quality of oil extracted with CO_2 from rapeseed press cake has been assessed in terms of the free fatty acid (FFA) content and peroxide value (POV) (65). The peroxide value, a measure of oxidative degradation, has values between 1 and 2 meq O_2/kg oil, lower than the values found for conventionally extracted oils. These low values are a consequence of the benign treatment implicit in cold prepressing and in the extraction step. The free fatty acid content in the extracted oil fluctuates between 0.5 and 6 wt% and decreases with extraction time. Phosphatides are practically insoluble in the supercritical gas, so the extracted rapeseed oil contains less than 5 ppm phosphatides. This confirms the advantage of SFE over the conventional extraction process, since degumming can be omitted from the subsequent refinement of the oil.

With respect to cake quality, protein solubility decreases only slightly in SFE from that of the starting press cake, whereas the sulfur content is increased from 0.96 to 1.24 wt%. The value for thioglucoside content rose from 0.04 wt% in press cake to 0.06% in extracted cake, within the limit prescribed by animal feed regulations (66).

Economic Aspects

The economics of processing oilseeds by SFE involve a considerable initial expenditure for capital equipment. Because this process must be conducted on a batchwise basis, the required pressure vessels can be quite numerous, large, and expensive compared to the costs associated with conventional extraction of oilseeds using hexane. As a consequence, current use of SFE has been in processing high-priced oils, such as evening primrose oil, that have medicinal value (62). Even an energy comparison of SFE with conventional liquid extraction can make SFE a more expensive process (65).

The specific energy requirement e_i (i.e., energy requirement per unit mass of starting material, rapeseed) of each process step i is given by the product of the specific solvent requirement (m_{CO_2}/m) and the enthalpy difference Δh_i involved in each process step:

$$e_i = (m_{CO_2}/m)\Delta h_i \qquad [3.5]$$

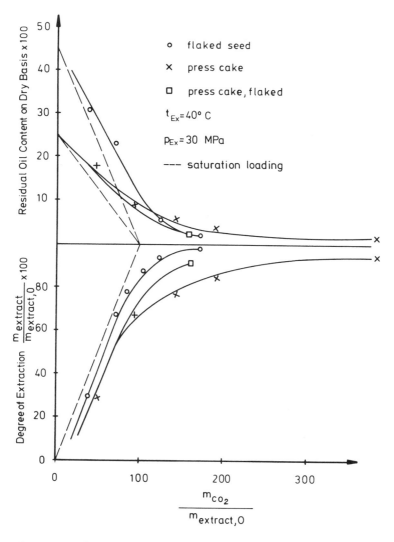

Fig. 3.20. High-pressure extraction of rapeseed oil.

The specific solvent requirement can be obtained from Figs. 3.20 and 3.21, since the ratio of the initial oil mass $m_{extract,0}$ to the mass of starting material m is known. For the energy calculation, the specific solvent requirement for high-pressure extraction, down to a residual oil content on dry basis of 2 wt%, is shown in Fig. 3.22. In the energy calculations, only the values for flaked rapeseed press cake of 35 kg CO_2/kg starting material at 300 bar and 40°C, and of 21.5 kg CO_2/kg starting material at 750

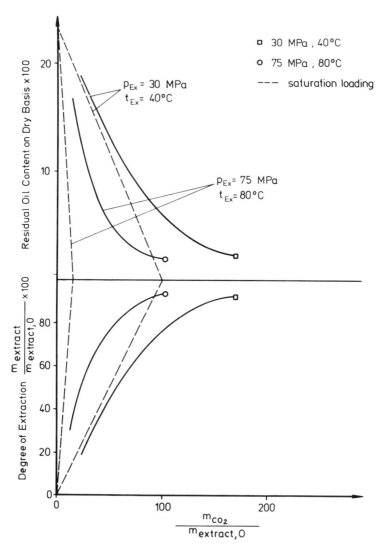

Fig. 3.21. High-pressure extraction of rapeseed press cake.

bar and 80°C, have been used, since the results for nonflaked rapeseed press cake and for flaked rapeseed were comparatively poor.

The relevant enthalpy differences are taken from the tables of the thermodynamic properties of CO_2 (34). The modes of operation noted in Table 3.6 have been computed and compared with results for conventional hexane extraction (68). The high-pressure extraction processes at 300 bar and at 750 bar are further subdivided according to the

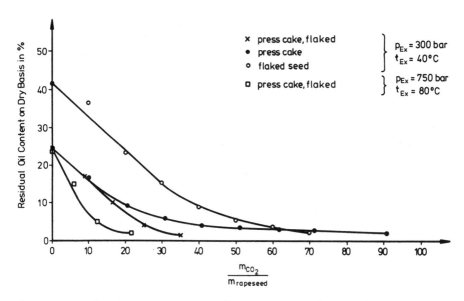

Fig. 3.22. Specific solvent requirement in the HP extraction of rapeseed.

TABLE 3.6 Energy Consumption in Various HP Extraction Cycles

	P_{Ex} (bar)	T_{Ex} (°C)	Separation	Operation	Mechanical energy (kWh/MT)	Heating (10^4kJ/MT)	Cooling (10^4kJ/MT)
A	Hexane extraction				75	75	42
B	300		Subcritical	Pump	277	431	518
C				Compressor	496	431	597
D	300		Supercritical	Pump	266	—	83
E				Compressor	278	—	87
F	750		Subcritical	Pump	438	160	305
G				Compressor	639	165	382
H	750		Supercritical	Pump	423	—	140
I				Compressor	469	—	156

methods of compression and of separation. The flowsheets for these process variants are described in Ref. 60.

The CO_2 can be compressed either in liquid form with a pump (assuming ηsv = 0.85) or in gaseous form with a compressor (assuming η_{sv} = 0.85). Separation can be achieved under subcritical conditions, 60 bar at 30°C, or supercritical conditions, 90 bar at temperatures exceeding 34°C (T_c = 31.06°C).

It should be noted that the table entries for mechanical energy requirement include the work involved in mechanical prepressing of the rapeseed. This quantity was determined to be 35 kWh/T seed (69) from a series of experiments on cold prepressing.

The calculated values per metric ton of rapeseed entering the screw press for prepressing lead to the following conclusions:

- High-pressure extraction of oilseed requires considerably more mechanical energy than does the conventional hexane extraction.
- The heat energy requirement favors supercritical separation, since the necessary heat energy may be introduced directly into the system via the temperature increase occurring during compression.
- The energy expenditure for cooling is higher than with hexane extraction, since the high-pressure cycles have been formulated without any consideration of heat recovery. Nevertheless, operating schemes using supercritical separation are favored here, too.

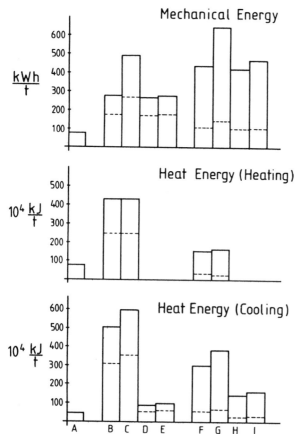

Fig. 3.23. Energy consumption in various HP extraction cycles.

In the ideal case, in which CO_2 is saturated with seed oil during the entire extraction period, the theoretical energy contributions, shown by dashed lines in Fig. 3.23, may be calculated for the operating schemes B through I. Operation at 750 bar, in particular then, becomes comparable to conventional hexane extraction. These energy considerations make it clear that high-pressure extraction of oilseed will become economical only if the extraction is conducted rapidly under conditions that maximize saturation solubility. This can only be achieved through continuous processing of oilseed. A proposal for such a process is presented in Ref. (70).

Continuous SFE of Oilseeds

A clear disadvantage of the deoiling of natural material by SFE using CO_2 is the necessity to produce large amounts of solid material in a batch process, filling and emptying the pressure vessels at atmospheric pressure. This results in

- Long production times, because of the need for additional process steps for filling, pressurizing, heating, recovery of solvent by depressurizing, and emptying of extracted seed.
- Extended times of production, due to decreasing solvent loading with time.

To overcome these disadvantages it is necessary to develop a device to achieve continuous transport of the solid material in and out of the extraction vessel.

A pilot plant for continuously deoiling rapeseed by SFE is shown in Fig. 3.24. The pressure in a cage screw press is normally used to separate the oil from the solid

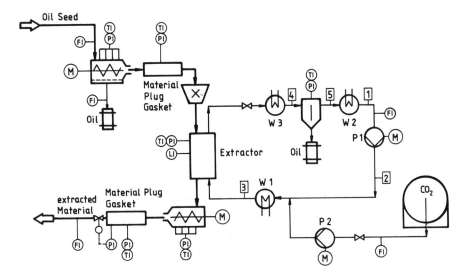

Fig. 3.24. Flow diagram of continuous SFE of oilseed.

mechanically. In the SFE plant the pressure in the solid material is used to ensure seal-ing on the pressure vessel. It is well known that the highest values of the pressure pro-files in cage screw presses are of the same order as the pressures required in the SFE process. The series of process steps as follows:

- Conditioning the material (cell disruption)
- Predeoiling by mechanical liquid-solid separation
- Production of a seed plug
- SFE of the predeoiled solid material
- Discharge of the solid material by an extruder

Figure 3.25 shows the system consisting of cage screw press, pressure vessel, and extruder.

The diagrams in Figs. 3.26 and 3.27 illustrate pressure profiles, measured in the pilot plant just described, for introducing rapeseed into the pressure vessel by cage screw press and for transporting extracted material out of the pressure vessel by the extruder machine. The vessel in this case is filled and pressurized with CO_2. The pres-sure profile in the cage screw press depends on the pressure in the connecting vessel. With increasing process pressures, the solid material will become more densified,

Fig. 3.25. Pilot plant for continuous transport of oilseed into and out of a pressure vessel.

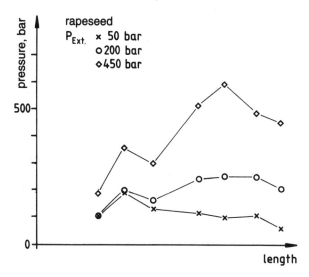

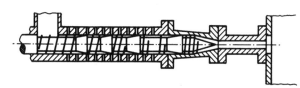

Fig. 3.26. Pressure profile for introducing the solid material into the high-pressure zone.

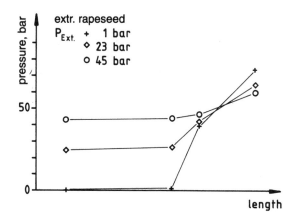

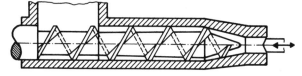

Fig. 3.27. Pressure profile for removing the solid material from the high-pressure zone.

resulting in smaller porosities and increasing pressure gradients in the extraction proper. This results in a continuous system to convey the solid particulate material into the pressure vessel in a controlled manner.

Porosities and diameters of capillaries in rapeseed, measured by mercury poros-imetry, are in the range of $0.15 < \varepsilon_p < 0.2$ and $20 \ \mu m < d_c < 50 \ \mu m$ depending on pressure difference and material used. The specific energy requirements increase from 350 kJ/kg at 50 bar up to 610 kJ/kg at 50 bar process pressure. These results are valid in the case of rapeseed, and the values represent overall energy demands, including mechanical prepressing and conveying the material into the pressure vessel.

The extracted solid material is removed out of the pressure vessel continuously in the direction of fluid flow. This fact is important in the design of the outlet system. A piston must be installed to act in opposition to the movement of the solid material, providing a pressure resistance. The piston may be controlled pneumatically, hydrauli-cally, or mechanically. The pressure exerted by the moving solid material must over-come the opposing piston pressure. It is possible to pump a bonding agent such as water through the moving solid material so that the conveying conditions are im-proved and the gradient of pressure is increased when the solid material is pushed through the gas-sealing section. This gas-sealing section keeps the pressure in the ves-sel constant.

Figure 3.26 shows pressure profiles measured when the extracted rapeseed is transported out of the vessel at different pressures. The profiles of pressure become flatter with high starting pressures. Compaction and development of high porosity in the solid material are avoided by transporting the solid mass through the extruder against the controlled pressure resistance. Otherwise, channeling will form in the material plug, or the plug will be thrown out of the gas-sealing section.

Porosities and diameters of capillaries of rapeseed in this case are in the range of $0.15 < \varepsilon_p < 0.4$ respective $20 \ \mu m < d_c < 60 \ \mu m$. They tend to have higher values compared to those recorded at the inlet system. The specific energy requirements decrease with higher process pressures and are in the range of 200 kJ/kg down to 100 kJ/kg, if extracted rapeseed is transported out of the CO_2 pressure vessel to atmo-spheric conditions.

References

1. Dickinson, J.T., U.S. Patent 2,660,590 (1947).
2. Palmer, G.H., and N.J. Fanwood, U.S. Patent 2,658,907 (1950).
3. Groll, H.P.A., German Pat. Appl. 1,079,636 (1953).
4. Zosel, K., German Pat. Appl. 1,493,190 (1964).
5. Vitzthum, O., and P. Hubert, German Pat. Appl. 2,127,596 (1970).
6. Schütz, E., Ph.D. Thesis, University of Saarbrücken (1979).
7. Stahl, E., E. Schütz, and H.K. Mangold, *J. Agric. Food Chem. 28:*153 (1980).
8. Friedrich, J.P., and G.R. List, *J. Agric. Food Chem. 30:*192 (1982).
9. Friedrich, J.P., and A.J. Heakin, *J. Am. Oil Chem. Soc. 59:*288 (1982).
10. Friedrich, J.P., and E.H. Pryde, *J. Am. Oil Chem. Soc. 61:*223 (1984).

11. Snyder, J.M., J.P. Friedrich, and D.D. Christianson, *J. Am. Oil Chem. Soc. 61:*1851 (1984).
12. Friedrich, J.P., U.S. Patent 4,466,923 (1984).
13. Stahl, E., K.W. Quirin, and R.J. Blagrove, *J. Agric. Food Chem. 32:*938 (1984).
14. List, G.R., and J.P. Friedrich, *J. Am. Oil Chem. Soc. 62:*82 (1985).
15. Peter, S., German Patent Appl. 3,429,416 (1986).
16. List, G.R., and J.P. Friedrich, *J. Am. Oil Chem. Soc. 66:*98 (1989).
17. Dakovic, S., J. Turkulow, and E. Dimic, *Fat Sci. Tech. 91:*116 (1989).
18. Gottschau, T., *GIT 33:*1133 (1989).
19. Eisenbach, W.O., in *Proceedings of the 2nd International Symposium on Supercritical Fluids,* Nice, 1989, p. 719.
20. Eggers, R., and W. Stein, *Fette Seifen Anstrichm. 86:*10 (1984).
21. Bulley, N.R., M. Fattori, and A. Meisen, *J. Am. Oil Chem. Soc. 61:*1362 (1984).
22. King, M.B., T.R. Bott, K. Kassim, M. Barr, and N. Sanders, in *Preprints of the International Symposium on High Pressure Chemical Engineering,* Erlangen, 1984, p. 301.
23. Brunner, G., Ber. Bunsenges. *Phys. Chem. 88:*887 (1984).
24. Eggers, R., U. Sievers, and W. Stein, *J. Am. Oil Chem. Soc. 62:*1222 (1985).
25. Lack, E., Ph.D. Thesis, University of Graz (1985).
26. Lack, E., G. Bunzenberger, and R. Marr, in *Preprints of the International Symposium on High Pressure Chemical Engineering,* Erlangen, 1984, p. 63.
27. Lee, A.K.K., N.R. Bulley, M. Fattori, and A. Meisen, *J. Am. Oil Chem. Soc. 63:*921 (1986).
28. King, M.B., T.R. Bott, M.J. Barr, and R.S. Mahmud, *Sep. Sci. Tech. 22:*1103 (1987).
29. Fattori, M., N.R. Bulley, and A. Meisen, *J. Am. Oil Chem. Soc. 65:*968 (1988).
30. Christianson, D.D., and J.P. Friedrich, *J. Food Sci. 49:*229 (1984).
31. List, G.R., J.P. Friedrich, and D.D. Christianson, *J. Am. Oil Chem. Soc. 61:*1849 (1984).
32. Quirin, K.W., Ph.D. Thesis, University of Saarbrücken (1984).
33. Quirin, K.W., D. Gerard, and J. Kraus, *Fat Sci. Tech. 89:*139 (1987).
34. Wilp, C., and R. Eggers, *Fat Sci. Tech. 93:*348 (1991).
35. List, G.R., and J.P. Friedrich, *J. Am. Oil Chem. Soc. 60:*719 (1983).
36. List, G.R., J.P. Friedrich, and J. Pominski, *J. Am. Oil Chem. Soc. 61:*1847 (1984).
37. Leiner, S., Ph.D. Thesis, University of Saarbrücken (1986).
38. Stahl, E., K.W. Quirin, and H.K. Mangold, *Fette Seifen Anstrichm. 83:*472 (1981).
39. Tolboe, O., J.R. Hansen, and V.K.S. Shukla, in *Proceedings of the 2nd International Symposium on Supercritical Fluids,* Nice, 1988, p. 685.
40. Favati, F., J.W. King, and M. Mazzanti in *Proceedings of the 3rd International Conference on Supercritical Fluids,* Boston, 1991, p. 2.
41. Taniguchi, M., and R. Tsuji, *Agric. Biol. Chem. 49:*2367 (1985).
42. Saito, M., and Y. Yamauchi, *J. Chrom. Sci. 27:*79 (1989).
43. Quirin, K.W., D. Gerard, and J. Kraus, *Gordian 9:*156 (1986).
44. Brannolte, H.D., H.K. Mangold, and E. Stahl, *Chem. Phys. Lipids 33:*297 (1983).
45. Goncalves, M., A.M.P. Vasconcelos, G. Azevedo, H.J. Neves, and M. Nunes da Ponte, *J. Am. Oil Chem. Soc. 68:*474 (1991).
46. Forssel, P., M. Cetin, G. Wirtanen, and Y. Maelkki, *Fat Sci. Tech. 92:*319 (1990).
47. Djarmati, Z., R.M. Jankow, E. Schwirtlich, B. Djulinac, and A. Djordjevich, *J. Am. Oil Chem. Soc. 68:*731 (1991).
48. Polak, J.T., M. Balaban, A. Peplow, and A.J. Philips, *ACS Series 406:*449 (1989).

49. Simoes, P.C., M. Nunes da Ponte, M. Caparica, H. Matos, and E. Azevedo, in *Proceedings of the 2nd International Symposium on High Pressure Engineering,* Erlangen, 1990, p. 409.

50. del Valle, J.M., and J.M. Aquilera, *Ind. Eng. Chem. Res. 27:*1551 (1988).

51. Klein, T. "Phasengleichgewichte in Gemischen aus pflanzlichen Ölen und Kohlendioxid," Ph.D. Thesis, University Dorfmund, 1988.

52. Chrastil, J., *J. Phys. Chem. 86:*3016 (1982).

53. Baltes, J., *Gewinnung und Verarbeitung von Nahrungsfetten,* Paul Parey, Hamburg, 1975.

54. Chien, J.T., and J.E. Hoff, *Chem. Eng. J. 43:*B103 (1990).

55. George, C.S., and D.G. MacDonald, *Can. J. of Chem. Eng. 64:*80 (1986).

56. Paulaitis, M.E., V.J. Krukonis, R.T. Kurnik, and R.C. Reid, *Rev. Chem. Eng. 1:*179 (1983).

57. Lee, A.K.K., N.R. Bulley, M. Fattory, and A. Meisen, *J. Am. Oil Chem. Soc. 63:*921 (1986).

58. Gil, M.G., M. King, and T.R. Bott, in Proceedings of the 2nd International Symposium on Supercritical Fluids, Nice, 1988, p. 651.

59. Wiebe, R., and V.L. Gaddy, *J. Am. Chem. Soc. 61:*31 (1939).

60. Eggers, R., Chem.-Ing.-Techn. 53:551 (1981).

61. Eggers, R., and U. Sievers, in *Proceedings of the 1st International Congress on Fluid Handling Systems,* Essen, 1990, p. 339.

62. Eggers, R., and U. Sievers, *ACS Series 406:*478 (1989).

63. Brunner, G., *Fluid Phase Equil. 10:*289 (1983).

64. Brunno, G., *Fette Seifen Anstrichm. 88:*464 (1986).

65. Eggers, R., and U. Sievers, *J. Chem. Eng. Japan 22:*641 (1989).

66. Marquard, R., *Fette Seifen Anstrichm. 83:*129 (1981).

67. Sievers, U., *Fortschr.-Ber. VDI-Z,* Reihe 6, Nr. 155, VDI-Verlag, Düsseldorf (1984).

68. Schumacher, H., *J. Amer. Oil Chem. Soc. 60:*417 (1983).

69. Homann, Th., M. Knuth, and W. Stein, *Fette Seifen Anstrichm. 83:*570 (1981).

70. Eggers, R., and Schade, G., German Pat. Appl., 332296806 (1983).

Chapter 4

Supercritical Fractionation of Lipids

S. Peter

Institute für Technische Chemie, University Erlangen-Nürnberg, Egerlandstraße 3, D-91058 Erlangen, Germany

Increasing concern over the use of chemical solvents in the manufacture of foods and pharmaceuticals has spurred the development of critical fluid extraction. Critical fluid extraction makes use of the unique behavior of compounds and mixtures near or above their critical points. These systems are characterized by dramatic changes in their solvent power and mass transfer characteristics with changes in temperature and pressure.

Two major benefits of such a process are the potential for low processing temperatures and ease of solvent recovery. These advantages make critical fluid extraction particularly applicable to food processing and for the production of essences and flavors. Several industrial plants for critical fluid extraction are operated batchwise for the extraction of solid materials using dense carbon dioxide (1). A solvent above its critical pressure is designated a *supercritical* fluid if the temperature is in the region below the solvent's critical temperature but above its critical pressure. This includes mixtures of a supercritical component and an entrainer that are used as the extracting phase where the pressure exceeds the critical pressure of the binary system.

Extraction by critical fluids is based on the observation that their solvent power is related to their density. In the critical region, the density of specific fluid substances dramatically change with pressure and temperature. This is shown in Fig. 4.1 for carbon dioxide. The operation conditions are preferably chosen in the range of the reduced pressure from 1 to 5 and a reduced temperature from 0.8 to 1.5 for CO_2. If a mixture of a supercritical component and an entrainer is used as extractant, the operation pressure must exceed the critical pressure of the binary system. For example, the critical curve of the system (carbon dioxide/propane) is shown in Fig. 4.2 (2) with the binodal curves at 40 and 60°C displayed. A mixture in a state above these critical loci is called a supercritical mixture.

A major task in designing effective critical fluid extraction is the optimization of the operational pressure and temperature. The procedure of critical fluid extraction consists of two steps:

1. Transfer of substances from a raffinate phase into an extractant

2. Separation of the dissolved substances from the extractant.

Step 2 is effected by reducing the density of the critical fluid solvent, by reducing the pressure, elevating the temperature, or both. The simplest method for this step, called the regeneration of the supercritical solvent, is to reduce the density by expansion. At

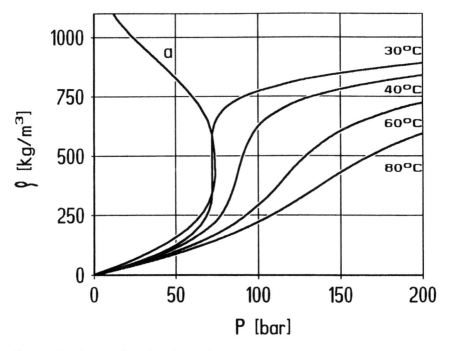

Fig. 4.1. Density of carbon dioxide as a function of temperature and pressure.

low densities the solvent capacity of the fluid disappears and the dissolved substances precipitate. By decreasing the pressure step by step, fractionation of the dissolved substances is possible.

Since recompressing of the circulating extractant is expensive, other regenerating procedures have been considered. For example, the adsorption of caffeine on activated carbon and absorption of caffeine by water have been proposed for the regeneration of the circulating gas. Another method for reducing the energy of compression consists of the partial expansion of the critical solvent, separation of the precipitated material, and cooling of the regenerated extractant so that it condenses to a liquid. The liquified solvent is then recycled via a heat exchanger in which it is heated to the process temperature (3).

The separation of mixtures that are in liquid state under operation conditions can be carried out in a countercurrent procedure. Usually a multistage extraction is necessary. In order to increase the solubility of the substances to be separated and to enhance the separation factor, an entrainer (or *cosolvent*) is often added to the dense gas (4). The addition of an entrainer complicates the thermodynamic behavior. This disadvantage must at least be compensated by the special benefits induced by the entrainer.

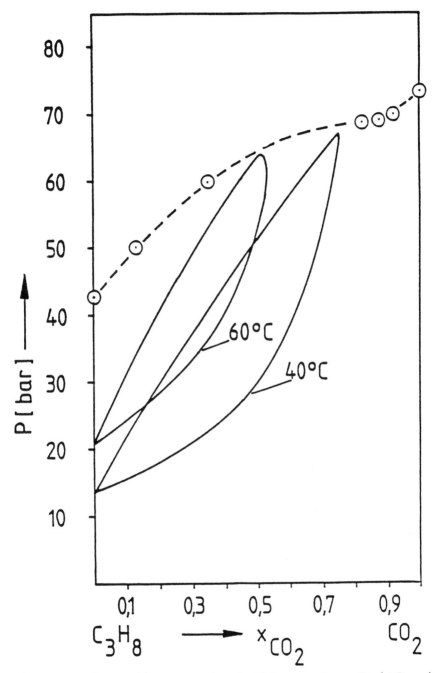

Fig. 4.2. Critical curve of the system carbon dioxide/propane. *Source:* Roof, J.G., and J.D. Baron, *J. Chem. Eng. Data 12*:292 (1967); Weidner, E., Ph.D. Thesis, Separation of Lecithin and Soybean Oil by a Supercritical Extractant, University of Erlangen-Nüremberg (1985).

An entrainer provides the following advantages:

1. Enhancement of the solvent power of the critical fluid extractant
2. Increasing the pressure and temperature dependence of the fluid's solvent power
3. Enhancement of the separation power by entrainers (or cosolvents) that are specific for the desired separation

In the region near the critical curve of the binary system of the supercritical component and the entrainer, the temperature dependence of the solvent power is so enhanced that regeneration of the critical extractant can be accomplished by a change of temperature.

The equipment for carrying out the separation by a countercurrent multistage extraction consists of at least two columns. The first acts as a separation column and the other serves as a precipitation column for the recovery of the solvent. Frequently a third column is used to complete the regeneration, as shown in Fig. 4.3. As the dynamic viscosity of the critical fluid extractants is about one to two orders of magnitude lower than that of the coexisting liquid phase, the critical fluid is usually used as the mobile phase in order to achieve a higher processing capacity.

In the case just described, the mixture to be separated is fed into the separation column at a specific concentration. The phase with the lower density, normally the critical fluid, flows upward through the column. During this process, the more soluble substances become enriched in the extractant and are removed from the separation column at the top, together with the circulating fluid. The less soluble part of the mixture flows downward, where it leaves the column as the raffinate, together with dissolved extractant (critical fluid).

The stream of extract that leaves the rectifying section of the main column at the top is fed to the regeneration column. By pressure reduction, temperature increase, or both, the dissolved components are precipitated and flow downward. The product withdrawn from the regeneration column is divided into the top product and the reflux for the extraction column. The circulating supercritical extractant leaves the regeneration column at the top and is returned to the bottom of the extraction column *via* a heat exchanger after recompression of the fluid. The bottom products from the regeneration are continuously withdrawn. If the solvent power of the critical fluid is enhanced by an entrainer, the regeneration of the extractant will require less of a pressure reduction. Hence, the costs for recycling the extractant are reduced.

In the selection of suitable operating parameters, the density of the coexisting phases must be taken into consideration. If an alkane is dissolved in an organic liquid phase, the density decreases, whereas dissolution of carbon dioxide in an organic liquid increases the density, as shown in Fig. 4.4. The capacity of an extraction column depends on the density difference between the raffinate and extract phase; therefore, densities of the coexisting phases should differ by more than 150 kg m^{-3}. This density difference is desirable in order to get a reasonable flooding point in the column.

The column flooding point also depends on the fluid phase viscosity. Therefore, some knowledge concerning the viscosity of the coexisting phases is required. The

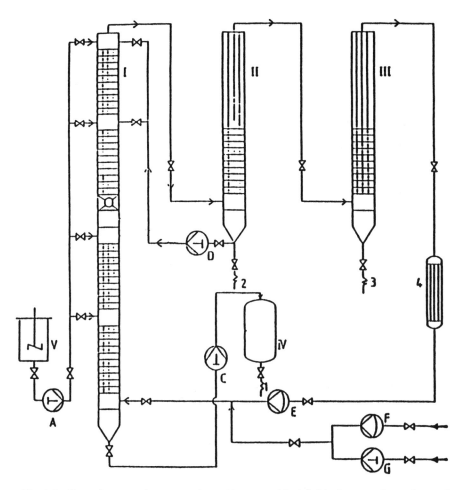

Fig. 4.3. Flow diagram of an extraction with supercritical fluids: I, separation column; II, III, regeneration columns; IV, product removal; V, feed storage; A, feed pump; B, solvent compressor; C, raffinate pump; D, reflux pump; F, gas compressor; G, entrainer pump; 1, raffinate discharge; 2,3, extract discharge; 4, heat exchanger. *Source:* Ender, U., and S. Peter, *Chemical Engineering Processes 26:*207 (1989).

dynamic viscosity of the liquid phase decreases dramatically with increasing content of the supercritical fluid extractant. The kinematic viscosity also changes considerably with the composition of the liquid phase. However, this depends on the nature of the supercritical extractant, as whether the kinematic viscosity will increase or decrease with increasing pressure (5). Solute mass transfer is enhanced if the dynamic viscosity of the liquid phase decreases. In Fig. 4.5 the mass transfer coefficient for the dissolution of oleic acid in dense carbon dioxide is plotted vs. the viscosity of the coexisting liquid phase. Remember that the concentration of the dissolved gas increases with

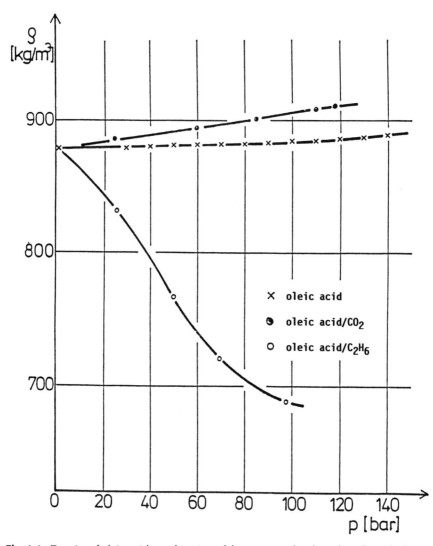

Fig. 4.4. Density of oleic acid as a function of the content of carbon dioxide and ethane. *Source:* Jakob, H., Ph.D. Thesis, Flow-Behavior of Coexisting Gas and Liquid Phases at High Pressures, University of Erlangen-Nüremberg (1988).

decreasing viscosity; that effect also contributes to the overall improvement of the mass transfer.

By means of a cascade of precipitation columns the top product of the separation column can be fractionated. Each of the columns is thereby operated under specific conditions. In this particular case, the rate of mass transfer may have remarkable influence on the result of the tandem separation process.

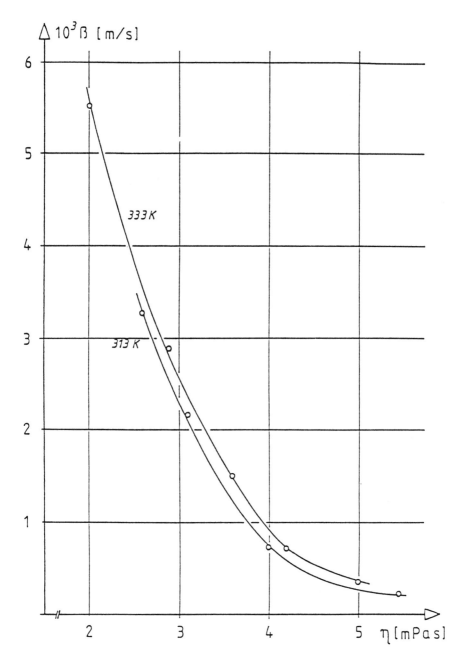

Fig. 4.5. Mass transfer coefficient for the dissolution of oleic acid in dense carbon dioxide as a function of the viscosity of the liquid phase. *Source:* Beyer, A., Ph.D. Thesis, Mass Transfer at the Dissolution of Substances of High Molecular Mass in Dense Gases, University of Erlangen (1990).

The interfacial tension between the liquid and gaseous phases also decreases dramatically with increasing system pressure. In order to obtain a supercritical extractant loading that is of commercial interest, it is necessary that the pressure be high enough that the interfacial tension amounts to only about 1 to 2 mN/m. Under these conditions the liquid phase disintegrates into fine droplets if a certain pressure is exceeded, as shown in Fig. 4.6. As a result, the liquid phase does not flow down as a film adhering to the packing of the column but moves down the column as fine droplets (6).

Under the conditions of critical fluid extraction, the behavior of the residence time has been investigated as a function of the volumetric flow. From these tracer-based measurements, characteristic parameters such as the axial dispersion coefficient and the Bodenstein number were determined:

$$\mathrm{Bo} = \frac{UL}{D_{ax}}$$

with U, gas velocity; L, height of the packed column; and D_{ax}, axial dispersion coefficient. The Bodenstein number corresponds to the ratio of convective to dispersive

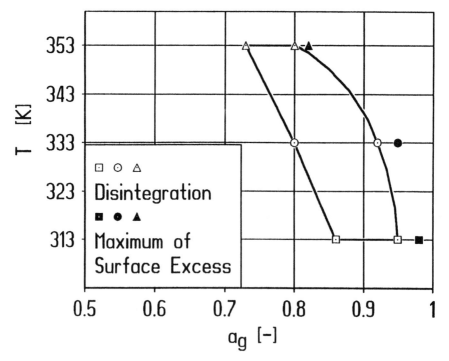

Fig. 4.6. Disintegration of a falling film of linoleic acid in presence of ethane as a function of pressure at 313, 333, and 353 K. *Source:* Hiller, N., et al., *Chemical Engineering Technology 16:*206 (1993).

forces. In Fig. 4.7, the Bodenstein number is shown as a function of the gas velocity for several packings. Obviously, the different packed columns exhibit different behavior under flow (7).

The influence of the flow behavior on the efficiency of the packings is shown in Fig. 4.8. The number of theoretical stages per meter is plotted as a function of the Bodenstein number for the following packings: Sulzer CY with drop dispenser; spray column; drop dispenser; and Sulzer SMV. With increasing Bodenstein number (i.e., decreased back-mixing), the number of theoretical stages per meter of packing increases. Note that within the accuracy of measurement, the data fall on the same curve. The specific surface of the packings is apparently of no significance (7).

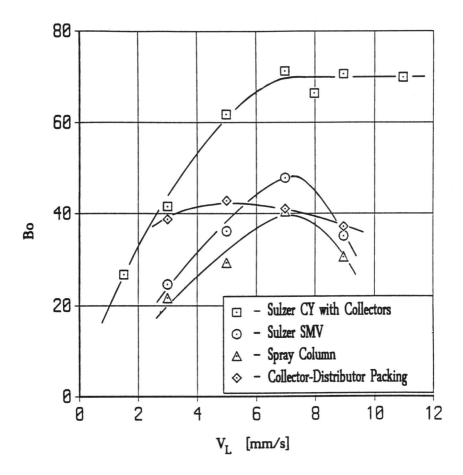

Fig. 4.7. Bodenstein number versus void pipe velocity at 12 MPa and 313 K for several packings. *Source:* Czech, B., Ph.D. Thesis, Behavior of a Countercurrent Column at the Production of Diglycerides By Near-Critical Extraction, University of Erlangen, 1991.

Several supercritical fluid–based processes have been developed that employ the principles just discussed. The following processes will be considered in more detail in the following section:

1. Removal of impurities from edible oils and fats
2. Deoiling of raw lecithin
3. Preparation of phosphatidylcholine from soybean lecithin
4. Preparation of mono- and diglycerides
5. Production of powders by rapid expansion
6. Isolation of polyunsaturated fatty acids from fish oil

Removal of Impurities from Edible Oils and Fats

The separation of mucilage-free fatty acids and flavoring compounds is normally accomplished in several steps. With supercritical dense gases it is possible to combine these two operations into one step, which can be carried out at moderate pressures and low temperature. Two processes have been proposed in the patent literature.

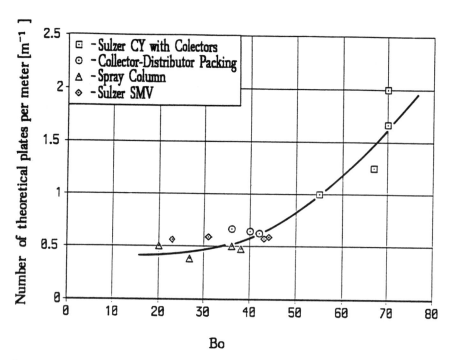

Fig. 4.8. Number of theoretical plates per meter packings vs. Bodenstein number. *Source:* Czech, B., Ph.D. Thesis, Behavior of a Countercurrent Column at the Production of Diglycerides By Near-Critical Extraction, University of Erlangen, 1991.

According to Zosel (8), the components dissolved by the gas can be removed from the supercritical extractant by adsorption on activated charcoal under nearly the same conditions as maintained in the extraction stage, which is conducted at 20.0 MPa and 423 K. The regeneration of the sorbent depends mainly on the adsorption equilibria; in this context the research of van Leer, et al. (9) should be consulted. The free fatty acid contents cited in the foregoing case were reduced from 0.4 to 0.02 wt%.

Similarly, the process as proposed by Coenen and Kriegel (10) has the advantage that the separation of fats and mucilage is carried out during the extraction stage. Triglycerides, fatty acids, aldehydes, ketones, terpenes, fragrances, and aromatic substances were all dissolved in dense propane at 15.0 MPa and 423 K and subsequently separated from mucilaginous materials and sludge. The latter substances are insoluble and form the raffinate. The separation of free fatty acids, odoriferous compounds, and fats occurs during a multiple-step depressurization process. The solvent is then liquefied and recycled via a pump into the extractor stage. The content of free fatty acids in the oil fraction is reduced to 0.04 wt%, aldehydes and ketones being no longer detectable.

The separation of free fatty acids from triglycerides of a natural oil is possible by using just a supercritical system. However, adding ethanol as an entrainer doubles the K- factors. A concentration between 0.5 and 5 wt% of the free fatty acids can be obtained in the gaseous phase at a pressure of about 14.0 MPa, whereas in the absence of an entrainer a pressure between 20 to 35 MPa is required to achieve the same concentration range.

The solubility of palm oil in carbon dioxide with ethanol as an entrainer is shown in Fig. 4.9 as a function of pressure and temperature. The ethanol content of the gas phase amounts to a constant value of about 10 wt%. At 13.0 MPa and 343 K the solubility of palm oil in the gas phase is only 2 wt%. Hence, by increasing the temperature to 383 K, the oil solubility is reduced to a negligible value (11).

Separation of free fatty acids from palm oil was also investigated. Carbon dioxide was employed as supercritical component and ethanol as entrainer. Operating conditions were as follows: separation column, 13.5 MPa, 353 K; regeneration column, 13.5 MPa, 383 K. The experiments were carried out on laboratory equipment with a fatty acid throughput of 1 kg/h as the top product. An increase in temperature reduced the amount of free fatty acids in the cycle gas from about 3 wt% to 0.1 wt%. The content of free fatty acids of the oil fraction was reduced to 0.04 wt%; aldehydes and ketones were found to be negligible.

The foregoing high-pressure methods have the potential of using a single step to replace several steps in the conventional technique, such as removal of fatty acids by alkali treatment or vacuum distillation, and deodorization by steam refining at high temperatures and low pressures.

Deoiling of Raw Lecithin

Heigel and Hüschens describe a method for the separation of soybean oil from raw lecithin by supercritical carbon dioxide at 20.0 to 50.0 MPa and 308 to 353 K (12).

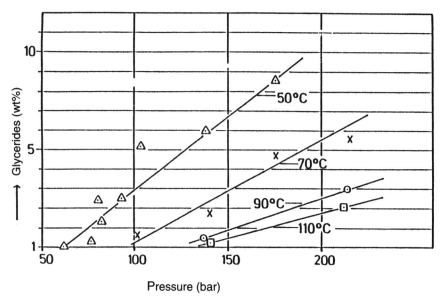

Fig. 4.9. Solubility of palm oil in carbon dioxide with ethanol as an entrainer. *Source:* Brunner, G., and S. Peter, *Separation Science and Technology, 17(1):*199 (1982).

The di- and triglycerides, fatty acids, hydrocarbons, and wax esters were all soluble in dense carbon dioxide. An autoclave is charged with the viscous raw lecithin and the dense gas is contacted with this substitute. As the lecithin is gradually deoiled, the process slows down, because of the increasing viscosity of the lecithin, and comes to a halt before complete removal of the oil.

The procedure suggested by Coenen and Hagen (13) used liquefied gases such as ethane and carbon dioxide. The operating pressure is between twice the critical pressure and 35.0 MPa; the operating temperature between 273 K and the critical temperature. The low temperature may be unfavorable as regards the viscosity of the starting material. The dissolved oil is removed from the extractant by decreasing the pressure to 1.0 to 3.0 MPa in the case of carbon dioxide and 3.0 to 4.0 MPa in the case of ethane. The purified lecithin was obtained as a powdery material and contained 90% acetone-insoluble material.

Stahl, et al. (14) suggested the dilution of the raw lecithin with a conventional solvent to reduce its viscosity. The most suitable solvent was found to be acetone. The mixture of raw lecithin and acetone was extracted with dense carbon dioxide, and the pressure and temperature chosen so that the binary acetone/carbon dioxide system was supercritical. However, removal of the acetone from the purified lecithin still had to be effected.

A semicontinuous and complete deoiling of raw lecithin is possible by means of the high-pressure jet extraction proposed by Stahl et al. (15). The viscous starting material is pressed through a capillary with an inner diameter of 0.2 mm into a mixing device, where it is brought into intensive contact with a stream of dense carbon dioxide. The mixing device is shown schematically in Fig. 4.10. The extraction conditions were 90.0 MPa and 363 K. The lecithin content of the gaseous phase was between 0.7 and 1 wt%. The mass flow of raw lecithin in the jet device was approximately 60 g/h, while that of carbon dioxide was 2.5 to 3 kg/h. A powdery lecithin with a residual oil content of 3.5 wt% was obtained. The average particle size of the lecithin was less than 0.1 mm. The CO_2 can be regenerated at a pressure of 15.0 MPa and temperatures of 333 to 353 K. The residence time of the lecithin in the jet device is about 0.05 s. Therefore, any temperature effect stress on the lecithin is negligible.

The above-mentioned processes require pressures between 35.0 and 90.0 MPa. The pure lecithin cannot be continuously withdrawn from the separator, because it does not form liquid solutions with carbon dioxide. Also, decomposition can occur at temperatures below the melting point.

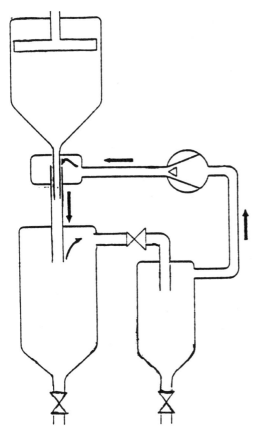

Fig. 4.10. Mixing device for removing soybean oil from lecithin by dense carbon dioxide. *Source:* Stahl, E., et al., *Berichte der Bunsengesellschaft für Physiklische Chemie 88*:900 (1984).

To achieve a continuous extraction process it is necessary to maintain the deoiled lecithin in a liquid state within the countercurrent column. This can be achieved by using propane as an entrainer. Lecithin and soybean oil are both very soluble in propane, whereas carbon dioxide dissolves only soybean oil with high selectivity. Therefore, the applicability of carbon dioxide/propane mixtures was investigated further (16).

The formation of a liquid phase in the quasi-ternary lecithin/carbon dioxide/ propane system was investigated as a function of pressure and temperature. The results are shown in Fig. 4.11. The pressure required to form a liquid lecithin solution increases with both increasing carbon dioxide concentration and temperature. Propane, in this case, acts as an entrainer. It increases the solvent power of the carbon dioxide and lowers the pressure drop required for the regeneration of the supercritical extractant.

The phase equilibria in the quasi-quaternary system lecithin/soybean oil/ propane/ carbon dioxide at 8.0 MPa and 323 K are shown in Fig. 4.12. The front side of the tetrahedron represents the phase equilibrium of the quasi-ternary system soybean oil/ propane/carbon dioxide, whereas the right-hand, rear side shows that of lecithin/ propane/carbon dioxide. The shaded area represents the binodal surface.

The operating point for a separation column must be within the two-phase region below the shaded area. The distance of the operating point from the binodal surface should be sufficient so that a suitable density difference between the coexisting phases is achieved. In Fig. 4.12 a proper operating line is represented by ZE. Z indicates a feed composed of 60% lecithin and 40% soybean oil, and E represents the composition of the extracting agent. The position of the overall operating point on line ZE depends on the solvent-to-feed ratio.

The distance between operating point and binodal surface can further be altered by changing the composition of the extractant. With an increasing carbon dioxide/propane ratio the distance between operating point and binodal surface is enhanced. However, to avoid the formation of solid lecithin, a maximum value of the carbon dioxide/propane ratio must not be exceeded. The conditions for continuous operation are inferred from the measured phase equilibria.

The process was tested in a plant essentially consisting of a separation and two regeneration columns. The plant has been shown schematically in Fig. 4.3. The separation column has an inner diameter of 65 mm and 6 m height. It was designed for operation at a pressure of 15.0 MPa and 373 K. The viscous raw lecithin can be fed into the separation column at several levels. In the continuous experiments at a pressure of 8.0 MPa and temperatures between 313 to 328 K, the critical extracting agent, consisting of 80 wt% propane and 20 wt% carbon dioxide, flowed upward through the column. Soybean oil and a small amount of lecithin dissolve in the extractant.

A mixture of propane, carbon dioxide, soybean oil, and lecithin leaves the separation column at the top and enters the first regeneration column. By a temperature increase to 348 K, the lecithin is selectively precipitated. Because of its higher density, liquid lecithin flows down through the packings of the first regeneration column and is drawn off and pumped back to the top of the separation column as reflux.

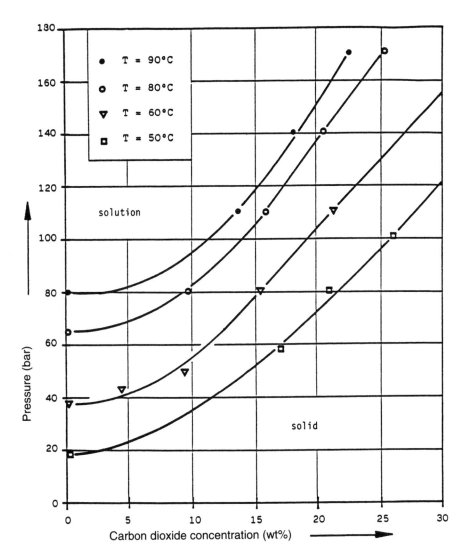

Fig. 4.11. Liquid formation in the system lecithin/carbon dioxide/propane. *Source:* Peter, S., et al., *Chemical Engineering Technology 10:*37 (1987).

The virtually lecithin-free extracting agent leaves the first regeneration column at the top and is expanded into the second regeneration column, which is operated at 6.0 MPa and 373 K. There the dissolved soybean oil is precipitated and continuously removed as bottom product. The regenerated extracting agent also leaves the second regeneration column at the top. It is recompressed to 8.0 MPa, cooled, and recycled to the bottom of the separation column. In the separation column lecithin flows down

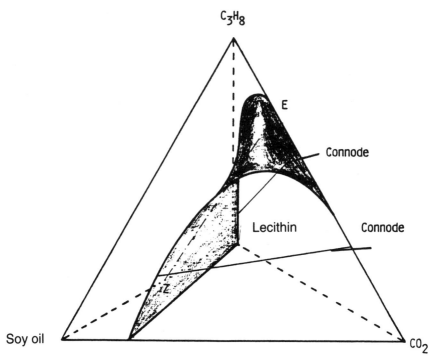

Fig. 4.12. Phase equilibria in the quasi-quaternary system of lecithin/soybean oil/carbon dioxide/propane at 323 K and 8 MPa. *Source:* Weidner, E., Ph.D. Thesis, Separation of Lecithin and Soybean Oil by a Supercritical Extractant, University of Erlangen-Nüremberg (1985).

through the packings, countercurrent to the extracting agent. On its way downward the lecithin comes in contact with an increasingly pure extractant and becomes completely deoiled. The liquid mixture of propane, carbon dioxide, and lecithin is collected at the bottom of the separation column and continuously removed through an expansion valve. Thereby a powdery, tasteless, yellow white, solvent-free lecithin is obtained.

A typical thin-layer chromatogram of the above product (line 1) is compared with a commercial acetone-extracted product (line 5) in Fig. 4.13. The thin-layer chromatogram demonstrates that an oil-free product was prepared by critical fluid extraction.

A temperature gradient along the separation column combined with partial recycle of the bottom product of the first regeneration column improved the process efficiency for lecithin up to 0.97. For continuous removal of pure lecithin from the separation column, the pressure drop in the product pipe must be low, in order to avoid precipitation of solid lecithin. Precipitation must occur only in the expansion valve. Therefore, the expansion valve should be mounted not far from the product vessel, and the pipe diameter downstream from the expansion valve must continuously

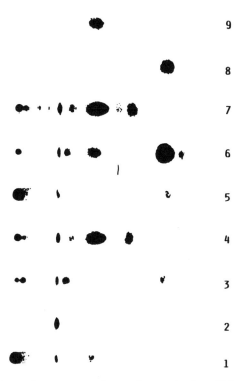

Fig. 4.13. Thin-layer chromatogram for product evaluation: (1) lecithin from fluid extraction 92%, (2) sitosterol, (3) oleates (mono, diglycerides), (4) acetone-soluble compounds from fluid extraction 8%, (5) lecithin from acetone extraction 96%, (6) oil from acetone extraction, (7) acetone-soluble compounds 8% (fluid extraction), (8) soybean oil, (9) oleic acid. *Source:* Peter, S., et al., *Chemical Engineering Technology 10:*37 (1987).

increase to obtain pure lecithin powder. The equipment has to be carefully grounded in order to avoid electrostatic charging.

The addition of propane to supercritical carbon dioxide decreases the required operation pressure to 8.0 MPa. To regenerate the extractant a pressure decrease to 6.0 MPa at a temperature of 373 K is sufficient. Under these conditions a centrifugal pump may be used for recycling the extracting agent if the plant capacity is about 5000 t/yr or more.

Preparation of Phosphatidylcholine from Soybean Lecithin

Soybean lecithin contains a number of different phosphatides and other compounds, such as glycerides. Aside from soybean lecithin, egg yolk lecithin is also available on the commercial market. The average composition of these products differ remarkably as can be seen from Table 4.1.

TABLE 4.1 Composition of Commercial Lecithin Products

Compound	Soybean Lecithin (wt%)	Egg Yolk Lecithin (wt%)
Phosphatidylcholine (PC)	31	78
Phosphatidylethanolamine (PE)	21	17
Phosphatidylserine (PS)	3	—
Phosphatidylinositol (PI)	18	0. 6
Phosphatidic acid	2	—
Other phospholipids	10	—
Phytoglycolipids	15	—
Plasmogen	—	1
Sphingomyelin (SPH)	—	2. 5

Lecithin from egg yolk is much more expensive than soybean lecithin; therefore, the separation of PC from soybean lecithin has been investigated by several authors. To the present author's knowledge, only chromatographic processes are used commercially for lecithin purification. Chromatographic separation processes, however, are expensive; therefore, the development of a cheaper isolation method is of interest.

Aliphatic alcohols have been utilized for the fractionation of phosphatides (17). Pardun (18) observed that with methanol as extractant the selectivity for phosphatidylcholine can be enhanced by addition of 15% water. By means of countercurrent extraction, it is possible to make a product consisting of 45 wt% phosphatidylcholine from soybean lecithin. Detailed work has been carried out on the fractionation of lecithin by means of preparative chromatography. Silica gel, alumina gel, and magnesium oxide exhibit a high selectivity for phosphatidylcholine (19). The chromatographic methods yield products with a PC content of 99 wt%.

In pursuit of an economical method for separating PC from soybean lecithin detailed experiments have been carried out on the fractionation of lecithin by critical fluid extraction. In the presence of water, a small but not negligible part of PC is converted into lysophosphatidylcholine. Therefore, water-free extraction systems have been extensively investigated (Peter, S., Zh. Dheng, B. Czech and W. Weidner, unpublished communication). Dense carbon dioxide mixtures, with various alkanols such as methanol, ethanol, and isopropanol as entrainers, were found to be efficient extracting agents. Only a few stages of a countercurrent extraction run yielded products having more than 90% PC. In the thin-layer chromatogram of PC-enriched and soybean or egg yolk lecithin is shown in Fig. 4.14.

Pressures of 20 MPa are sufficient to achieve adequate loading of the extractant in the supercritical fluid (1–3 wt%). By adding propane to the dense carbon dioxide, it is possible to gain commercial loadings at decreased pressures. With a mixture of 90% carbon dioxide and 10% propane as the supercritical extractant, a loading of 4 wt% lecithin could be achieved at 15 MPa and 328 K. The separation factor remained unchanged under these conditions. A decrease of the separation factor occurs only at elevated propane contents.

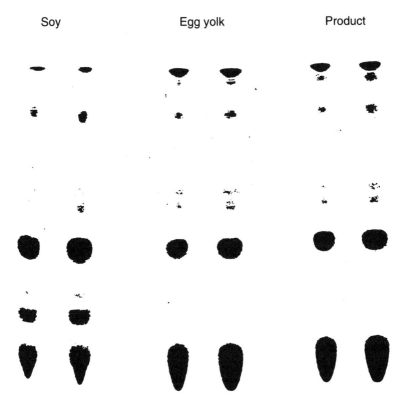

Fig. 4.14. Thin-layer chromatogram of PC-enriched soybean lecithin and egg yolk lecithin.

Preparation of High-Grade Mono- and Diglycerides

Glycerol esters of one or two fatty acids are an important source of fatty nonionic surface-active agents (surfactants) that are versatile and appear to have excellent commercial potential. These products are used as emulsifiers, wetting agents, lubricants, plasticizers, and detergents. They are mild and have no toxicity.

Industrial mono- and diglycerides contain quantities of free glycerol. Depending on the ratios of fat to glycerol employed, the product mixtures are called commercially 40% monos or 60% monos. A 90% mono can be prepared by short-path distillation of the optimum reaction product resulting from glycerolysis of a fat or oil. When pure monoglycerides are desired, special methods are required for their preparation. The 90% monoglycerides of stearic acid are favored for use as emulsifiers in the food industry.

The manufacture of commercial mono- and diglyceride mixtures represents a special case of alcoholysis in which either fatty acids or a fat are reacted with an excess of glycerol. The reaction is carried out on a large scale as a preliminary process in the preparation of edible, oil-soluble surface-active agents and also for the production

of detergents and alkyd resins. Mixtures of mono- and diglycerides are useful interme-
diates for acetostearins, lactopalmitins and lactostearins, diacyltartaric acid esters, suc-
cinylated monoglycerides, stearylmonoglyceride citrate, and other products for use in
the food and detergent industries.

Mono-, di-, and triglyceride mixtures are prepared by the direct condensation of
fats or fatty acids with glycerol at elevated temperatures. The reaction temperature is
usually about 250°C. At these conditions the mixture of glycerol and glycerides is
generally one phase. Patents have been issued on both batch and continuous processes
and on a variety of catalysts. The reaction products usually contain 40 to 60%
monoester, 30 to 45% diester, 2 to 8% triester, 1% free fatty acids, 5 to 14% glycerol,
0.1% catalyst, and 1% impurities.

If the mixture of glycerides is used as primary reactant in the manufacture of
monoglycerides, the glycerol content must be adjusted to values below 5%. This is
done by distillation. Distillation for removing glycerol occurs at 30 to 50 mbar at temperatures ranging from 453 to 483 K. Unfortunately, a part of the monoglycerides is
lost with the glycerol.

For use in food industry, pharmaceuticals, and cosmetics the monoester content
(generally monostearates) is enhanced by short-path distillation to values of 90 to
95%. High-vacuum short-path distillation is preferred to keep the temperature as low
as possible in order to avoid interesterification reactions. No interesterification catalyst
should be present during the heating of a monoglyceride. In cosmetics, diglycerides of
high purity (about 90%) are required. For these purposes, glycerol is removed in a
second distillation step to values between 0.2 to 0.5%. Approximate values reported
for vapor pressures of monoglycerides are shown in Table 4.2 (20).

Monoglyceride concentrations above 96% cannot be achieved by distillation
because of interesterification and degradation to glycerol and free fatty acids at the
required temperatures. Also, isomerization occurs from a 1-monoglyceride to a 2-
monoglyceride. This isomerization must be avoided, because 1-monoglycerides have
higher emulsifying capacity.

For further improvement of the monoglyceride content, the temperature of the
fractionation must be reduced. In critical fluid extraction, both separation and extractant removal are feasible at low temperatures. Monoglycerides of 99% purity can be
prepared by means of critical fluid extraction using mixtures of carbon dioxide and

TABLE 4.2 Vapor Pressure Values for Monoglycerides (20)

Monoglyceride	Pressure (mm)	Temperature (°C)
Monocaprin	1	175
Monolaurin	1	186
Monomyristin	1	199
Monopalmitin	1	211
Monostearin	0.2	190
Monoolein	0.2	186

propane as an extractant (21). In order to ascertain favorable operating conditions, comprehensive knowledge of the phase equilibria of the system under consideration is required as a function of pressure, temperature, and concentration.

The phase behavior of the quasi-quaternary system consisting of carbon dioxide/ propane/monoglycerides/di-, triglycerides is shown in Fig. 4.15 at 12.0 MPa and 313 K in a tetrahedron graph. The quasi-ternary monoglycerides/di-, triglycerides/carbon dioxide system forms the base of the tetrahedron. Within this system a mixing gap exists between the di-, triglycerides and carbon dioxide, as well as between the monoglycerides and carbon dioxide. The quasi-ternary system di-, triglycerides/ propane/ carbon dioxide is of type I with one binary mixing gap between di-, triglycerides and carbon dioxide. The binodal curve for the system consisting of monoglycerides/ propane/carbon dioxide is similar. However, the two-phase region of the latter is larger than in the system di-, triglycerides/propane/carbon dioxide. Under the chosen conditions, the monoglycerides/di-, triglycerides/propane system is single-phase. By adding an adequate amount of carbon dioxide, a mixing gap is induced.

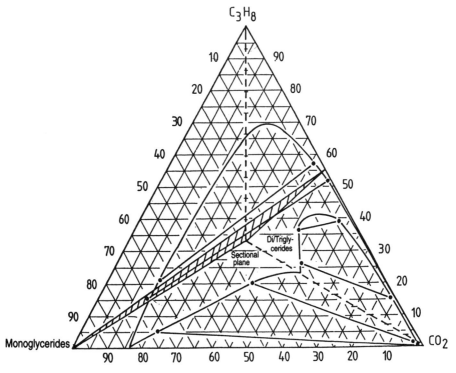

Fig. 4.15. Phase behavior of the quasi-quaternary system carbon dioxide/ propane/monoglycerides/di-, triglycerides of oleic acid at 12.0 MPa and 313 K. *Source:* Peter, S., and U. Ender, *Fat Science and Technology 91(7):*260 (1989).

A two-phase region of the quasi-quaternary system is defined by a plane that intersects the side surfaces of the tetrahedron in the binodal curves. For an extractant composed of 55 wt% propane and 45 wt% carbon dioxide the intersectional plane is shown in Fig. 4.16 at 12.0 MPa and 313 K. From the phase behavior it follows that monoglycerides can be obtained as an almost pure raffinate.

The phase equilibria in the systems just described gave surprising results concerning the effectiveness of the entrainer. In Fig. 4.17, the separation factor between monoglycerides and diglycerides is shown at constant pressure and temperature (as a function of the propane concentration of the extractant). At low propane concentrations the monoglycerides are enriched in the extract phase; that means that the monoglycerides would be the top product of a separation column. With increasing propane concentration the separation factor turns to values smaller than 1, then passes through a minimum, and again approaches 1 at the plate point of the system. Monoglycerides of high purity can be obtained as bottom product using compositions of the extracting agent in the range of the minimum.

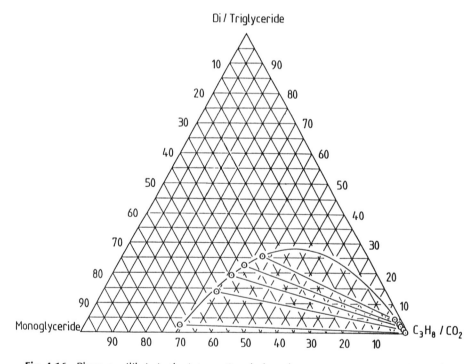

Fig. 4.16. Phase equilibria in the intersectional plane for an extractant composition of 55% propane and 45% carbon dioxide of the quasi-quaternary system carbon dioxide/propane/monoglycerides/di-, triglycerides of oleic acid at 12.0 MPa and 313 K. *Source:* Peter, S., and U. Ender, *Fat Science and Technology 91(7):*260 (1989).

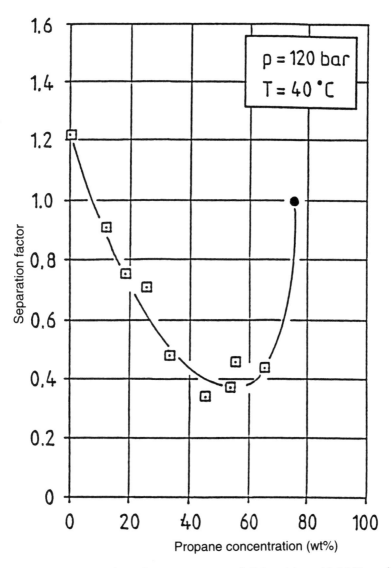

Fig. 4.17. Separation factor between mono- and diglycerides at 12.0 MPa and 313 K as a function of the propane concentration of the extractant. *Source:* Peter, S., and U. Ender, *Fat Science and Technology 91(7)*:260 (1989).

The flooding point of a countercurrent column depends on the density difference between raffinate and extract phase; therefore the densities of carbon dioxide/propane mixtures were measured. The results are shown in Fig. 4.18 at 8.0 and 12.0 MPa and different temperatures. The calculated values are obtained by means of the Schmidt-

Wenzel equation of state (22). Experimental and calculated data correspond very well; they differ clearly from those provided by ideal mixing rules.

In Fig. 4.19 the density of the coexisting liquid phase is plotted as a function of the propane content of the gas phase. At 12.0 MPa and 313 K the density of the coexisting liquid phase amounts to 800 kg-m^{-3} for an extracting agent composed of 55 wt% propane and 45 wt% carbon dioxide. The density of the extracting agent is 580 kg-m^{-3} under these conditions, as can be seen from Fig. 4.18.

The pure monoglyceride preparation was tested using equipment described earlier in the section describing the deoiling of raw lecithin. The experiments were carried out at 12.0 MPa and 313 K in a separation column that was equipped with Sulzer CY packings. The propane concentration in the extractant amounted to 55 wt%. The feed was kept at 1.6 kg/h. At these conditions the mixture of carbon dioxide and propane was supercritical (2).

The monoglyceride concentration of the bottom product increases with the solvent-to-feed ratio at a given height in the packings. With a concentration of monoglycerides in the feed of 60 wt%, a bottom product with 99 wt% monoglycerides (glycerol content deducted) was obtained at a solvent-to-feed ratio of 40 and a packing height of 3.2 m.

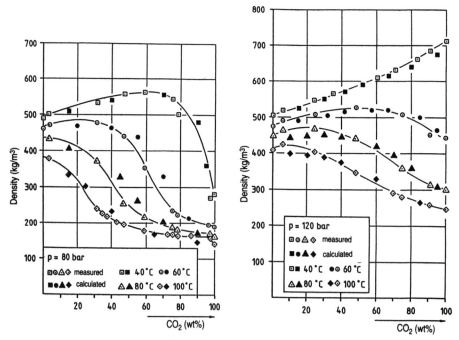

Fig. 4.18. Densities of carbon dioxide/propane mixtures at 80 and 120 bar as a function of composition at several temperatures. *Source:* Peter, S., and U. Ender, *Fat Science and Technology* 91(7):260 (1989).

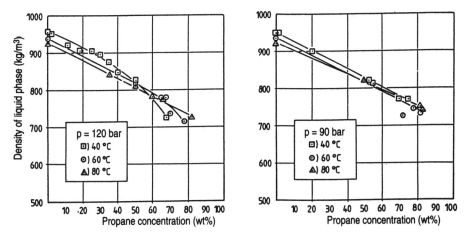

Fig. 4.19. Density of the coexisting liquid phase in the system carbon dioxide/propane/mixture of oleic acid glycerides as a function of the propane content at 120 bar and 313 K. *Source:* Peter, S., and U. Ender, *Fat Science and Technology* *91(7):*260 (1989).

The gas chromatograms of feed, bottom, and top product are shown in Fig. 4.20. The bottom product is nearly free of di- and triglycerides. The top product contains about 10 wt% monoglycerides and 70 wt% diglycerides (23).

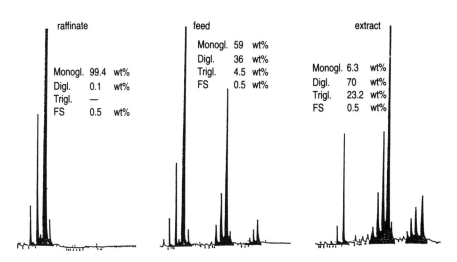

Fig. 4.20. Gas chromatogram of feed, raffinate, and extract, *Source:* Ender, U., and S. Peter, *Chemical Engineering Processes 26:*207 (1989).

A very promising method to produce high-grade monoglycerides proceeds from the reaction product of the glycerolysis. This product has the following approximate composition:

Glycerol 5–14 wt%

Free fatty acids 1 wt%

Glycerol monoesters 45–55 wt%

Glycerol diesters 30–40 wt%

Glycerol triesters 2–8 wt%

Catalyst 0.1 wt%

Impurities 1 wt%.

The mixture is extracted with propane at 5.0 MPa and temperatures above 323 K. The glycerol-diesters and glycerol-triesters are extracted by the propane using a counter-current process. Glycerol monoesters and glycerol form the raffinate. The loading of the extractant is high (about 15 to 20 wt%), and therefore the solvent-to-feed ratio is relatively favorable, with values between 2 and 4.

The mixture of propane, a small amount of monoglycerides, and di- and triglycerides leaves the separation column and enters the regeneration column. Reducing the pressure to between 2.0 and 3.0 MPa, at the same temperature as in the extraction column, causes the dissolved nonvolatile material to precipitate. The regenerated extractant is then recycled to the extraction column. A part of the precipitated glycerides is also pumped back to the extraction column as reflux. The glycerol monoester yield of this step amounts to about 95%.

The raffinate is divided into two liquid phases. It is drawn off and the phases are separated by decanting. The heavy phase consists of glycerol with a very small portion of glycerides and impurities (e.g., residues of catalysts). This phase is then recirculated to the reactor. The light phase is expanded to ambient pressure, releasing the dissolved propane. The degassed light phase contains the glycerol monoesters and about 15% glycerol.

The light phase containing the monoglycerides is than mixed with a hydrocarbon (for example, C_3 to C_6) and an alkanol (perhaps ethanol) in a ratio of about 1:1:1. Subsequently, the glycerol is extracted from the solution with water in a countercurrent column. The glycerol content of the monoglycerides is thereby reduced to values smaller than 0.4 wt%. The addition of hydrocarbons to the solution of the monoglycerides prevents the forming of an emulsion; hydrocarbons between C_2 and C_{10} are suitable. The hydrocarbon and the alkanol are removed from the monoglycerides by vacuum distillation or stripping. If glycerol is removed from the glycerides, a distillation temperature of 473 K is necessary, whereas with use of an alkanol (C_1 to C_3) and the hydrocarbons already mentioned, fractionation can be achieved at substantially lower temperatures. This process for removing glycerol from glycerides is especially profitable if glycerol has to be separated from monoglycerides, that are manufactured from coconut oil or palm kernel oil. The yield of glycerol monoesters by this step amounts to 90% (Czech, B., S. Peter and E. Weidner, patent pending).

Production of Powders by Rapid Expansion

The rapid expansion of solutions in supercritical media through a nozzle in order to produce fine powders of nonvolatile, thermally labile compounds has been thoroughly investigated by several authors, including Debenedetti (24) and Matson et al. (25). The resultant particle sizes are determined by competitive processes such as nucleation, particle growth, cooling due to the Joule-Thomson effect, and supersaturation.

By rapid expansion of supercritical solutions thin films, or crystalline or amorphous powders with narrow and controllable particle size distribution can be produced. Usually, crystallization or precipitation from the solution in a dense supercritical medium is utilized; however, this procedure requires relatively high pressures in order to generate solutions with high concentrations. The processing of raw lecithin in a countercurrent column by near-critical fluid extraction at 9.0 MPa and 330 K results in an extract containing the soybean oil and a raffinate containing the lecithin. The raffinate consists of a mixture of 64 wt% lecithin, 30 wt% propane and 6 wt% carbon dioxide. When the liquid raffinate is drawn off from the extraction column through an expansion nozzle, a fine powder of lecithin is obtained. Carbon dioxide and propane vaporize in the expansion valve. The temperature decreases, and lecithin precipitates as a solid. When the pipe diameter increases continuously downstream from the expansion valve, a powdery lecithin with bulk density between 150 and 200 kg/m^3 is obtained. By changing the opening angle, the resultant particle size can be controlled.

In the course of investigations on the dissolution of fatty acids in dense supercritical carbon dioxide and ethane using a falling film, disintegration was observed if a certain pressure was exceeded (Fig. 4.7), producing a mist of tiny droplets. The interfacial tension of such systems decreases with increasing activity of the supercritical component as indicated in Fig. 4.21.

Because the boundary between a liquid and gas is in a state of tension, one can indirectly study the interfacial adsorption by means of the Gibbs adsorption equation:

$$\Gamma 1 = -\left(\frac{1}{RT} \frac{d\sigma}{dl\,na_1} \right)_{P,T}$$

where Γ = surface excess. The activity of the pure gas is used as a first approximation, because the solute concentration does not exceed 5 wt%. A virial equation with four parameters relating interface tension σ and gas activity a_1 can be applied to represent the measuring results. From the differentiation of this equation, the numerical values of the change in σ with activity can be obtained.

The interfacial excess of the dense gas in the linoleic acid/carbon dioxide system at 313, 333, and 353 K is shown in Fig. 4.22 as a function of gas activity. The surface

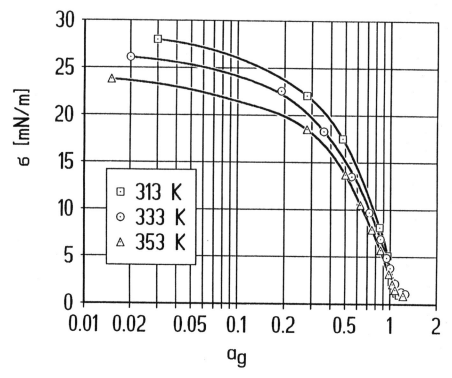

Fig. 4.21. Interfacial tension of the system linoleic acid/carbon dioxide as a function of pressure at 313 K. *Source:* Hiller, N., et al., *Chemical Engineering Technology 16*:206 (1993); Peter, S., and A. Beyer, *Abstract Handbook of the Second International Symposium on High Pressure Chemical Engineering,* Erlangen, 1990, pp. 159–164.

excess increases with increasing activity, runs through a marked maximum, and decreases rapidly to small values. The region of film disintegration is marked by black points. Disintegration occurs near the maximum value of the interfacial excess of the supercritical component. At the lower boundary of the disintegration process, the formation of droplets begins. The size of the droplets in this phase is in the mm region. With further increase in the gas activity, the size of the droplets decreases. Finally at the upper boundary limit, a fine mist of droplets emerges.

Because the input of mechanical energy into the film is small and constant, it is obvious that in the region of film disintegration, the interfacial energy and the energy of adsorption become equal. Thus, the spontaneous collapse of the film into very tiny droplets can be rationalized. This phenomenon is the basis for producing fine powders of solid materials (26).

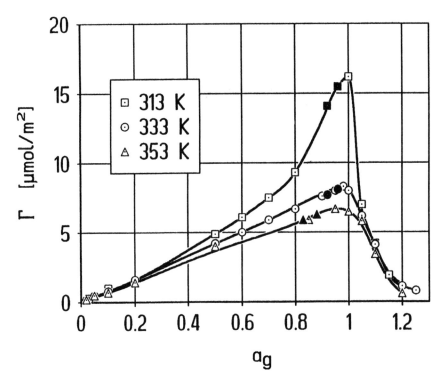

Fig. 4.22. Interfacial tension in the system linoleic acid/carbon dioxide at 313, 333, and 353 K as a function of gas activity. *Source:* Hiller, N., et al., *Chemical Engineering Technology* 16:206 (1993); Peter, S., and A. Beyer, *Abstract Handbook of the Second International Symposium on High Pressure Chemical Engineering,* Erlangen, 1990, pp. 159–164.

In the course of deoiling of soybean lecithin by critical fluid extraction, a liquid raffinate consisting of lecithin, propane, and carbon dioxide is obtained. The raffinate is in a state beyond the maximum of the surface excess. Therefore, the region of spontaneous disintegration is passed when the raffinate is withdrawn under expansion. By consequence of the Joule-Thomson effect, temperature and particle growth decrease simultaneously, and a powdery lecithin with bulk densities between 150 and 200 kg/m^3 is obtained. In this procedure, the concentration of the primary solution is much higher than that of a solution in the supercritical solvent at equal pressure. The pressures necessary to get high concentrations of the material to be pulverized are considerably lower.

The same characteristics are observed in the preparation of high-grade glycerol monostearates. When the liquid raffinate, consisting of propane, carbon dioxide, and monoglycerides of stearic acid, is rapidly expanded from 8.0 MPa and 333 K to ambient pressure, a very fine powder emerges. The temperature decreases due to the Joule-Thomson effect by about 50 to 70 K, depending on the ratio of propane to carbon dioxide.

Polyunsaturated Fatty Acids from Fish Oil

Polyunsaturated fatty acids are of major interest for their therapeutic potential. Consumption of polyunsaturated fatty acids is associated with lower incidences of heart attacks and absence of autoimmune or inflammatory diseases, such as rheumatoid arthritis and ulcerative colitis. The characteristic feature of fish oils is their high content of polyunsaturated fatty acids of the ω-3 type. Fish and marine-mammal oils contain substantial amounts of fatty acid moieties with 20 or 22 carbon atoms and four to six double bonds.

Toothed-whale oils are not promising for the preparation of polyunsaturated fatty acids, because they often contain glycerol ethers, wax esters, or hydrocarbons such as pristane and squalene. About 1 million tons of commercial fish oils are available in triglyceride form as a primary product for the preparation of high-grade polyunsaturated fatty acids.

The fatty acids of most interest in marine oils are eicosapentaenoic acid, 20:5 (EPA) and docosahexaenoic acid, 22:6 (DHA). Both EPA and DHA are ω-3 in structure. In enrichment processes arachidonic acid (20:4) usually concentrates with EPA. It amounts to 1/10 to 1/20 of EPA present. Commercially available marine oils have iodine values ranging from about 95 to 195 (pilchard oils). The totals for unsaturated acids correspond then to 10 and 43%. These values define the range that can be recovered at maximum from marine oils. EPA and DHA usually total about 80% of all polyunsaturated fatty acids in marine oils.

The fish body oils are divided in two groups: those with more than 10% (20:1 and 22:1) and those with less. If there is little or no (20:1 and 22:1), then EPA and DHA are usually high, perhaps about 20% of total fatty acids. If much 20:1 and 22:1 are present, they dilute the EPA and DHA to as little as 8 to 10%. A relatively high proportion of EPA and DHA is in the 2 position of glycerides. The fish liver oils usually consist of more 22:1 than 20:1, and 20:5 approximately equal to 22:6 fatty acid composition. Fish oils contain about 0.5% dissolved water, even in refined oils, unless they are vacuum-stripped at high temperatures. All fish oils contain 0.5 to 1.0% sterols, and usually significant levels of cholesterol.

The content of unsaturated fatty acids can be enhanced by slow cooling to about 5°C, thereby crystallizing the saturated fatty acids. The unsaturated fatty acids remain in the liquid phase (27). The only other method that can enrich oils without using solvents is short-path distillation, which requires high vacuum; however, some degradation of EPA and DHA is unavoidable because of the high temperatures used in the distillation (200°C).

Separation of these desired fatty acids is accomplished by conversion to their monoesters. These can be produced from triglycerides by transesterification with methanol or ethanol. Alkaline alcoholysis, according to Gauglitz and Lehmann (28), has been found to be suitable for manufacturing methyl esters or ethyl esters from fish oils. The Solexol process with supercritical propane yields some enrichment of fish oils (29). Rough calculations suggest that EPA plus DHA can be increased from 28 to 47% using this process.

Eisenbach (30) investigated the separation of codfish ethyl esters by means of supercritical carbon dioxide as selective solvent. A fraction of C_{22} ester with a purity of 98% could be achieved in a two-stage batch extraction but at very low throughput. Stout and Spinelli (31) patented a process that comprises transesterifying fish oil glycerides with a lower alkanol and extracting the esters with carbon dioxide under supercritical conditions at 17.0 MPa and 74 to 75°C. The extraction occurs in fewer steps than in the process of Eisenbach and is therefore more economical.

In the quest for a continuous process, the phase equilibria were investigated in systems composed of the ethyl ester of a fish oil and carbon dioxide. The fish oil used utilized the following composition:

C_{14}	1.4%	$C_{20:0}$	6.8%	
C_{15}	0.3%	$C_{20:5}$	3.1%	(EPA)
C_{16}	25.9%	C_{21}	4.0%	
C_{17}	2.0%	$C_{22:0}$	7.3%	
C_{18}	38.9%	$C_{22:6}$	3.4%	(DHA)
C_{19}	1.3%	$>C_{22}$	5.6%	

In Fig. 4.23 the concentration of the ethyl esters in the coexistent phases is plotted vs. pressure for the quasi-binary system of fish oil ethyl ester and carbon dioxide at 50, 70, and 90°C. The ester concentration of the gaseous phase is shown on an extended scale. Pressures above 20.0 MPa are required to generate loadings in the gaseous phase of more than 3 wt%. The loading of the gas phase may be enhanced by the addition of propane as an entrainer. At concentrations up to 20 wt% the separation factor does not change greatly.

The continuous fractionation of fish oil ethyl esters with supercritical carbon dioxide was also investigated in a countercurrent extraction column. A schematic diagram of the apparatus is given in Fig. 4.24. A sufficient amount of ethyl ester is prepared from an acid-refined and bleached fish oil. Tocopherol is added to the fish oil ethyl ester in a concentration of 0.1 to 0.5% for stabilization. In addition, the feed vessel is flooded with carbon dioxide in order to prevent oxidation of the oil.

The preheated feed is then pumped in the column. The extractant enters the column at the bottom. The extractant leaves the column at the top and passes through a heat exchanger and pressure control valve. A portion of the dissolved esters precipitates in the separator, A1, and is then recycled to the top of the extraction column as reflux. The regenerated extractant is recycled to the extraction column *via* a heat exchanger and a gas ballast.

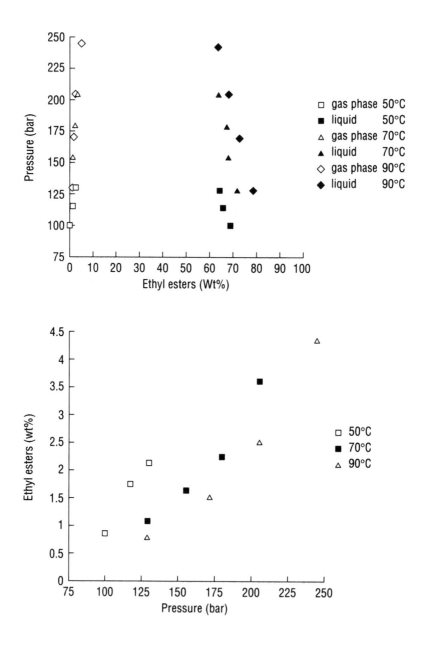

Fig. 4.23. Phase equilibria in the fish oil ethyl ester/carbon dioxide system at 50, 70, 90°C.

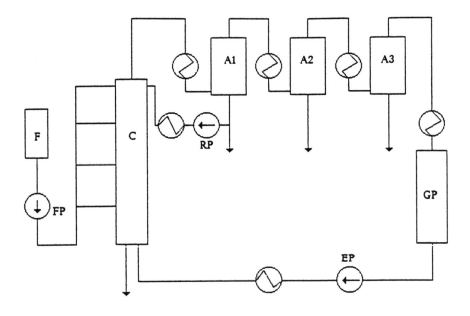

F feed vessel
C extraction column   FP feed pump
A1–A3 separator   RP reflux pump
GP gas buffer   EP extractant pump

Fig. 4.24. Flow diagram of supercritical extraction of fish oil ethyl esters.

The fish oil feed had the following composition:

C_{14}	11.1%	$C_{20:5}$	12.2%
C_{16}	38.8%	$C_{22:1}$	1.3%
C_{18}	24.5%	$C_{22:5}$	1.5%
$C_{20:1}$	3.1%	$C_{22:6}$	5.7%
		$C_{24:1}$	0.8%

A feed rate of 3.35 kg/h yielded a product at 0.76 kg/h. A solvent-to-feed ratio of 29 was used at a pressure of 31.0 MPa and temperature at 309 K, resulting in a loading of about 3.7%. The resultant raffinate composition was

C_{14}	6.8%	$C_{20:5}$	19.0%
C_{16}	29.7%	$C_{22:1}$	2.0%
C_{18}	23.6%	$C_{22:5}$	2.1%
$C_{20:1}$	3.6%	$C_{22:6}$	11.0%
		$C_{24:1}$	2.7%

The composition of the extract was

C_{14}	11.6%	$C_{20:5}$	12.4%
C_{16}	43.7%	$C_{22:1}$	1.1%
C_{18}	21.5%	$C_{22:5}$	1.0%
$C_{20:1}$	2.5%	$C_{22:6}$	5.1%
		$C_{24:1}$	1.0%

The ratio of 20:5 to 20:1 amounted to 5.3 in the raffinate and to 4.96 in the extract; the ratio of 22:6 to 22:1 in the raffinate was 5.5, and 4.63 in the extract. The enrichment of the polyunsaturated fatty acid is higher than that of the other fatty acids of the same chain length. Thus, a separation appears to occur not only with respect to molecular weight but also with regard to the number of double bonds in the fatty acids.

As the column had about four theoretical stages, the separation efficiency leaves much to be desired. An optimization between loading and separation factor may yield results of commercial interest.

Conclusions

The solubility of organic substances in dense gases depends on their vapor pressure—that is, their molecular weight, if interaction forces with the dense fluid are absent. The separation of the monoesters of long-chain fatty acids from those of shorter chain length is easily achieved by means of dense gases.

Substances that differ by the number of OH groups (e.g., a mixture of mono-, di-, and triglycerides) can be separated by a supercritical extractant composed of carbon dioxide and propane. The triglycerides are more soluble in the nonpolar supercritical extractant than the diglycerides are the monoglycerides are less soluble than the diglycerides. The more OH groups a substance contains, the less its solubility in the supercritical nonpolar solvent. The interaction forces between the OH groups outweigh the influence of the molecular weight. Therefore the preparation of high-grade monococoates is feasible without changing the pattern of the fatty acids.

The formation of hydrogen bonding is a characteristic feature of OH groups; therefore, the distribution between the coexisting phases obviously depends on extent of hydrogen bonding of the different compounds in both phases.

The polarity of dense carbon dioxide can be increased by addition of water. Thus, the solubility of the weak base caffeine in dense carbon dioxide can be doubled by the addition of water. At 293 K, saturation with carbon dioxide lowers the pH of water from 6 at a pressure of 3×10^{-4} bar to about 4 at 100 bar.

High-grade phosphatidylcholine was prepared from soybean lecithin by using a supercritical mixture of carbon dioxide and ethanol as an extractant. Association between phosphatidylcholine and ethanol is stronger than that between ethanol and the remaining phosphatides. As a result, its distribution between the coexisting phases is enhanced by the ethanol, and it becomes enriched in the gaseous phase.

References

1. Bruno, Th.J., and J.F. Ely, *Supercritical Fluid Technology,* CRC Press, Boca Raton, FL, 1991; Brunner, G., *Intl. Chem. Eng. 30:*191 (1990); Brunner, G., and S. Peter, *Ger. Chem. Eng. 5:*181 (1982); Paulaitis, M.E., V.J. Krukonis, R.T. Kurnik, and R.C. Reid, *Rev. Chem. Eng. 1(2):*179 (1983); *Chemical Engineering at Supercritical Fluid Conditions,* edited by M.E. Paulaitis, J.M.L. Penninger, R.D. Gray, Jr., and P. Davidson, Ann Arbor Science, Ann Arbor, MI 1983; *Supercritical Fluid Technology,* edited by J.M.L. Penninger, M. Radosz, M.A. McHugh and V.J. Krukonis, Elsevier, Amsterdam (1985); Peter, S., *Ber. Bunsenges. Phys. Chem. 88:*875 (1984); *Angew. Chem. Intl. Ed. Engl. 17(10):*701 (1978).
2. Roof, J.G., and J.D. Baron. *J. Chem. Eng. Data 12:*292 (1967).
3. Zosel, K., Angew. *Chem. Intl. Ed. Engl. 17(10):*702 (1978).
4. Peter, S., G. Brunner and R. Riha, *Ger. Chem. Eng. 1:*26 (1978); Peter, S., G. Brunner, and R. Riha, *Fette Seifen Anstrichm. 78:*45 (1976); Peter, S., *Ber. Bunsenges. Phys. Chem. 88:*875 (1984).
5. Peter, S., and H. Jakob, *J. Supercritical Fluids 4:*166 (1991).
6. Peter, S., M. Seekamp, M. and A. Bayer in *Proceedings of the First International Symposium on Supercritical Fluids,* Nice, Oct. 17–19, 1988, Vol. 1, edited by M. Perral, p. 99.
7. Czech, B., , Ph.D. Thesis, Performance of a Counter-Current Column of the Production of Diglycerides by Near-Critical Extraction, University of Erlangen, 1991.
8. Zosel, K., German Patent DAS 2 332 038 (1973).
9. Van Leer, R.A., T.R. Bergstresser and M.E. Paulaitis, *Spring Meeting of the Electrochemical Society, St. Louis, 1980, Extended Abstracts Vol. 80–1,* Abstract No. 549.
10. Coenen, H., and E. Kriegel, German Patent DOS 2 843 920 (1978).
11. Brunner, G., and S. Peter, *Sep. Sci. Technol. 17(1):*199 (1982).
12. Heigel, W., and R. Hüschens, German Patent DE 3 011 185 (1980).
13. Coenen, H., and R. Hagen, German Patent DE 32 29 041 A1 (1982).
14. Stahl, E., Quirin, K.W., and A. Hübgen, German Patent DE 33 29 249 (1983).
15. Stahl, E., K.W. Quirin, A. Glatz, D. Gerard, and G. Rau, *Ber. Bunsenges. Phys. Chem. 88:*900 (1984).
16. Peter, S., M. Schneider, E. Weidner, and R. Ziegelitz, *Chem. Eng. Technol. 10:*37 (1987).
17. Scholfield, C.R., H.J. Rutton, F.W. Tanner, Jr., and J.C. Cowan, *J. Am. Oil Chem. Soc. 25:*386 (1948). Liebing, H., and J. Lau, *Fette Seifen Anstrichm. 78:*123 (1976).

18. Pardun, H., *Fette Seifen Anstrichm. 86:*55 (1984).
19. Betzing, H., and H. Eikermann, German Patent DE A 16 17 679 (1967); Napp, W., German Patent DE A 32 27 001 C1 (1982); Fujiwara, S., T. Mukohara and Y. Koh, *Shokuhin Sangyo Senta Gijutsu Kenkyu Hokuku 17:*61 (1991).
20. Kuhrt, N.H., E.A. Welch and F.J. Kowarik, *J. Am. Oil Chem. Soc. 24:*49 (1947).
21. Peter, S., and U. Ender, *Fat Sci. Technol. 91(7):*260 (1989).
22. Schmidt, G., and H. Wenzel, *Chem. Eng. Sci. 35:*1503 (1980).
23. Ender, U., and S. Peter, *Chem. Eng. Process. 26:*207 (1989).
24. Debenedetti, P.D., *AIChE J. 36:*1289 (1990).
25. Matson, D.W., K.A. Norton and R.D. Smith, *Chemtech 1989,* p. J80; Matson, D.W., J.L. Fulton, R.C. Peterson, and R.D. Smith, *Ind. Eng. Chem. Res. 26:*2298 (1987); Matson, D.W., R.C. Peterson, and R.D. Smith, *J. Mat. Sci. 22:*1919 (1987); Matson, D.W., J.L. Fulton, and R.D. Smith, *Mat. Lett. 6:*31 (1987).
26. Hiller, N., H. Schiemann, E. Weidner, and S. Peter, *Chem. Eng. Technol.,* in press; Peter, S., and A. Beyer in *Abstract Handbook of the Second International Symposium on High Pressure Chemical Engineering,* Erlangen, 1990, pp. 159–164.
27. Bailey, R.E., *Marine Oils with Particular Reference to Those of Canada,* Bull. 89, Fisheries Research Board of Canada, Ottawa, 1952.
28. Gauglitz, E.J., Jr., and L.W. Lehmann, *J. Am. Oil Chem. Soc. 40:*197 (1963).
29. Passimo, H.J., *Ind. Eng. Chem. 41:*280 (1949).
30. Eisenbach, W., *Ber. Bunsenges. Phys. Chem. 88:*882 (1984).
31. Stout, V.F., and J. Spinelli U.S. Patent 4,675,132 (1986).

Chapter 5

Oilseed Solubility and Extraction Modeling

J.W. Goodrum[a], M.K. Kilgo[a], and C.R. Santerre[b]

[a]Biological and Agricultural Engineering Department, University of Georgia, Athens, GA and [b]Food Science Department, University of Georgia, Athens, GA.

Introduction

It is believed that the oil produced from oilseeds by supercritical carbon dioxide (SC-CO$_2$) extraction may be superior to products of hexane extraction or an expeller, due to selective extraction of desirable oil fractions from the seed. Because the solvent passes through a stationary bed of seeds, SC-CO$_2$ oil recovery is not susceptible to mechanical damage as is the screw press, which may be damaged by rocks and other foreign material in the feed being forced through the metal screws. Hexane extraction presents a severe fire hazard to the equipment operator and is associated with solvent residues in products and with energy-intensive solvent recycle distillation operations.

Several groups have studied SC-CO$_2$ extraction of many types of oilseeds. Some work concentrated on improving flavor and shelf life of edible oil (2,3). Pressure, temperature, particle size, and moisture were reported as key extraction parameters by Stahl et al. (4) and by Snyder et al. (5). List and Friedrich (6) investigated oil recovery from corn and cottonseed. Goodrum and Kilgo (7,8) focused on oil recovery from peanuts via CO$_2$.

The potential for selective recovery of materials from oilseeds continues to be studied. Early studies cited above reported that less phospholipid ("gums") was extracted by SC-CO$_2$ than by the conventional hexane method. In general, it has been found that CO$_2$ and other supercritical solvents may be optimized for selective recovery of desired components by adjusting temperature, pressure, or extraction time. By combining co-solvents with CO$_2$, the selectivity has been increased in some instances. It seems likely that selective triglyceride recovery will be useful in the future for high-value triglyceride fractions, perhaps from novel oilseeds.

Studies of the Solubility of Peanut Oil in CO$_2$

A typical fatty acid composition of peanut oil is given in Table 5.1 (9,10). The relative fatty acid concentration in peanut oil varies considerably between peanut varieties (9,11); however, the major triglycerides are present in all varieties, and it is expected that the general solubility patterns for oil in CO$_2$ will be very similar for most varieties.

Equipment and General Procedure

Goodrum and Kilgo (8) sought to identify variables that lead to efficient oil extraction using CO_2 solvent. They also included a study of methods to reduce extraction of gums and other undesirable chemical fractions. They determined the solubility of peanut and rapeseed oil by a dynamic method in carbon dioxide as a function of pressure, temperature, particle size, moisture, solvent flow rate, and batch size.

The experiments were intended to aid in the design and testing of solvent extraction equipment and controls suitable for efficient large-scale operation. The large 2 in. (5 cm) id × 15 in. (38 cm) id extraction chamber permitted data collection on flow uniformity and thermal gradient control in the extraction chamber.

The equipment used for oil extraction with CO_2 includes the basic units described by Stahl et al. (2) and Friedrich et al. (3). Figure 5.1 shows a schematic of the extraction chamber and temperature control system. An air-driven two-stage compressor (Haskel Model #54488, Burbank, CA) supplied SC-CO_2 without pressure fluctuations. Solvent from the compressor entered a heated 2 L extraction vessel filled with peanut meal. The pressure of the exiting CO_2 was reduced by a micrometering valve (Autoclave Engineers, Erie, PA). This drastic reduction in pressure leads to cooling effects. Therefore, the decompressed solvent/oil mixture was heated to avoid oil and CO_2 solidification. As the mixture entered the receiver, the oil separated from the CO_2 and was drained out of the bottom.

Each experiment was performed on a batch basis. The extractor was filled with peanuts and then connected to the system. In order to minimize small observed thermal gradients in the fixed meal bed, the bed was preheated with three heating tapes connected to temperature controllers (Oven Industries, Inc., Mechanicsburg, PA) for several hours before beginning the extraction. After the preheat period, CO_2 solvent was allowed to flow through the bed at the chosen flow rate. During the extraction, heating of the feed was enhanced with a section of tightly coiled stainless steel wire at

TABLE 5.1 Typical Composition of Peanut (9) and Rapeseed (10) Oils

Fatty Acid		Peanuts	Rapeseed[a]
16:0	Palmitic	9.9	3.7
18:0	Stearic	2.5	1.1
18:1	Oleic	51.0	56.1
18:2	Linoleic	32.6	24.3
18:3	Linolenic	—	12.4
20:0	Arachidic	1.1	—
20:1	Eicosenoic	1.0	2.3
22:0	Behenic	1.4	—
22:1	Erucic	—	—
24:0	Lignoceric	0.5	—

[a]Ref. (10), p. 20

the inlet of the pressure vessel. This provided a large amount of metal surface area for conduction of heat from the vessel walls to the incoming CO_2.

The oil was collected from the separator and weighed at intervals. The corresponding weight of CO_2 used was recorded by passing the exhaust CO_2 gas through a flow meter (Sprague, Model 22, Owenton, KY) before being vented to the atmosphere. As a check on mass balance, the change in weight of CO_2 supply cylinders was recorded.

The temperature profile inside the heated extraction vessel was monitored and the heat rate adjusted to obtain the desired isothermal temperature conditions. In initial experiments surface thermometers measured the temperature at each end of the extractor. Later, semiconductor thermosensor probes were attached at four locations on the outside of the extractor. Two additional thermocouples were installed in the inlet and outlet solvent streams. A microcomputer displayed the temperatures, made a printed record, and recorded data to disk (see Fig. 5.1).

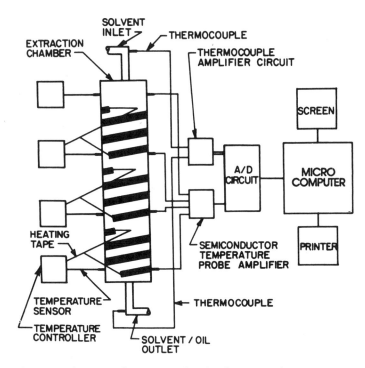

Fig. 5.1. Schematic of extraction chamber heating and temperature sensing system.

Oilseed Grinding Methods and Sample Analysis. Several methods of particle size reduction were tested for their ability to produce high oil yields. In one method whole peanuts were fed by batches into a rotary cutter and ground to various sizes. Ranges of particle size were determined by sieving the particles with wire screens. Another method of size reduction was to pass the peanuts through a rolling mill (6″ × 6″ Rollmaster with corrugated rolls, Roskamp Mfg., Waterloo, IA). Particles were classified according to the thickness as well as the size ranges of the sieves.

For some experiments the oil was partially recovered from oilseeds in a screw press (Hander New Type 52, CeCoCo, Hander Corp., Osaka, Japan) prior to SC-CO$_2$ extraction. Rittner (12) examined this pretreatment approach in some depth. By bringing the press to its maximum production point, which is characterized by continuous flakes of brown, very thin meal, and then backing off the auger by a fixed number of turns, the amount of oil that was recovered in the press was controlled. Samples were extracted with hexane in a Soxhlet extractor (Fisher Scientific, Pittsburgh, PA) using AOCS Standard Method A63-49 (13) to determine the exact amount of oil remaining in them.

Oilseed moisture contents reported throughout this paper were determined in a Steinlite moisture meter (Stein Laboratories, Atcheson, KS) unless otherwise stated. Appropriate calibration charts were used for peanuts and rapeseed.

Extractor Trials

Horizontal Extractor Trial. The first extraction experiments were performed using a horizontal extractor. Samples, 1 kg each, of peanuts were ground to less than 1.68 mm diameter in a rotary cutter. Ninety percent by weight was between 0.86 and 1.68 mm in diameter, giving an average diameter of 1.27 mm. The CO$_2$ flow rate was maintained at 20 L/min as measured at STP (standard temperature and pressure, 0°C and 1 atm). Perforated baffles were tested as a means of obtaining a uniform cross-sectional flow of solvent through the extractor unit.

Vertical Extractor Trials. Experiments were performed with the extractor oriented vertically. Samples, 500 g each, of peanuts were ground to less than 0.86 mm and processed. A lower limit of the particle size could not be obtained because the oily peanut particles stuck together and would not go through the screen. The CO$_2$ flow rate was 40 L/min as measured at STP. Trials were performed with solvent flowing either upward or downward through the bed.

Extractor Trial Results

Horizontal Extractor Trials. With the horizontal unit, estimates of the solubility were made from the initial maximum extraction rate at pressures of 275 to 680 bar and temperatures of 25 to 100°C. The results are in qualitative agreement with those obtained by Stahl et al. (4) and Friedrich et al. (14) for soybean oil. The highest oil recovery was 50% after 3 h of extraction.

The peanut meal removed from the extractor was uneven in color, with the bottom section remaining yellow and oily. This apparent solvent channeling may have been largely due to meal settling in the extractor, leaving a lower-resistance flow path along the top of the meal. Also, the solvent CO_2 density was less than that of peanut meal. Therefore, the peanut meal near the bottom was exposed to less fresh solvent. In order to increase the recovery from the meal sample, vertical baffles were placed inside the horizontal extractor to direct the CO_2 flow downward. The baffles permitted a maximum oil recovery of 70%, but visible solvent channeling remained.

Vertical Extractor Trials. With the extractor oriented vertically and CO_2 flowing upward through the bed, the extraction remained at the maximum rate for longer periods of time and higher oil recoveries were obtained than with the extractor oriented horizontally. Although the maximum extraction rates did not significantly vary from values obtained with the horizontal reactor, the total oil yield increased to 99% at 550 bar and 95°C during a 3 h extraction.

With downward solvent flow through the vertical extractor, the initial extraction rates and overall yields were the same or slightly better than those of the upward flow experiments (Figs. 5.2 and 5.3). Increases in the extraction rates were attributed to improvements in the uniformity of the extractor interior temperatures. Because of higher temperatures induced by convection effects at the top of the extractor, it was easier to heat solvent entering at the top than at the bottom. More uniform temperatures throughout the extractor produced more reliable results than experiments in which the first part of the bed was cooled by the incoming solvent.

Physical Description of Oilseed Meals

Since the product of the vertical extractor had uniform, reproducible visual characteristics, these are briefly described as follows. Each sample was removed from the extractor by pushing the material out onto a table, maintaining the cylindrical shape of the meal. The meal located at the solvent inlet was extracted, whereas the meal at the solvent outlet was not completely extracted. The change from extracted to unextracted meal could be seen as a gradual color change.

In general, the color of the meal product was unaffected by pressure of extraction. Unextracted zones of the sample were somewhat darker and more oily than the extracted areas.

Higher temperatures, with constant pressure, produced darker meal. As temperatures reached 75°C, meal from the solvent inlet changed from chalk white to light brown. At 95°C, the meal product was dark brown.

The extracted meal occupied less volume as extraction pressure and temperature increased. This volume change was noted by measuring the volume of untreated ground peanuts compared to the volume of the meal product. At 25°C, the meal crumbled when touched. At 100°C, residual meal volume decreased with increasing CO_2 pressure. There was a visual similarity to sintering of ceramic powders. At 100°C and

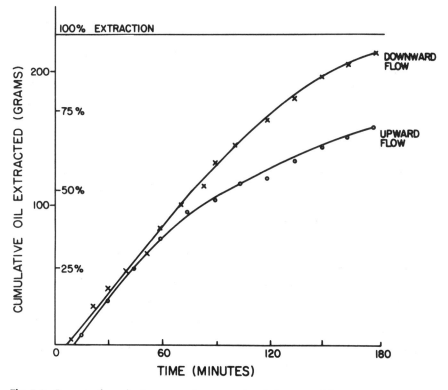

Fig. 5.2. Peanut oil production curve for upward and downward flow at 415 bar and 50°C.

8000 psi (550 bar), the residual meal was a coherent mass that was difficult to fracture manually. This bulk volume reduction of the ground peanuts was attributed to compaction of the meal by the high pressure of the solvent and to volume reduction from the elimination of oil from the meal.

Crude oil products contained 1/2 to 1% filterable solids and varied in color with changing extraction conditions. In general the oil color ranged from yellow to dark brown as the temperature of extraction was increased from 25 to 120°C. None of these physical changes affected the oil recovery.

Experimental Designs

When investigating a new type of process or trying to improve an existing one, one finds that a large number of process conditions may affect the production rate or yield; however, usually only a few of these factors have a significant effect (14). The effect of these critical parameters on the process is known as the Pareto principle (15). *Fractional factorial* experimental designs will in many cases quickly identify these factors.

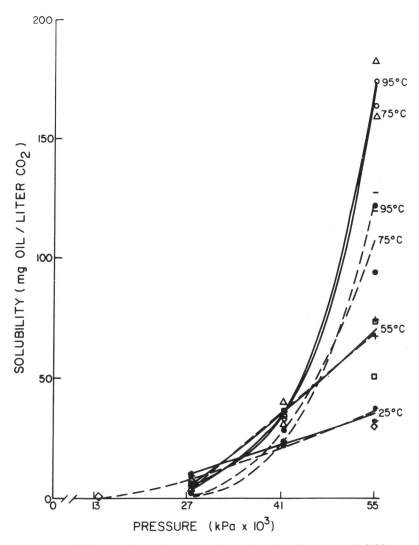

Fig. 5.3. Solubility isotherms for peanut oil as a function of pressure. Solid lines indicate downward CO_2 flow; dotted lines indicate upward CO_2 flow.

There are several reasons for designing an experiment statistically. First, the effects of two or more variables are studied simultaneously. Second, by performing the smallest number of experiments that will provide unambiguous results, the materials and time devoted to the investigation are reduced. Third, designed experiments are the only way to detect interactions between variables when they exist. Finally, by measuring experimental error, the significance of main variables and interactions can be estimated (17).

In large designs the number of runs necessary to perform a complete factorial experiment (in which every variable is tested in a separate series) is sometimes beyond the resources of the investigator. The main advantage of fractional factorial design is that it enables five or more factors to be studied simultaneously in an experiment of practical size, so the investigator can quickly discover which factors have an important effect on the process.

To use a *half-factorial* design without replication, two assumptions must be made. First, each run measures the combination of two effects, known as *aliases* (18,19). In the case of a design of five factors, the aliases are either a main effect with a four-factor interaction or a two-factor interaction with a three-factor interaction. Second, since only a "vital few" of the factors being studied are expected to be significant, it is assumed that third- and higher-order interactions probably will not occur; therefore, each run actually measures only a main effect or a two-factor interaction. To analyze the results, an estimate of the normal variation of the process is needed.

Half-Factorial with Ground Peanut. The effect of pressure, temperature, moisture, particle size, and flow rate on the extraction were studied with a half-factorial experimental design. The conditions that were used are shown in Table 5.2. The desired particle diameters were obtained by grinding whole, raw, runner-type peanuts with a rotary cutter and then sieving to divide the particles into size ranges. Moisture levels were obtained either by adding water to ground peanuts or by drying whole peanuts in an oven before grinding. 526 g of the 5% moisture peanuts and 579 g of the 15% moisture peanuts were used in order to keep the amount of available oil constant at 250 g.

TABLE 5.2 Half-Factorial Design for Ground Peanuts

Exp.#	Pressure (bar)	Temperature (°C)	Moisture (% by wt)	Flow rate (L/min)	Size (mm)
1	415	25	5	40	0.864–1.70
2	550	25	5	40	3.35–4.75
3	415	95	5	40	3.35–4.75
4	550	95	5	40	0.864–1.70
5	415	25	15	40	3.35–4.75
6	550	25	15	40	0.864–1.70
7	415	95	15	40	0.864–1.70
8	550	95	15	40	3.35–4.75
9	415	25	5	60	3.35–4.75
10	550	25	5	60	0.864–1.70
11	415	95	5	60	0.864–1.70
12	550	95	5	60	3.35–4.75
13	415	25	15	60	0.864–1.70
14	550	25	15	60	3.35–4.75
15	415	95	15	60	3.35–4.75
16	550	95	15	60	0.864–1.70

Pressure-Temperature Study. The half-factorial experiments were supplemented with a further study of pressure and temperature effects with moisture, particle size, and flow rate held constant. The moisture level, 9%, represented normal storage conditions for the oilseed. Particle size (<1.18 mm) and flow rate (40 L CO_2/min as measured at STP) were chosen for optimal solubility measurements. Detailed solubility curves and kinetic data were obtained with the vertical extractor and downward solvent flow. For each experiment, 500 g of peanuts were extracted. Each combination of the following temperatures and pressures was used: 25°C, 55°C, 75°C, 95°C; 275 bar, 415 bar, and 550 bar.

Particle Size Study. Reducing the particle size of the oilseed material, thus increasing the surface area and rupturing the cell walls, is necessary for high oil recovery (5). Essentially no diffusion takes place through the cell walls; therefore, the total amount of oil available for extraction depends on the amount of surface area. Small particles have a high surface-area-to-volume ratio; therefore, much of the oil is located on the surface rather than inside the particles. Since more oil is exposed to the solvent, the total oil yield is higher with small particles than with large ones.

A series of detailed tests examined the effect of particle size. Temperature and pressure were fixed at 25°C and 550 bar to yield maximum experimental precision and accuracy. Moisture and CO_2 flow rate were also constant at 9% and 40 L/min, respectively. Five size ranges were used: 0.864 to 1.18 mm, 1.18 to 1.70 mm, 1.70 to 2.36 mm, 2.36 to 3.35 mm, 3.35 to 4.75 mm. Peanut halves were also extracted. After extraction the particles were sieved to check the size distribution. The batch size was 500 g. This data was also used to model the extraction kinetics.

CO_2 Extraction of Prepressed Peanuts. A $2 \times 2 \times 2 \times 2$ factorial design was used to study the effect of preparing the peanuts for SC-CO_2 extraction by pressing some of the oil from them with a screw press. Two levels of pressure (415 and 550 bar) and two levels of temperature (25°C and 95°C) were used in the SC-CO_2 extraction. The amount of oil that was recovered in the press was controlled by adjusting the position of the feed auger. Two samples were chosen for the supercritical CO_2 extractions: 25.6% and 56.4% recovery from the press, determined by AOCS Method Ab3-49 (13). Because it was not known whether or not the action of the screw press would be sufficient to break open the oil cells or to provide enough surface area to give complete extraction with the SC-CO_2 unit, some of the meal was also flaked on the rolling mill. 742 g of low-oil-content peanuts and 527 g of high-oil-content peanuts were used in order to keep the amount of oil available for SC-CO_2 extraction constant at 225 g.

Flaked Peanut Experiments. Half of a $2 \times 2 \times 2 \times 2$ factorial was performed with flaked peanuts, with temperature, pressure, particle size, and moisture as the variables. The pressures were 415 and 550 bar; the temperatures were 25 and 95°C; the moisture levels were 5 and 15%; and two particle sizes were used. The smaller size consisted of particles sieved to the 0.86 to 2.38 mm size and flaked at 1.27 mm thickness, whereas the larger size was sieved to 3.35 to 4.75 mm and flaked at 3.05 mm thickness. The batch size was 526 g for the 5% oil peanuts and 575 g for the 15% oil peanuts in order to keep the amount of oil available for extraction constant at 240 g.

Experimental Design Results

Three sets of data were analyzed: first, a half-factorial design studying the effects of solvent pressure, solvent temperature, solvent flow rates, peanut moisture and peanut particle size; second, various sizes of both ground and flaked peanuts; third, a factorial design using four temperatures and three pressures.

Several methods were used to analyze the data. For the half-factorial experiments both an *F*-test and a graphical analysis of the effects (19) were used. For the particle size experiments the correlation between particle size and each category (fatty acid distribution, residual oil, and moisture) was estimated by linear regression. Graphs of the *P-T* data were used to visually determine trends.

The dependence of the moisture content of the oil on both the solvent temperature and initial peanut moisture was significant ($P < 0.01$). Also, there was a significant ($P < 001$) effect due to the interaction between moisture and temperature.

The residual oil content of the inlet meal was used as a measure of the total amount of oil that can be extracted at various solvent and oilseed conditions. The amount of residual oil in the extracted meal was dependent on particle size ($P < 0.01$). Numerical analysis also indicated that pressure affected the oil content ($P < 0.01$), but graphical analysis indicated that this effect was due to the noise of the system. For the particle size experiments, the relationship between the size effects and the residual oil content had a high correlation. For the pressure-temperature factorial, there was a difference in residual oil content between 4000 psi (275 bar) and 6000 psi (415 bar) but not between 6000 psi (415 bar) and 8000 psi (550 bar). This pressure effect occurred because the slow extraction rates at 4000 psi meant that the assumption that the inlet meal was completely extracted of oil was incorrect. Table 5.3 summarizes the results.

Results for Half-Factorial with Ground Peanuts. The effect of the process variables on solubility, total oil recovery, and amount of volatiles lost in the vented solvent was examined (Table 5.4). For the solubility, the pressure, temperature and the pressure-temperature interaction effects were significant ($P < 0.01$). (All significance levels were calculated as discussed in Ref. 18.) Oilseed moisture and particle size effects were significant to a lesser degree ($P < 0.10$). Neither flow rate nor any other two-factor interaction was obtained from the statistical results:

$$Y = 55.0 + 49.3 \frac{X_1 - 7000}{1000} + 51.8 \frac{X_2 - 60}{35}$$

$$+ 40.1 \left(\frac{X_1 - 7000}{1000}\right)\left(\frac{X_2 - 60}{35}\right) \tag{5.1}$$

TABLE 5.3 Summary of Statistically Significant Effects for the Fatty Acids in Extracted Peanut Oil (Based on Half-Factorial Experimental Design)

Fatty acid number	Graphical parameters	Numerical ($P < 0.01$)	Particle-size Ground	Flaked
14:0	Size	Size, moisture	$r = -.32$	$-.08$
16:0	—	moisture	$-.04$	.16
18:0	Flow rate	flow rate	.70	$-.15$
18:1	Moisture	Moisture	.17	$-.03$
18:2	—	Size	$-.33$	.99
18:3, 20:0, and 20:1	—	—	$-.03$	$-.93$
22:0	—	Flow rate	.70	$-.22$
24:0	Size	Size	.15	$-.71$
	Temp × moisture		$-.29$	$-.92$

where

$$Y = \text{solubility (mg oil/L CO}_2)$$
$$X_1 = \text{pressure (psi)}$$
$$X_2 = \text{temperature (°C)}$$

For percent oil recovered during a 3 h test, the particle size and temperature effects were significant ($P < 0.01$). The pressure effect was less significant ($P < 0.10$). Higher temperatures may promote breakdown of the oil cells and increase the diffusion rate of the oil through the oilseed particle. Also, since the CO_2 density is dependent on pressure and temperature, it appears that more oil inside the particles is extracted. Considering only the particle size and temperature effects to be significant, the following equation predicts the total yield:

$$Y = 54.25 + 44.5 \, \frac{X_1 - 0.555}{0.145} + 19.75 \, \frac{X_2 - 60}{35} \qquad [5.2]$$

where,

$$Y = \% \text{ yield}$$
$$X_1 = \text{particle size (mm)}$$
$$X_2 = \text{temperature (°C)}$$

A mass balance for the extraction process consistently shows a loss of material, ranging from 2% to 20%. We attribute this to loss of volatile light organics and water vapor in the exhaust CO_2 stream. The amount of volatiles lost was affected by the amount of moisture ($P < 0.05$) and the moisture-temperature and moisture-pressure interactions ($P < 0.10$). The higher the moisture level in the ground peanuts, the greater the mass balance loss. At higher temperatures, volatile components would have a

TABLE 5.4 Half-Factorial Results for Ground Peanuts.

Exp #	Solubility (mg oil/L)	Yield (in 3 h)	Volatiles lost (q)
1	29.2	63%	34
2	23.0	21	18
3	37.0	36	35
4	139.7	99	11
5	23.3	24	48
6	38.3	66	58
7	42.6	71	103
8	141.36	54	120
9	22.4	23	96
10	37.2	74	29
11	31.3	80	45
12	48.6	33	26
13	22.9	63	58
14	36.2	21	58
15	33.6	44	—
16	172.6	96	99

greater vapor pressure, so a greater mass would be lost in the exhaust CO_2 stream. The positive moisture-pressure is due to water's enhanced solubility in $SeCO_2$ under pressure. The following equation can be used to predict the amount of volatiles lost, considering only the moisture effect to be significant:

$$Y = 55.5 + 37.5 \frac{X_1 - 10}{5}$$

[5.3]

where

$$Y \;=\; \text{amount of volatiles (grams)}$$
$$X_1 \;=\; \%\ \text{moisture}$$

Since only a few two-factor interactions appear to have an effect, our assumption that all third- and higher-order interactions are negligible is a good one. Flow rate was found to have no effect on the solubility, percent oil recovered, or volatiles lost. However, the extraction can be completed more quickly at higher flow rates, because more CO_2 passes through the extractor per unit of time.

Results for Pressure-Temperature Study. The dependence of oil recovery rate on CO_2 pressure is given in Fig. 5.4. The initial solubility is greater at higher pressure, and therefore the initial recovery rates are much greater at higher pressures. Note that for the fixed-bed configuration used the rate of oil recovery at 550 bar approximated the oil recovery rate, the rate at 275 bar near the end of the three-hour test. The dependence of oil recovery rate on temperatures may be seen in Figs. 5.5 and 5.6. By comparison of these figures, the dominant influence of pressure on oil recovery becomes evident. Also note that at 275 bar the initial oil recovery rate of 7.6 mg oil/L CO_2 decreased to 0.5 mg oil when bed temperature was increased from 27°C to 100°C. The overall yield after a 3 h run decreased from 24% to 1%. This temperature crossover occurred at approximately 415 bar and is attributed to variations in the density of CO_2.

The pressure-temperature solubility study is summarized in Fig. 5.3. It can be seen from the graph that increasing the pressure at a constant temperature can increase the solubility by as much as 150 mg oil/L CO_2. Increasing the temperature at constant pressure decreases the solubility below about 350 bar but increases it at higher pressures. Figure 5.3 also shows the results from the previous study with upward flow. For most levels of the factors, the flow direction did not greatly affect the initial extraction rate. The largest difference occurred for 550 bar at 75°C and 95°C. With upward flow a temperature gradient of up to 50°C developed between the top and bottom of the ex-

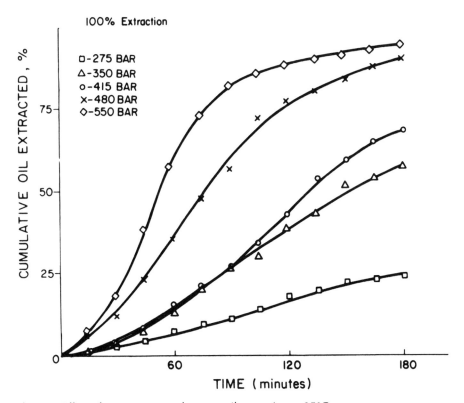

Fig. 5.4. Effect of pressure on total peanut oil extraction at 25°C.

J.W. Goodrum et al.

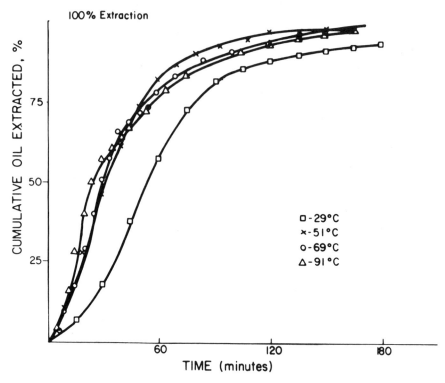

Fig. 5.5. Effect of temperature on total peanut oil extraction at 550 bar.

tractor as the incoming CO_2 cooled the inside of the extractor. With downward flow it was possible to maintain the temperature throughout the experiment. Consequently, the extraction rate remained constant longer and declined more quickly near the end of the extraction (Fig. 5.2).

Results of Particle Size Study. In general, decreasing the particle size of the ground peanuts from a range of 3.35 to 4.75 mm to a range of 0.86 to 1.19 mm increased the total oil recovery from 36% to 82%. This relationship agrees with the hypothesis that only the surface oil is easily extracted with SC-CO_2. The relationship between oil recovery and particle size fits the quadratic equation

$$Y = 103.4 - 21.8X + 1.17X^2 \qquad [5.4]$$

with a correlation R-squared of 0.9697, where Y is the percent oil recovered and X is the average particle size (mm). This result is for 550 bar, 25°C, 40 L/min for three-hour tests.

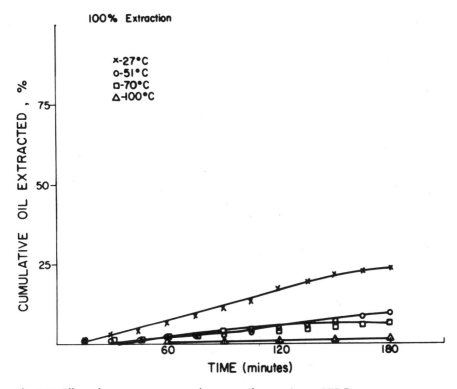

Fig. 5.6. Effect of temperature on total peanut oil extraction at 275 Bar.

The extraction rate is limited by the solubility of oil in $SC\text{-}CO_2$. For all of the particle sizes studied, the extraction began at a constant rate and then leveled off at a total yield that depended on the size (Fig. 5.7). Even with only 90% confidence, the null hypothesis (that there was no difference in initial extraction rates) could not be rejected. Therefore, the particle sizes tested do not appear to affect the maximum solubility.

After the residual peanut meal was sifted, it was found that some of the particles broke into smaller pieces. When the extraction vessel was opened immediately after letting the pressure out, the particles jumped and a popping could be heard, suggesting that most of the breakup occurred as the CO_2-saturated particles were depressurized.

Results of CO_2 Extraction of Prepressed Peanuts. For the initial extraction rate, pressure, temperature and the pressure-temperature interaction effects were all significant (P < 0.025). Neither percent oil remaining after pressing nor treatment (screw press only or additional crushing in the rolling mill) had an effect on the initial extraction rate.

The total yield from a batch of peanuts from the combination of pressing and $SC\text{-}CO_2$ extraction depended only on pressure (P < 0.01). In other experiments,

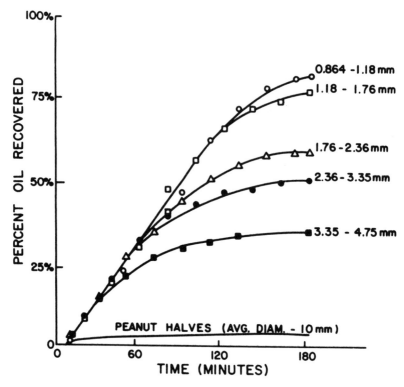

Fig. 5.7. Effect of particle size on peanut oil extraction with SC-CO_2, 550 bar, 25°C, 40 liters CO_2/min.

temperature and size were the significant effects. Since this design did not study size effects but instead the effect of treatment (screw press only or press plus rolling mill), it can be concluded that further preparation by rolling the pressed meal was not necessary to break open oil cells to expose the oil to the solvent. Temperature did not have an effect in these experiments, but pressure did. The cause of this contradiction between these and previous results is unknown. The high degree of oil cell fracturing in the screw press may reduce the need for temperature-controlled oil diffusion in the oilseed particle.

For the amount of volatiles lost, no effects were found to be significant. This result agrees with previous studies in which only the amount of moisture in the peanut meal was found to affect the amount of volatiles lost. In this study, the initial meal moisture was constant for each experiment.

From the results of the experiments, equations describing the system were derived. The initial extraction rate can be predicted from the following equation:

$$Y = 54.3 + 34.3 \; \frac{X_1 - 7000}{1000} + 32.2 \; \frac{X_2 - 60}{35}$$

$$+ 31.5 \left(\frac{X_1 - 7000}{1000} \right) \left(\frac{X_2 - 60}{35} \right) \qquad [5.5]$$

where

Y = initial extraction rate (mg oil/L CO_2)
X_1 = pressure (psi)
X_2 = temperature (°C)

The equation describing the total yield under these conditions is as follows:

$$Y = 83.6 + 7.2 \left(\frac{X_1 - 7000}{1000} \right) \qquad [5.6]$$

where

Y = total % yield per batch of peanuts
X_1 = pressure (psi)

Results of Flaked-Peanut Experiments. The results were very similar to those of the experiments with ground peanuts. Solubility was affected by temperature, pressure, and the temperature-pressure interaction. Particle size and moisture level did not have any effect in this experiment as was found with ground peanuts. The yield was affected only by the particle size, although temperature also had an affect with ground peanuts. The amount of volatiles lost was affected only by the amount of moisture, although both the moisture-temperature and moisture-pressure interactions were significant ($P < 0.01$) with ground peanuts.

Rapeseed Solubility Studies

There has been increasing interest in recovery of rapeseed oils in recent years, particularly the canola class of rapeseed varieties. A listing of major fatty acids in rapeseed oil is given in Table 5.1 (10). Bulley et al. (7) applied a general thermodynamics approach to evaluation of mass transfer coefficients and other characteristics of the rapeseed oil recovery process using SC-CO_2. Goodrum and Kilgo (20) conducted extraction studies of two varieties of rapeseed. They examined both solubility and extraction kinetics over wide ranges of CO_2 solvent and oilseed conditions. The experimental equipment, procedure, and experimental design approach were similar to those already described for peanuts.

Experimental Design for Oil Recovery from Rapeseed

Half-Factorial with Dwarf Essex Rapeseed. Half of a $2 \times 2 \times 2 \times 2$ factorial experimental design (half-factorial design) was used to study SC-CO_2 extraction of the Dwarf Essex rapeseed variety. 6000 and 8000 psi (415 and 550 bar) were used as the two extraction pressures, and 25°C and 95°C were the two temperatures. The two moisture levels were 6% and 15%. The samples with smaller average particle size had the following distribution:

Smaller than 0.503 mm	10%
0.503–0.864 mm	30%
0.864–1.19 mm	35%
1.19–1.68 mm	25%

The mass average was 0.974 mm. The larger particle size distribution was as follows:

Smaller than 0.503 mm	2%
0.503–0.864 mm	5%
0.864–1.19 mm	10%
1.19–1.68 mm	35%
1.68–2.38 mm	48%

The average was 1.62 mm. The two flow rates were 40 and 60 L CO_2/min (measured at STP). For the experiments at the higher moisture level 36 g water were added to 424 g ground rapeseed at 6% moisture to make 460 g of material at 15% moisture. The seeds were thoroughly mixed with the water and allowed to equilibrate for several days at 4°C. (See the preceding discussion of the experiments with peanuts for additional experimental design description.)

CO_2 Extraction of Pre-pressed Rapeseed. Dwarf Essex rapeseed was processed in the screw press to partially recover its oil content. Residual oil content of the meal was found by a standard hexane extraction method (13). A $2 \times 2 \times 2$ factorial design was used to study the effects of pressure, temperature, and percent oil recovered in the press on the solubility and volatiles lost. The two SC-CO_2 pressures used were 6000 and 8000 psi (415 and 550 bar); the two temperatures were 35°C and 100°C; and the two levels of oil recovery were 27% and 41%. The moisture was not altered from the storage level of 4.5%. Batches of 500 g each were used.

Rapeseed Variety Comparison. A $2 \times 2 \times 2$ factorial design was used to determine the differences between two varieties of rapeseed: Dwarf Essex and Cascade. The pressures and temperatures used were 6000 and 8000 psi (415 and 550 bar) and 35°C and 100°C. Samples with small particle size were obtained from seeds flaked on the rolling mill at 0.75 mm thickness and sieved to remove all particles larger than 1.18 mm; the large-particle-size samples were obtained from seeds flaked at 1.65 mm thickness and sieved to obtain only particles in the range of 1.18 to 2.38 mm. The moisture was not altered from that of storage: 6.15% for Cascade and 5.75% for Dwarf Essex. The Dwarf Essex seeds contained 46.9% oil, and the Cascade 43.57%. 145 g of Dwarf Essex and 135 g of Cascade were used to keep the amount of oil recovered at 63 g.

Experimental Results

Half-Factorial with Dwarf Essex Rapeseed. The half-factorial experiments with Dwarf Essex rapeseed were completed with results that were very similar to those of the peanut half-factorial experiments. The solubility was affected by the pressure, the temperature, and the pressure-temperature interaction ($P < 0.01$). Particle size ($P < 0.05$) and the pressure–particle size interaction ($P < 0.10$) may also be significant. The yield was affected by both the particle size and the temperature ($P < 0.01$). No other effects were significant. Both moisture and temperature affected the amount of volatiles lost ($P < 0.01$).

A few further experiments were performed to determine more closely the effect of pressure and temperature on the solubility. Figure 5.8 shows the solubility isotherms for Dwarf Essex rapeseed and peanuts. It can be seen from the graph that oil from finely ground rapeseed does not dissolve as readily in SC-CO_2 as does oil from ground peanuts. Differences in fatty acid and triglyceride composition between peanuts and rapeseed (Table 5.1) may account for this difference. There also may be differences in the oil cell structure in the two seeds, making it more difficult for the rapeseed oil to diffuse into the SC-CO_2 solvent.

Observed characteristics of rapeseed products are similar to those of peanuts. Rapeseed residual meal samples followed the same trends as those for peanuts, except that the dark brown skins of the rapeseed made the entire sample appear darker than peanut meal extracted at the same conditions. The oil samples seemed to be more viscous than peanut oil samples extracted at the same conditions.

Prepressed Rapeseed. The meal from which 50% or more oil had been recovered in the screw press was shaped like potato chips. The samples that had a low percentage of oil recovered in the press crumbled rather than retaining the chip shape.

Pressure, temperature, and the pressure-temperature interaction effects were all found to be significant ($P < 0.01$), which agrees with all previous experiments. No conclusions could be drawn about the maximum yield, because the extraction rates were too slow to extract all of the oil from the sample completely in 4 h.

The following equation estimates the initial solubility of oil in CO_2 for prepressed Dwarf Essex rapeseed:

$$Y = 18.2 + 18.75 \frac{X_1 - 7000}{1000} + 19.5 \frac{X_2 - 67.5}{32.5} + 15.65 \left(\frac{X_1 - 7000}{1000} \right) \left(\frac{X_2 - 67.5}{32.5} \right) \quad [5.7]$$

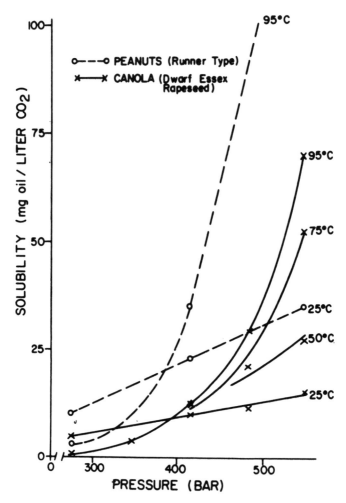

Fig. 5.8. Comparison of temp. dependent solubility of peanuts and dwarf essex rapeseed.

where

$$Y = \text{solubility (mg oil/L CO}_2)$$
$$X_1 = \text{pressure (psi)}$$
$$X_2 = \text{temperature (°C)}$$

Rapeseed Variety Comparison. For this set of experiments, no factors had a significant effect on the equilibrium oil solubility in CO_2. These results differ from all other experiments, where pressure, temperature, and the pressure-temperature interaction were significant factors. In this experiment, as in the previous ones, lower yields were obtained for rapeseed than those previously found for peanuts. Yields varied from only 4 to 59% for rapeseed, whereas for smaller-sized peanut particles the yields varied from 35 to 95%. The lower yields indicate that not enough oil was available for the extraction to reach the maximum rate. In other words, the extraction had already reached the exponential decay part of the extraction kinetics curve. Temperature and particle size had a significant effect on the total yield ($P < 0.01$), as had been found in other experiments with peanuts and rapeseed. Increasing the temperature and decreasing the particle size both increased the yield. No difference between the two varieties of rapeseed was seen.

Oil Recovery Models for Peanuts and Rapeseed

Extraction Kinetics Model: Linear and Second-Order Kinetics (Region-Two Model)

For the effect of particle size on the extraction, several classes of curves were fitted to the peanut extraction data. Both a quadratic function

$$Y = A_0 + A_1\theta + A_2\theta^2 \tag{5.8}$$

where

$$Y = \text{\% oil yield (total weight of oil collected/total weight of oil in the sample before extraction)}$$
$$\theta = \text{time (minutes)}$$
$$A_0, A_1, A_2 = \text{constants}$$

and an exponential function of the form

$$Y = A_0[1 - \exp(-A_1\theta)] \tag{5.9}$$

where the variables are the same as those defined in Eq. (5.8), fit the data well for values of from 0 to 180 min ($P < 0.001$). However, after both equations were differentiated to obtain the maximum points, unreasonable maximum yields were obtained at the smaller particle sizes. For example, a maximum oil yield of 154% was obtained with Eq. (5.8), and 270% with Eq. (5.9) for the particle size 0.864 to 1.18 mm.

We suggest that the extraction consists of two major processes and is best described by a combination of two curves. The initial saturated solvent condition is represented by a straight line. Then an asymptotic approach to maximum oil recovery occurs as the surface oil is depleted from the particles, and oil slowly diffuses out to complete the extraction. The maximum recovery is inversely dependent on particle diameter and thus depends on the amount of surface area that is available for extraction. The data fit well ($P < 0.001$) with the following equation, which is a combination of a straight line and Eq. (5.9):

$$Y = (A + B\theta)Z_1 + C1 - \exp(-D(\theta - \theta') + Y')Z_2 \tag{5.10}$$

where

Y = total oil produced (g)
θ = time (min)
(θ', Y') = transition point between the straight line and the exponential curve
Z_1 = 1 up to the transition point and 0 after
Z_2 = 0 up to the transition point and 1 after $(= 1 - Z_1)$
A, B, C, D = constants

Table 5.5 shows the values of the constants and transition points for each particle size, and one example of the fit of the curve is shown in Fig. 5.9.

For each constant (A, B, C, D) the 95% confidence intervals overlap for each of the five particle sizes. Thus, the data could also be represented by one equation with only the transition point (q', Y') dependent on particle size, for example,

$$Y = (-6.5 + 1.3\theta)Z_1 + 53(1 - \exp(-0.0215(\theta - \theta') + Y'))Z_2 \qquad [5.11]$$

where the variables are the same as those described for Eq. (5.10). The average value of each constant was used.

Fitting Oil Recovery Data to a Chromatographic Model

A desorption model of a chromatography column was selected to describe SC-CO_2 extraction of peanut oil in a fixed bed (21). Other models have been described in Ref. 22. Particle size studies (2,5,8) suggest that only surface oil is readily available for

TABLE 5.5 Constants in the Oil Extraction Kinetics Equation, Eq. (5.10)

Particle diameter (mm)	A	B	C	D	θ'^a	Y'^a
0.364–1.18	−9.64	1.36	57.88	0.0213	110	143
	(−15.8, −3.5)[b]	(1.27,1.45)[b]	(27,89)[b]	(0.0013,0.041)[b]		
1.18–1.70	−11.94	1.49	64.29	0.0228	90	119
	(−19.3,−4.6)	(1.36,1.63)	(47,82)	(0.0098,0.036)		
1.70–2.36	−3.45	1.24	56.75	0.0256	70	80
	(−19.6,12.7)	(0.88,1.60)	(36,77)	(0.0027, 0.049)		
2.36–3.35	−4.06	1.32	45.02	0.0206	60	75
	(−10.6,2.5)	(1.15,1.49)	(36,54)	(0.011, 0.030)		
3.35–4.75	−3.25	1.17	45.25	0.0171	40	43
	(−11.2,4.7)	(0.87,1.46)	(35,55)	(0.0092,0.025)		
0.864–1.18	−0.46	1.36	39.89	0.0312		
(1000 g batch)	(−9.7,10.6)	(1.3,1.43)	(28,51)	(-0.0026,0.065)	250	330
0.503–1.18	−7.68	1.07	71.60	0.0140	250	258
(flaked batch)	(−12.5,-2.8)	(1.04,1.11)	(61,82)	(.0092,.0187)		

[a]Transition point coordinates.
[b]95% confidence limits are shown in parentheses.

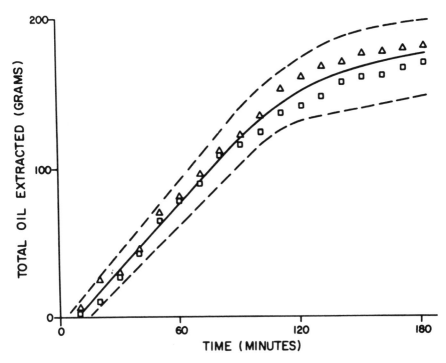

Fig. 5.9. Oil extraction rate for 1.18–1.70 mm particles of ground peanut (the squares and triangles represent two replications).

extraction. If this is true, the mechanism of SC-CO_2 peanut oil extraction is similar to the solid/liquid stripping or desorption process of a chromatography column.

In desorption processes solute species are dissolved into the fluid phase from the solid phase. Assuming that equilibrium between solvent and oil is attained, batch extraction of oil from peanuts using supercritical CO_2 may be described as a desorption process in a porous, fixed bed. Peanut meal is the *particulate phase* or *bed packing,* peanut oil is the *solute,* and CO_2 is the *solvent.*

Figure 5.10 describes the desorption wave and shows how the concentration of the solute in the exiting solvent changes with time. In (a) the inlet layer of solid is in contact with fresh solvent. The solute concentration in the solvent reaches saturation in the first part of the bed; therefore, no solute is desorbed further down the column, and the solvent at the exit is saturated with solute. As additional solvent flows through the bed (b), a zone forms in which the concentration of solute in the solid phase is changing. This zone moves through the column as a wave, usually at a much slower rate than the linear velocity of the solvent through the bed.

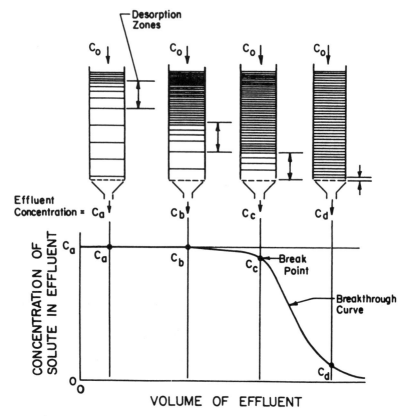

Fig. 5.10. The desorption wave. (Adapted from Treybal, 1980).

When a decrease in the amount of solute in the exiting solvent is first detectable (c), this is known as the *breakpoint*. At (d), the bed is essentially exhausted of solute. The curve of the solute concentration in the solvent versus time between the breakpoint and total depletion of the solute is known as the *breakthrough curve*. If the desorption were infinitely rapid, the curve would be a vertical line. In practice, it is generally S-shaped, although it may be a steep, flattened, or distorted S.

The model includes several assumptions (21):

1. The process is isothermal. Usually, desorption is an endothermic process, and the heat absorbed by the reaction alters the rate of desorption. Because the temperature in the extractor is held constant with heating tapes, this is a good analogy to the SC-CO_2 process.

2. The mixture of solute in the solvent is dilute. This is true for oil in CO_2, because the solubility at 25°C and 8000 psi (550 bar) is less than 2% by weight.

3. The equilibrium desorption isotherm is concave to the solution-concentration axis. This condition is necessary for desorption to occur.

4. The desorption zone is of a constant height as it moves down the bed. It should be noted that if not enough time is given for the zone to reach its equilibrium length, it will still be changing at the point that it reaches exhaustion, giving incorrect results.

5. The length of the bed is large compared to the length of the zone.

Figure 5.11 shows both an experimental and an idealized breakthrough curve. In the ideal case, the mass transfer rate is infinitely rapid and the curve is reduced to a vertical line. This line is used to estimate the breakpoint on the experimental curve by drawing the line so that the shaded areas are equal. θ_B is the time at which the breakpoint occurs, whereas θ_T is the time of exhaustion. The total length of the bed, Z, is equal to Z_s, the length of the bed that is free of oil, plus the length of unused bed, L_{UB}. If V is the velocity of advancement of the zone, then at any time θ, $Z_s = V\theta$. At θ_s, $Z_s = Z = V\theta$, and at breakthrough, $Z_s = V\theta_B$; therefore,

$$
\begin{aligned}
L_{UB} &= Z - Z_s \\
&= V\theta_s - V\theta_B \\
&= V(\theta_s - \theta_B) \\
&= Z(\theta_s - \theta_B)/\theta_s
\end{aligned}
\qquad [5.12]
$$

where

L_{UB} = length of unused bed (section containing partially extracted oilseed) = length of the desorption zone

Z = length of the bed

θ_B = time at which the drop in the solute concentration of the exiting solvent first becomes detectable

θ_T = time at which the solute concentration in the exiting solvent becomes negligible

θ_s = approximately $1/2\ (\theta_T - \theta_B)$

Experimental Procedure for Chromatographic Model Verification: Breakthrough Curves. The extraction vessel had an internal length of 86 cm and an internal diameter of 5.72 cm (2 $1/4$ in.). The data for each breakthrough curve were obtained from 1 kg batches of peanuts. The peanuts were ground in a rotary cutter and sieved to obtain particles in the range of 0.864 to 1.18 mm diameter. Each batch was extracted at 8000 psi (550 bar), 25°C, and CO_2 flow of 40 L/min (at STP). The extraction was stopped after five consecutive readings of less than 2 g/15 min were obtained. Solubility for each reading was calculated by dividing the amount of oil collected (mg) by the amount of CO_2 used (L).

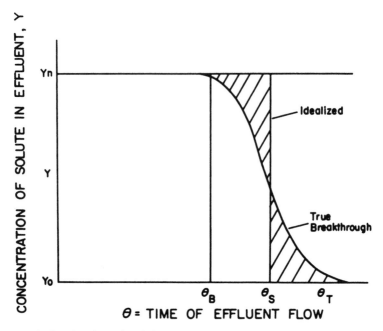

Fig. 5.11. Idealized and true breakthrough curves, where Yn and Yo are the maximum and minimum solute concentration in the effluent, respectively. (Adapted from Treybal, 1980.)

Two additional runs were made with meal from the rolling mill. The peanuts were flaked to 1.27 mm thickness and then sieved to remove the pieces larger than 1.18 mm. For each run, 860 g of this material was placed loosely in the extractor to prevent packing.

Results: Chromatographic Model. Figures 5.12 and 5.13 show the fit of the peanut oil extraction data to the chromatographic column model. The inverse tangent function was used to approximate the breakthrough curve because it closely approximates the shape of the chromatographic extraction curve. This function has an asymptotic approach to both a maximum (representing the initial extraction rate or true solubility) and a minimum (representing total oil depletion). Constants were added to scale the function to the extraction data, yielding the following equation:

$$Y = A_0 \arctan\left(\frac{A_1 - X}{A_2}\right) + \frac{A_0}{2} \qquad [5.13]$$

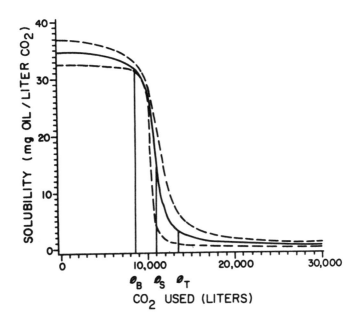

Fig. 5.12. Breakthrough curve for peanut oil extraction from ground peanuts at 550 bar and 25°C. (Dotted lines show 95% confidence level).

where

Y = extraction rate (mg oil/L CO_2)

X = CO_2 used ($L \times 1000$) [(time = $X/(40$ L/min)]

A_0, A_1, A_2 = constants

Table 5.6 shows the values of A_0, A_1, A_2, θ, Z, and L_{UB} for ground and flaked peanuts. This breakthrough curve model gives an approximate value for L_{UB} and shows that the depletion of oil in a particular small segment of the bed occurs rapidly. This suggests that a relatively short column would be adequate for a continuous extraction system.

The model assumes that the length of the bed is large compared to the length of the gradient zone. Because the experimental bed length was only about 4 times the length of the gradient zone, the zone length may have still been changing at the exhaustion point, limiting the accuracy of the model. A better model of the concentration gradient would be more complex and require additional data that is not generally known for complex systems such as peanut meal–peanut oil–CO_2. Such additional

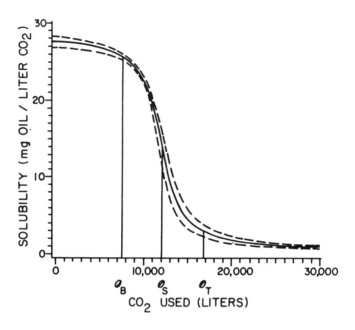

Fig. 5.13. Breakthrough curve for flaked peanuts extracted at 550 bar and 25°C. (Dotted lines show 95% confidence level).

data includes the diffusivity of oil in CO_2 at high pressures and the mass transfer coefficients of oil in peanut meal and in compressed CO_2.

It should be noted that the results computed from Eq. (5.13) apply to a column of specific dimensions. For other SC-CO_2 extraction columns, it is essential that the chemical extraction parameters inside the extractor unit—particularly temperature, pressure, particle size, and solvent linear velocity—be analogous. A simple extrapolation of extractor dimensions could lead to significant differences between the computations and extractor performance. This model was partially described in Ref. 23.

Summary and Conclusions

Using SC-CO_2, peanut oil was extracted from ground peanuts at pressures from 2000 to 10,000 psi (135 to 680 bar) and temperatures from 25 to 120°C. Above 6000 psi (415 bar), increasing the temperature to the maximum possible without heavily charring the peanuts (120°C) significantly increased the initial extraction rate. Increasing the pressure at constant temperature increased the rate.

TABLE 5.6 Concentration Gradient Model Parameters

Parameter	Ground peanuts	Flaked peanuts
A_0	32.46	28.81
A_1	10.96	12.09
A_2	0.743	1.559
θ_B (minutes)	217	182
θ_S (minutes)	274	300
Z (cm)	57	66
L_{UB} (cm)	11.9	28.2

The vertical extractor produced higher total oil recovery and more uniformly extracted meal samples than did the horizontal extractor. The vertical extractor with downward solvent flow produces more uniform extraction from peanuts and a greater percent oil recovery than either vertical extraction with upward flow or horizontal extraction.

At higher temperatures (75°C and above) roasting begins to occur; however, this was not detrimental to the extraction rate or overall oil recovery.

Decreasing the particle size increases the overall yield per batch of peanuts, as seen in both the half-factorial and particle size experiments. The factorial experiments suggested that particle size may also have a small effect on the solubility, although the particle size experiments did not confirm this effect. Increasing the moisture increases the amount of volatiles lost.

The flow rate does not affect the solubility, percent oil recovered, or volatiles lost for flow rates of 40 to 60 L CO_2/min (at STP). However, the extraction can be completed more quickly at a faster flow rate, because the amount of CO_2 that passes through the extractor per unit time is greater.

CO_2 extraction of peanut oil from prepressed peanuts is a viable alternative to other extraction methods. The highest total yield was 98.1% at 95°C and 8000 psi (550 bar). The two factors—percent oil recovered in the press and treatment of the meal—did not have any effects on the process.

A combination of a straight line and an exponential curve was found to describe the experimental oil production kinetics accurately. The straight-line portion is associated with the initial saturated solvent condition, whereas the exponential curve describes the unsaturated and diffusion-limited processes associated with the last stage of oil recovery from an oilseed particle.

The breakthrough curve model provides a relatively simple basis for future analysis of vegetable oil extraction processes with SC-CO_2 solvent. Data for supercritical CO_2 extraction of peanut oil in a fixed-bed extractor was fitted to an inverse tangent function. This function was taken from the shape of the breakthrough curve of a solid/liquid chromatographic extraction. The length of the concentration gradient in the extractor was calculated to be 2.1 times the column diameter of 5.72 cm for ground peanuts and 4.9 times the diameter for flaked peanuts in a cylindrical column. The analogy proposed between the chromatographic process and the SC-CO$_2$ process is in good agreement with fit of the data to the model for a chromatographic column. The results indicate that continuous extraction at the maximum extraction rate is possible in a short column within the solvent flow rate, temperature, and pressure ranges examined.

None of the effects that were studied definitely affected the fatty acid distribution in the resultant oil. Temperature, moisture, and a temperature-moisture interaction effect all influence the amount of moisture in the extracted oil. Size, and possibly pressure, affect the residual oil content of the inlet meal and, therefore, the total oil yield that can be obtained from a sample of peanuts.

Recovery of peanut and rapeseed oil with a combined process of partial recovery in a screw press plus extraction of the remaining oil with SC-CO$_2$ is technically a viable alternative to other oil recovery methods. Oil recoveries of 95% (peanuts) and 75% (rapeseed) have been demonstrated.

The initial extraction rate for rapeseed was consistently lower than the rate for peanuts at the same extraction temperature and pressure.

No differences in SC-CO$_2$ extraction rates of yields were found between Dwarf Essex and Cascade varieties of rapeseed.

References

1. Bulley, N.R., and M. Fattor, *J. Amer. Oil Chem. Soc. 61(8):*1362 (1984).
2. Stahl, E., E. Schutz, and H.K. Mangold, *J. Agric. Food Chem. 28:*1153 (1980).
3. Friedrich, J.P., G.R. List, and A.J. Heakin. *J. Amer. Oil Chem. Soc. 59(7):*288 (1982).
4. Stahl, E., K.W. Quirin, A. Glatz, D. Gerard, and G. Rau, *Ber. Bunsenges. Phys. Chem. 88:*900 (1984).
5. Snyder, J.J., J.P. Friedrich, and D.D. Christianson, *J. Amer. Oil Chem. Soc. 61(12):*1851 (1984).
6. List, G.R., and J.P. Friedrich. *J. Amer. Oil Chem. Soc. 61(12):*1847 (1984).
7. Goodrum, J.W., and M.B. Kilgo, *Energ. Agr. 6:*265 (1987).
8. Goodrum, J.W., and M.B. Kilgo, *Trans. ASAE 30(6):*1865 (1987).
9. *Peanut Oil,* Georgia Peanut Comm., Tifton, GA, 1980.
10. *Canola Oil, Properties and Performance.* Monograph published by the Canola Council, Winnipeg, Manitoba, Canada, 1987.
11. *Peanut Science and Technology,* American Peanut Research and Education Soc., Yoakum, TX, 1982.
12. Rittner, Hernan, *J. Amer. Oil Chem. Soc. 61(7):*1200 (1984).
13. American Oil Chemists' Society, *Official Methods,* American Oil Chemists' Society, Champaign, IL, 1969.
14. Friedrich, J.P., and E.H. Pryde, *J. Amer. Oil Chem. Soc. 61(2):*223 (1984).
15. Kilgo, M.B., *Qual. Eng. 1(1):*45 (1989).
16. Juran, J.M., *Managerial Breakthrough,* McGraw-Hill, New York, 1964.
17. Hahn, G.J., *J. Qual. Technol. 9(1):*13 (1977).

18. Davies, O.L., *The Design and Analysis of Industrial Experiments,* Hafner, New York, 1954.
19. Box, G.E.P., W.G. Hunter, and J.S. Hunter. *Statistics for Experimenters,* Wiley, New York, 1978, pp. 329–334.
20. Goodrum, J.W., and M.B. Kilgo, *Trans. ASAE 32(2):*727 (1989).
21. Treybal, R.E., *Mass Transfer Operations,* 3rd edn., McGraw-Hill, New York, 1980, pp. 623–641.
22. Lee, A.K.K., N.R. Bulley, M. Fattori, and A. Meisen. *J. Amer. Oil Chem. Soc. 63(7):*921 (1986).
23. Goodrum, J.W., and M.B. Kilgo, *Trans. ASAE 31(3):*926 (1988).

Chapter 6

Modeling of the Supercritical Fluid Extraction of Oilseeds[1]

Ki-Pung Yoo and In-Kwon Hong+

Department of Chemical Engineering, Sogang University, C.P.O. 1142, Seoul, Korea.
+Department of Chemical Engineering, Dankook University, Seoul, Korea

Introduction

During the past decade, supercritical fluid extraction (SFE) technology has experienced many rapid advances. So far, SFE has been applied in areas as diverse as petrochemicals, food substances, polymers, surfactants, cosmetics, and pharmaceuticals and to hazardous wastes. As a result, SFE technology has become interdisciplinary and has been studied extensively as a viable separation process alternative (1,2).

Design engineers, however, have been apprehensive about the high capital costs as well as the safety problems of using high pressure. In addition, the mathematical models for the design and scale-up of such processes are limited. Future applications of SFE will be highly dependent on the ability of engineers to model mass transfer and scale-up characteristics in addition to phase equilibria. One must be able to identify and reject many speculative schemes that, based on valid process design and economic feasibility studies, show little promise of economic success. To accelerate the acceptance of SFE, we need first to achieve a sufficiently detailed understanding of the mass transfer and extraction behavior and control aspects of the process, and then to use this knowledge to make reliable empirical correlations for the economical design of such SFE processes.

For the food industry the emphasis of research and developing SFE technology up to the 1980s was mainly on accumulating experimental data on the solubility of natural food substances in CO_2 in order to evaluate roughly whether SFE technology could be a viable alternative process. In recent years however, attention has shifted toward scrutinizing SFE quantitatively as a potential candidate to replace current extraction methods. Thus, reliable mathematical representation of mass transfer and extraction rate data are crucial information for the practical design of an SFE process. The intent of this chapter is to provide a chemical engineering approach to modeling mass transfer rates for food processing with supercritical fluids. The discussion is limited to the extraction of edible oils from soybeans; however, the details presented here could be applied to other similar SFE processes (3–6).

[1]The authors are grateful to the Korea Science and Engineering Foundation for long-term financial support. We also thank our colleagues, Professor C.S. Lee at Korea University and Professors K.S. Lee and W.H. Lee at Sogang University, for their valuable input.

Process development for the SFE of edible oils is currently being carried out by several German and U.S. researchers who have been active in this field (8–24). Also, in the 1980s there was considerable interest in SFE technology at the U.S. Department of Agriculture's Northern Regional Research Center on the extraction of vegetable oils with SC-CO_2 (14–18). Studies done at that Center concerned the SC-CO_2 extraction of oils from soybeans, sunflower seeds, corn germ, wheat germ, safflower seeds, peanuts, and rapeseed. In all these cases, the SFE-extracted oils were compared with those obtained by traditional hexane extraction. According to these investigations, oil yields by SC-CO_2 extraction were comparable to those of hexane-extracted oil. The SC-CO_2-based oil showed significantly lower refining loss and phosphorous content, was lighter-colored, and was essentially degummed. As a result, data on triglyceride solubility in SC-CO_2 for SFE and on the oil composition of soybean oil have been described in a number of journals, but mass transfer rates (19–24) have not yet been widely studied.

In recent years the present authors also have been studying the extraction of vegetable oils with SC-CO_2, with an emphasis on the potential use of SFE technology in the edible oil industry in Korea (25–32). The work has been stimulated by the interest shown by a number of edible oil companies in Korea and by the Korea Science and Engineering Foundation. This chapter reports an overview of these efforts to establish an SFE process model for soybean oil recovery on fixed-bed equipment, together with other engineering information, such as optimization of extraction yield and operation energy demand for a large-scale SFE process.

Experimental Apparatus and Procedures

In order to judge the feasibility of the SFE process for edible oil from seeds, information about phase equilibria and mass transfer rates are needed. Also, for the SFE process to be economical, a proper type of separator with scale-up information is needed. Toward that goal, three types of experimental apparatus were originally constructed: a dynamic flow equilibrium cell, for the solubility measurement, a prototype and a pilot-plant scale SFE apparatus for the acquisition of mass transfer engineering data. The design specifications and the experimental procedures for each type of equipment in this section are briefly presented next.

Apparatus for Phase Equilibria

According to an approximate analysis of soybean-extracted oil, the major constituents are triglycerides containing palmitic, stearic, oleic, linoleic, and linolenic acids. To measure the solubilities of crude soybean oil and each of its constituents in supercritical CO_2, a dynamic flow–type equilibrium cell was used as shown in Fig. 6.1.

Liquid CO_2 at ambient temperature is charged to the system with a high-pressure pump and compressed to the desired operating pressure. The pump is used to deliver CO_2 continuously at flow rates slow enough to ensure that equilibrium is obtained

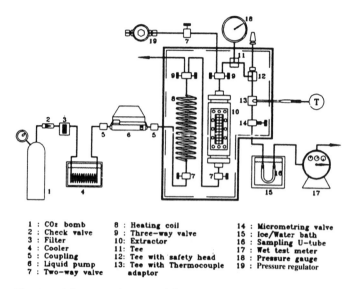

1 : CO₂ bomb	8 : Heating coil	14 : Micrometring valve
2 : Check valve	9 : Three-way valve	15 : Ice/Water bath
3 : Filter	10: Extractor	16 : Sampling U-tube
4 : Cooler	11: Tee	17 : Wet test meter
5 : Coupling	12: Tee with safety head	18 : Pressure gauge
6 : Liquid pump	13: Tee with Thermocouple	19 : Pressure regulator
7 : Two-way valve	adaptor	

Fig. 6.1. Schematic diagram of dynamic flow type extraction apparatus.

between the heavy solute and SC-CO₂ as it flows through the system. The CO₂ flows slowly through the column and becomes saturated with solute prior to exiting. After the saturated CO₂-rich phase exits, it is expanded to atmospheric pressure across a heated metering valve; the solute falls out of solution and is collected in a cold trap. The amount of solute collected in the cold trap for a given amount of time is determined gravimetrically, and the corresponding volume of CO₂ is measured with a wet test meter. The equilibrium pressure of the system is measured at the exit with a Heise gauge having ±15 psi (1.0 atm) accuracy. The temperature of the air bath is measured with a thermocouple, and is usually maintained with ±0.1°C. For determining the solubility of soybean oil, the sample of soybeans was washed in water, dried for 24 h at 50°C, dehulled, and crushed to a powder. The individual sample solutes, such as triglycerides and fatty acids, were directly purchased from chemical companies.

Prototype Fixed Bed

A flow diagram of the small-scale fixed-bed extractor for the measurement of extraction rates is shown in Fig. 6.2. The extraction section consists of two identical fixed beds (11 mm i.d. and 260 mm height). While one of the extractors is being used to carry out a rate experiment, the other extractor is disassembled, charged with a new sample and then assembled in order to save time. Commercial-grade CO₂ from a cylinder is cooled, liquefied, pumped to a preheater located in an air bath, and then converted to CO₂ in its supercritical state. The accuracy of the temperature maintained in the air bath is within ±0.1°C up to 120°C.

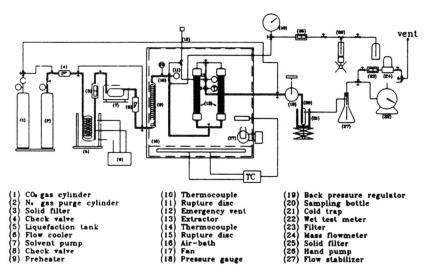

(1) CO₂ gas cylinder	(10) Thermocouple	(19) Back pressure regulator
(2) N₂ gas purge cylinder	(11) Rupture disc	(20) Sampling bottle
(3) Solid filter	(12) Emergency vent	(21) Cold trap
(4) Check valve	(13) Extractor	(22) Wet test meter
(5) Liquefaction tank	(14) Thermocouple	(23) Filter
(6) Flow cooler	(15) Rupture disc	(24) Mass flowmeter
(7) Solvent pump	(16) Air-bath	(25) Solid filter
(8) Check valve	(17) Fan	(26) Hand pump
(9) Preheater	(18) Pressure gauge	(27) Flow stabilizer

Fig. 6.2. Experimental apparatus for supercritical extraction of crude soybean oil.

After measurement of extraction pressure and temperature, the SC-CO₂ solvent was passed through one of the extractors charged with a 10-g sample of soybeans. The soybeans, obtained from the domestic market, were washed with water, dried for 24 h at 50°C, then dehulled and prepared as three types of sample dry-milled powders, which were shown to have mean particle diameters of 0.05 cm, 0.15 cm, and 0.28 cm, respectively, by sieve test. The SC-CO₂ stream with extracted oil is expanded to atmospheric pressure by a back-pressure regulator. The amount of extracted oil is collected into a preweighed sample bottle, which is located in a flow cooler. The sample bottle containing the extracted oil is then detached and weighed at regular time intervals. The solute-free CO₂ stream passes through a mass flow meter, where the flow rate is measured.

Parts for the system were obtained from the following manufacturers: Milton Roy (Riviera Beach, FL), 450 atm double-ended diaphragm pump (Constametric III); HIP Instrument Co. (Erie, PA), high-pressure on-off valves, unions, metering valves, ¹/₄ in (6.35 mm) tube, thermocouples; Dresser Industries (Gaithisburg, MD), Heise gauge; Techno Co. (Stratford, CT), a cooling bath up to –40°C. The air bath (800 mm high × 650 mm wide × 310 mm deep) is especially designed to maintain the temperature accurately using a controller, electric heater, and air circulation fan.

The whole apparatus is rated for operation at a pressure of 450 atm. All measured temperatures and pressures are recorded on a data logger at regular time intervals. The mass transfer rates were measured as functions of flow rate, pressure, temperature and particle size. The range of system parameters examined in this study is shown in Table 6.1.

TABLE 6.1 Operating Conditions in Mass Transfer Rate Experiment

System	Soybean–SC-CO_2
Particle size of soybean powder (cm)	0.025~0.2828
Height of bed (cm)	20.0
Diameter of bed (cm)	1.1
Temperature (°C)	50.0~70.0
Pressure (atm)	270.0~375.0
Flow rate (STD cm^3/sec at 0°C and 1 atm)	0.02~0.20
Reynolds number	1.0 < Re < 100.0
Schmidt number	10.0 < Sc < 15.0
Sherwood number	0.03 < Sh < 5.0

Pilot-Scale SFE Equipment and Control System

A rational and economic development of SFE process may be categorized by two steps, in the second of which the engineering data for scale-up can be collected. This step is based on fundamental studies of fluid-phase equilibria and of mass transfer obtained in the initial evaluation. In the second step such characteristics of the SFE process as the operating pressure, the pressure variations between unit processes, the recycling of fluids to prevent pollution problems, and the sensitivity of extraction rates with respect to operating pressure and temperature must be taken into account. The safety, operability, flexibility, and economic feasibility of the process must also be addressed at this stage. The final process system embodying these concerns is shown in Fig. 6.3. Also, the design detail for the two extractors and fractionators is shown in Fig. 6.4. The specification of the other parts is documented in Table 6.2.

A make-up CO_2 gas stream from the gas cylinder is added to the recycle solvent stream from separator 2 through an on-off control valve. This low-pressure CO_2 gas stream is sent to a gas booster. The pressurized gas stream from the gas booster is supplied to one of the extractors after the supercritical temperature and pressure are adjusted by a heater. The fluid extracts oil as it passes up through a bed packed with dry-milled soybean powder. The supercritical CO_2 containing the extracted oil stream is reduced in pressure by passing a control valve. The dissolved oil is partly separated in the first extractor and completely separated in the second separator at an even lower pressure. To prevent the accumulation of residue in the separators, the pressure in the second separator is maintained at a constant level, and the CO_2 stream in the second separator is often vented.

Filtering devices are installed at the exits of all extractors and separators to prevent entrainment in the extractor and the first separator. In addition, a check valve is installed between the CO_2 cylinder and the gas booster to prevent back-flow. The operational parameters for each of these units are summarized in Table 6.3.

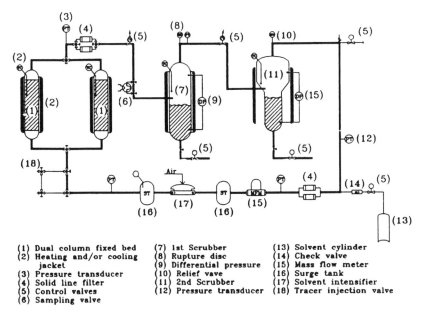

Fig. 6.3. Schematic pilot-scale SFE process.

(1) Dual column fixed bed	(7) 1st Scrubber	(13) Solvent cylinder
(2) Heating and/or cooling jacket	(8) Rupture disc	(14) Check valve
	(9) Differential pressure	(15) Mass flow meter
(3) Pressure transducer	(10) Relief vave	(16) Surge tank
(4) Solid line filter	(11) 2nd Scrubber	(17) Solvent intensifier
(5) Control valves	(12) Pressure transducer	(18) Tracer injection valve
(6) Sampling valve		

A Fuzzy Expert Control System for a Pilot-Scale SFE Process

The pilot-scale SFE process described in the previous section is constructed so as to operate semi-continuously via computer control. In considering the SFE process from the point of computer control, it can be assumed that the process temperatures in individual subunits can be controlled independently. However, there are still two important problems to be overcome:

1. Interaction among operational control variables such as the pressures at the extractor and the first separator (e.g., if the material balance differs from unit to unit)
2. The difficulty inherent in process modeling, due to high nonlinearities among process variables

In addition, the process is operated in batch mode, and a proper protocol for SFE startup is critical. These factors suggest the need for a robust and efficiently designed control system for the SFE process.

To solve the problem of operational control of the SFE process, one can consider decoupling techniques to reduce the degree of interaction among process variables or the model reference adaptive control (MRAC) method to overcome the difficulty in modeling the process. However, these recent control techniques are based on complicated mathematical theories, thus making it difficult to give a clear understanding of the underlying concepts to both plant engineers and operators.

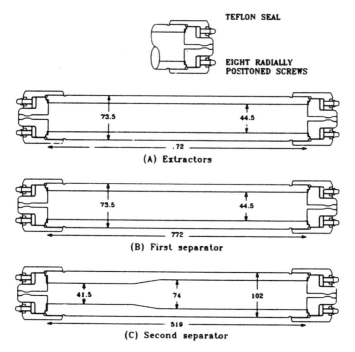

Fig. 6.4. Design details of pilot-scale extractors and fractionation separators.

TABLE 6.2 The Specification of Major Units used in a Typical Pilot-Scale SFE Process

Unit	Specification
Extractor (mm)	i.d.: 44.5, o.d.: 73.5, L: 772.0, SS306
Separator 1 (mm)	i.d.: 44.5, o.d.: 73.5, L: 772.0
Separator 2 (mm)	i.d. of upper part: 41.5, lower part 74.0
	o.d.: 102.0, L: 519.0
Control valve 1 (CV1)	ATO, Linear, C_v = 0.00018, 1/4" NPT,
	with air-operated positioner, Badger Co. (USA)
CV2	C_v = 0.0006, others are same as CV1
CV3	ATO, EQ%, C_v = 1.5, 1/2" Flange, with positioner,
	Hokushin Co. (Japan)
Gas booster	Model AGT-15/75, two-stage, air-driven motor, max.
	P-12,000 psig, Haskel Co. (USA)
Pressure transducers	0~1000 kg/cm^2 absolute pressure to 4~20 mA Valcom Co.
	(Tokyo, Japan)
Temperature sensors	K(CA)-type thermocouple

TABLE 6.3 Operation Ranges for Extraction and Fractionation in an SFE Process

Condition	Extractor	Separator 1	Separator 2
Temperature (°C)	30~80	35~50	Room temperature
Pressure (atm)	200~400	80~200	20~30
Flow rate (mol/min)	0.25~1.5	0.25~1.5	0.25~1.5

For this reason the present authors have introduced a new control method, a fuzzy expert system, to the SFE process (30,33,34). Based on the experience of manual operation of the process, the SFE process is represented by a highly interacting 3×3 MIMO system having the CO_2 recycling rate and the extractor and separator pressures as output variables and the air pressure to the gas booster, and to the extractor and separator control valves as manipulated variables. This simplified SFE process for the control is shown in Fig. 6.5. The fuzzy expert control system consists of three steps (Fig. 6.6):

1. Fuzzification of the sensor signals representing process variables
2. Fuzzy logic execution using control rules and membership functions on the basis on the experience of manual operation
3. Defuzzification of the fuzzified outputs

Three control pairs between manipulated and output variables are selected. Then seven membership functions, from "big negative" to "big positive," are defined for control and time rate of change of error, giving 49 fuzzy control rules. The membership functions are also defined in two steps as "coarse" and "fine," to enhance control performance. Fuzzy inference is performed using the max-min composition rule, and defuzzified control outputs are calculated based on the principle of the center of gravity. A few of the fuzzy control rules follow:

RULE 1: IF E = PB AND CE = NB, THEN, Control valve position 1
RULE 2: IF E = PB AND CE = NM, THEN, Control valve position 2
.
.
.
RULE 48: IF E = NB AND CE = PM, THEN, Control valve position 48
RULE 49: IF E = NB AND CE = PB, THEN, Control valve position 49

where E denotes the error between set point and process variable; CE, rate of change of the error; NB, big negative; NM, medium negative; NS, small negative; ZO, zero; PS, small positive; PM, medium positive; and PB, big positive.

In Fig. 6.7, the membership functions such as NB, NM, and PB are shown. By defining the two different types of membership function as coarse and fine, the control performance can be enhanced. Similar types of membership functions are also constructed for the pairings between extractor pressure and control valve 1 and between separator 1 and control valve 2. The proposed fuzzy control scheme has been found to be successful in experiment and numerical simulation.

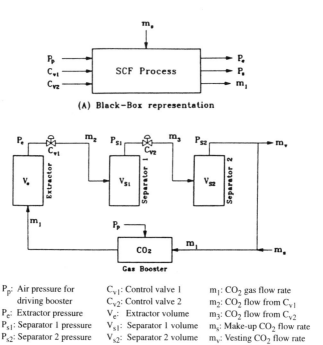

P_p: Air pressure for driving booster
P_e: Extractor pressure
P_{s1}: Separator 1 pressure
P_{s2}: Separator 2 pressure

C_{v1}: Control valve 1
C_{v2}: Control valve 2
V_e: Extractor volume
V_{s1}: Separator 1 volume
V_{s2}: Separator 2 volume

m_1: CO_2 gas flow rate
m_2: CO_2 flow from C_{v1}
m_3: CO_2 flow from C_{v2}
m_s: Make-up CO_2 flow rate
m_v: Vesting CO_2 flow rate

(B) Schematic diagram of SFE process

Fig. 6.5. Simplified representation of control loop for the SFE process shown in Fig. 6.3.

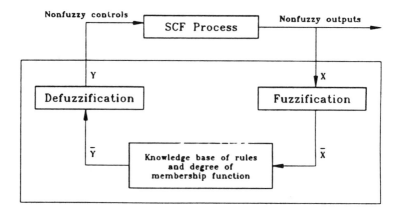

Fig. 6.6. General fuzzy control system for the SFE process (Fig. 6.3).

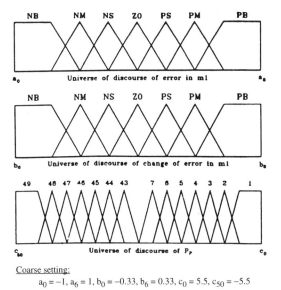

Coarse setting:
$a_0 = -1, a_6 = 1, b_0 = -0.33, b_6 = 0.33, c_0 = 5.5, c_{50} = -5.5$

Fine setting:
$a_0 = -0.5, a_6 = 0.5, b_0 = -0.03, b_5 = 0.03, c_0 = 1.1, c_{50} = -1.1$

Fig. 6.7. An example of membership functions between the error of CO_2 flow rate (m_1), rate of change of error of m_1 and air pressure for driving gas booster (P_p).

Modeling of Mass Transfer Rates

Phase Equilibria

Oil from seeds such as soybeans can actually be considered as a continuous mixture consisting of an infinite number of components. However, for modeling and correlating phase equilibrium data between SC-CO_2 and oil, we need to assume that the oil consists of a mixture of one or two representative major components. Thus, based on literature data (11–18), and a proximate analysis carried out in the present study, we assume the oil to be a mixture consisting of fatty acids and triglycerides. The general structure and representative composition of the oil is shown in Tables 6.4 and 6.5. To model the experimental solubilities of each component in SC-CO_2, we assume the supercritical fluid phase to be a compressed gas, and the solubility of the heavy component in the supercritical phase can be calculated by conventional thermodynamic models given in Eq. (6.1) as:

$$y_2 = \frac{P_2^{sat}}{P} \frac{\phi_2^s}{\phi_2} \exp\left[\frac{V_2^s}{RT}(P - P_2^{sat})\right] \qquad [6.1]$$

Here the nonideal behavior of the SC phase is characterized by the *fugacity coefficient* ϕ, which is evaluated using an equation of state and other specific thermodynamic relationships. The reliability of the correlation of the solubility depends on the

TABLE 6.4 Composition of Soybean Triglyceride

R, R', R''	Name	Wt% composition[a]
16:0	Palmitic acid	2.3~10.6
18:0	Stearic acid	2.4~6
18:2	Linoleic acid	23.5~30.8
18:3	Linolenic acid	4.9~51.5

[a]Trace amount of phospholipids, sterol, antioxidants, pigments etc., are neglected in phase equilibria.

TABLE 6.5 Physical Properties of Soybean Triglycerides

Property	Values
Density	$\rho = 0.8475 + 0.00030$ (Saponification value) $+ 0.00014$ (Iodine value)
Specific heat	$C_p = 1.93 + 0.0025t$ ($t = 15~60°C$, in kJ/K-g)
Vapor pressure	0.133 MPa at 254°C, 6.66 MPa at 308°C
Thermal conductivity	C_a 167 mW/m–K at 20°C, 163 mW/m–K at 100°C
Viscosity	$\log \eta = -0.73 + 46.6 \times 10/T$ (T in °C, η in cP)

accuracy of the determination of fugacity coefficients. For this reason, equations of state that can predict phase behavior in all regions are of the utmost importance for solubility determination. A new, statistical mechanics–based lattice equation of state (EOS) developed by coworkers of the present authors (28,30) is used in the calculation of the fugacity coefficient for the heavy component in the supercritical phase. The workability of the EOS for complex molecules has been demonstrated elsewhere (28,30), so here the resulting equation is simply presented in Table 6.6.

The pure-component parameters for some representative ingredients of soybean oil for the lattice EOS and the characteristic binary interaction parameter λ_{12}, between SC-CO$_2$ and solutes making up the oil, are presented in Tables 6.7 and 6.8, respectively. Also, some correlation results of experimental solubility data based on EOS parameters are given in Fig. 6.8 for the CO$_2$–triglyceride and in Fig. 6.9 for CO$_2$-triolein. On scrutiny of the correlation it can be concluded that the solubility of soybean oil in SC–CO$_2$ can be modeled reasonably well by the EOS, assuming that the soybean oil is a pseudo-mixture consisting of triglycerides and triolein.

Mass Transfer Rates

For the extraction of oil from pretreated soybean seeds at several pressures and temperatures two distinctive features characterized the extraction rates:

1. The extraction rate is constant during the initial extraction period.
2. Gradually the extraction rate shifts to an unsteady-state mode.

TABLE 6.6 Multicomponent Form of the Lattice Model–Based Equation of State and Chemical Potential

Equation of state:

$$\frac{\tilde{P}}{\tilde{T}} = -\ln(1 - \tilde{\rho}) + \frac{Z}{2}\ln\left[1 + \left(\frac{q_M}{r_M} - 1\right)\tilde{\rho}\right] - \frac{\theta^2}{\tilde{T}}$$

$$- \frac{Z}{2}\beta^2\left[\Sigma_j\Sigma_k\Sigma_l\,\theta_j\theta_k\theta_l\theta_m\varepsilon_{jk}(\varepsilon_{jk} + \varepsilon_{lm} - \varepsilon_{jm} - \varepsilon_{km}) - \Sigma_j\Sigma_k\theta_j\theta_k\,\varepsilon_{jk}(\varepsilon_{jk} - \varepsilon_{kl})\right]$$

where

$$\tilde{P} = P/P^*, \quad \tilde{T} = T/T^*, \quad \tilde{\rho} = V^*/V, \quad Pv_H = RT^* = (Z/2)\,\varepsilon_M$$

$$\varepsilon_M = \left[\Sigma_i\Sigma_j\,\theta_i\theta_j\varepsilon_{ij} + \frac{1}{2}\beta\Sigma_i\Sigma_j\Sigma_k\Sigma_l\theta_i\theta_j\theta_k\theta_l;\varepsilon_{ij}\,(\varepsilon_{ij} + \varepsilon_k - \varepsilon_{ik} - \varepsilon_{jk})\right]\bigg/\theta^2$$

$$\varepsilon_{ij} = (1 - \lambda_{ij})(\varepsilon_{ii} \cdot \varepsilon_{jj})$$

$$\theta = (q_M/r_M)/[\tilde{v} + (q_M/r_M) - 1]$$

$$\theta_i = (q_i/r_i)/[\tilde{v} + (q_M/r_M) - 1]$$

$$v_H = 9.75 \times 10^{-6}\ m^{-3}mol^{-1}, \quad Z = 10$$

Pure-component parameters, ε_{ij}, v^*

$$v^* = 0.9871 + 0.0213249T + [1.061526 - 6.002 \times 10^{-4}\,T + 0.107263\ \ln\ T] \bullet V_{vdw}$$

(T is in K; V_{vdw} is van der Waals volume in cm³/mol)

Chemical potential:

$$-\frac{\mu_i}{RT} = \gamma_i(T) - \frac{\tilde{r}_iP}{\tilde{T}} + \ln(q_i) - \ln(\theta_i) - r_i\left(1 - \frac{Z}{2}\right)\ln\left[1 + \left(\frac{q_M}{r_M} - 1\right)\tilde{\rho}\right]$$

$$- \frac{q_i\theta^2}{\tilde{T}} + \frac{Z}{2}q_i\beta\ \{2\Sigma\theta_j\varepsilon_{ij} - \beta\Sigma_j\Sigma_k\Sigma_m\theta_j\theta_k\theta_m\varepsilon_{jk}(\varepsilon_{jk} + \varepsilon_{lm} - \varepsilon_{jm} - \varepsilon_{km})$$

$$+ \beta\Sigma_i\Sigma_j\Sigma_k\theta_j\theta_k\theta_l[\varepsilon_{ij}\,(\varepsilon_{ij} - \varepsilon_{ik} + 2\varepsilon_{kl} - 2\varepsilon_{jk}) + \varepsilon_{jk}\,(\varepsilon_{jk} - \varepsilon_{kl})]\}$$

These characteristics can be modeled by combining a model of steady-state mass transfer at high oil concentrations with unsteady-state diffusion mass transfer at lower oil content in the solid matrix, as in a normal drying process. Thus, the extraction is assumed to proceed at a constant rate up to a critical oil content in the solid and then shifts to an unsteady-state governed by diffusion rate.

In the present authors' work, the two types of mass transfer calculations were carried out separately, and these results were combined to describe the overall observed

TABLE 6.7 Lattice Mode–Based Equation of State Pure-Component Parameters

Chemicals	v^a,cm/mol	ε_{11}/k, K	T, K
CO_2	36.470	84.747	325.0
Soybean triglycerides	889.430	121.759	335.2
Palmitic acid	282.084	117.517	360.0
Stearic acid	289.484	120.622	360.0
Oleic acid	278.130	128.306	360.0
Triolein	901.405	112.212	313.15

[a]In the parameter estimation, other pure properties such as vapor pressure and molar volume at saturation conditions are calculated by the methods suggested in the data book (Ref. 3).

TABLE 6.8 The Lattice Equation of State Binary Interaction Parameters

System	ND^a	λ_{12}	AAD^{b}%	T, K
CO_2–triglyceride	20	0.061	18.20	323.15
CO_2–palmitic acid	17	0.066	27.42	313.15
CO_2–stearic acid	30	0.087	29.34	313.15
CO_2–oleic acid	38	0.066	14.92	313.15
CO_2–triolein	43	0.082	16.05	313.15
CO_2–tristearin	9	0.096	23.42	313.15

[a]Number of experimental data points.
[b]AAD% = $\Sigma \ |(X_{i,exp} - X_{i,calc})/X_{i,exp}| \ / \ ND$

mass transfer data. During the constant-rate period, where steady-state mass transfer prevails, the extracted oil quantity is expressed as

$$m = k_g A_s V_t \nabla C_m \tag{6.2}$$

$$A_s = 6(1 - \Psi)/d \tag{6.3}$$

where k_g represents the external mass transfer coefficient; A_s the specific mass transfer area; V_t the volume of the fixed bed; ∇C_m the concentration difference of oil between the mass transfer interface and bulk mean concentration, which can be calculated from phase equilibria; Ψ the void volume fraction; and d the mean diameter of the particle.

After the experimental extraction rate data under supercritical conditions is fitted to Eq. (6.2), the mass transfer coefficient can be described as functions of dimensionless hydrodynamic variables, such as the Sherwood number, Sh ($KgM_{av}d/\rho D_g$); Reynolds number, Re ($\mu/\rho u d$); Schmidt number, Sc ($\mu/\rho D_g$); and Grashof number, Gr ($d^3 g\rho \nabla \rho/\mu^2$). These dimensionless numbers represent hydrodynamic properties and are commonly used to design and scale up separation processes in chemical engineering practices. The mass transfer between a fluid and a packed bed of solid can be described by a correlation, e.g., Sh $= f$ (Re,Sc,Gr), depending on the degree of convection. When both natural and forced convection are present, the following correlation fits the observed data well (36):

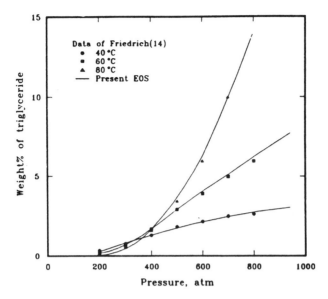

Fig. 6.8. Comparison of experimental and calculated P-x data for CO_2-triglyceride system at various temperatures; $\lambda_{12} = 0.066$ (40°C), 0.061 (60°C), and 0.057 (80°C).

$$Sh/(Sc \cdot Gr)^{1/4} = 1.05 \times 10^{-4}(Re/Gr)^{0.3986} \tag{6.4}$$

The diffusion coefficient D_g, in dimensionless numbers such as Sh and Sc, is calculated based on the following correlation as (1):

$$D_g = 3.3531 \times 10^{-4} \times 10^{-0.6860\rho} \tag{6.5}$$

in cm²/sec. Here, the density ρ of the CO_2–solute mixture is calculated based on the EOS described earlier. For calculating the other properties, such as viscosity, η, the supercritical phase was represented as pure CO_2 and the properties were estimated by a method recommended in the literature (35). The correlation of experimental data used to find the parameters in Eq. 6.4 is shown in Fig. 6.10.

As mentioned previously, the constant rate of extraction proceeds up to a certain critical concentration $(m_{shift}, \theta_{shift})$, after which the constant-rate mass transfer is succeeded by mass transfer that is unsteady-state with respect to the time θ. The extraction process for this unsteady state can be described by a set of coupled differential equations based on Fick's second law, which takes into account the time-dependent concentration flux and, accordingly, a time-dependent driving force and mean mass transfer coefficient, as well as the influence of the flow of the supercritical fluid phase. General solutions to such a set of equations will be quite difficult to obtain analytically; however, an approximate solution can be obtained by a quasi-steady-state analysis of the coupled equation (9).

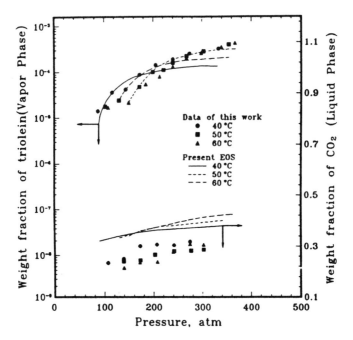

Fig. 6.9. Comparison of experimental and calculated P-x data for CO_2-triolein system at various temperatures, $\lambda_{12} = 0.082$ (40°C), 0.083 (60°C), and 0.084 (80°C).

$$\frac{m(\theta)}{m_{shift}} = A[1.0 - \exp(-K \cdot Fo)] \qquad (6.6)$$

The approximation usually takes the form of the first term of a Fourier series, where Fo represent Fourier's number, $D_g\theta/d^2$, and A and K are defined as parameters. The remaining quantity, m, of the extracted components is related to the initial mass by an exponential function of dimensionless time, expressed as the Fourier number, and the coefficients A and K, which have been calculated with the observed rate data. For the case where mass transfer is independent of the flow rate of SC-fluid, the parameter A has the value 0.6079, and the parameter K is related to Biot number, B_i ($k_g \cdot d/D_g$), which represents the mass transfer from the solid into gas relative to the mass diffusion in the solid phase represented by D_g. The values of K are then obtained from the observed rate data. The unsteady diffusion mass transfer is correlated with the fitted parameter K, resulting in the fits shown in Figs. 6.11 through 6.13.

Finally, the operating and hydrodynamic conditions (Table 6.9), and overall mass transfer correlation of the extraction rate data with respect to time, $m(\theta)$, are shown in Figs. 6.14 and 6.15. The mass transfer rate for the SFE of soybeans is similar to that for the drying of a solid, as mentioned previously. The extraction rate process is split

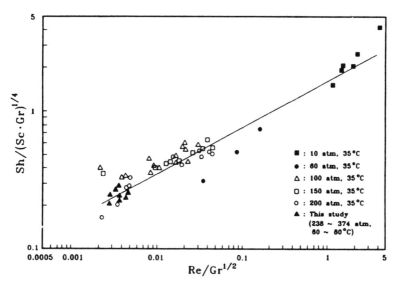

Fig. 6.10. Correlation of hydrodynamic variables for CO_2-oil system [literature data; one report in the literature is for CO_2-naphthalene (1)].

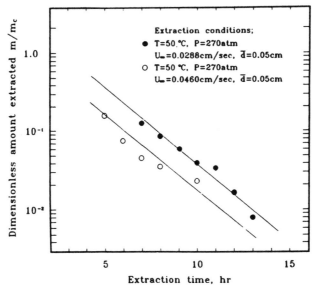

Fig. 6.11. Correlated mass transfer coefficients for the diffusion-controlled period.

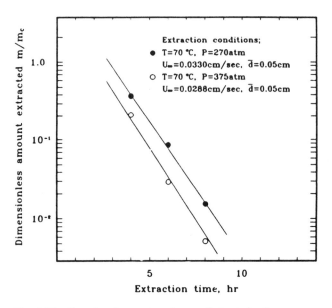

Fig. 6.12. Correlated mass transfer coefficients for the diffusion-controlled period.

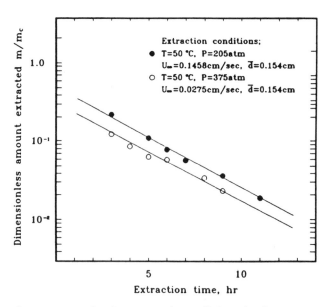

Fig. 6.13. Correlated mass transfer coefficients for the diffusion-controlled period.

TABLE 6.9 Operating and Hydrodynamic Conditions for Mass Transfer Experiments

Run No.	T (°C)	P (atm)	d (cm)	U (cm/s)	C_{sat} (g/cm³)	Re	Sc	Sh	K_g (cm/s)
1	50	272.11	0.0500	0.02888	0.00717	1.562	11.56	0.0471	7.63E-5
2	50	272.11	0.0500	0.04632	0.00717	2.500	11.56	0.0532	8.52E-5
3	50	374.15	0.0500	0.02629	0.01877	1.327	14.01	0.0265	3.65E-5
4	50	374.15	0.0500	0.06250	0.01877	3.145	14.02	0.0436	6.17E-5
5	70	272.11	0.0500	0.03290	0.00650	1.919	10.71	0.0374	5.99E-5
6	70	374.15	0.0500	0.02877	0.02020	1.599	12.28	0.0338	4.79E-5
7	50	204.08	0.1540	0.03356	0.00247	5.757	10.03	0.2316	1.35E-4
8	50	204.08	0.1540	0.09092	0.00247	15.600	10.03	0.5174	3.01E-4
9	50	204.08	0.1540	0.14580	0.00247	25.010	10.03	4.7390	2.75E-3
10	50	374.15	0.1540	0.02758	0.01877	4.287	14.01	0.5274	2.42E-4
11	50	204.08	0.2828	0.15328	0.00247	48.280	10.03	3.8000	1.20E-3
12	50	374.15	0.2828	0.05380	0.01877	15.360	14.01	0.5123	1.28E-4
13	50	374.15	0.2828	0.01706	0.01877	4.869	14.01	0.4208	1.05E-4
14	50	374.15	0.2828	0.05620	0.01877	16.040	14.01	0.4471	1.12E-4
15	50	374.15	0.1540	0.05340	0.01877	8.300	14.01	0.4039	1.85E-4

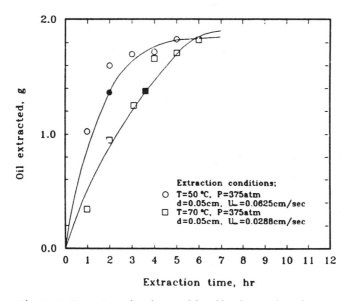

Fig. 6.14. Extraction of soybean oil fitted by the combined mass transfer models, at a constant pressure of 375 atm.

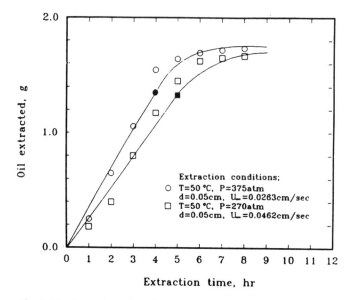

Fig. 6.15. Extraction of soybean oil fitted by the combined mass transfer models, at a constant temperature of 50°C.

into two steps. In the first region, the oil is removed from the surface of the solid and the macropores. In the second region, mass transfer is controlled by the diffusion of extractants from the micropores of the solid. The extraction rate of oil with respect to solvent consumption is summarized in Fig. 6.16, with data from literature (16). As one can see from that Figure, the extraction rate increases with the density of the solvent.

The hydrodynamic correlation of constant-rate mass transfer proposed in this work was obtained for a fixed geometry (e.g., diameter and height of the fixed bed). Taking into account this limitation, the correlation proposed here can be used for the purpose of process scale-up calculations.

Yield and Operation Energy for a Large-Scale SFE Process

The process parameters to be considered in the planning of a large-scale plant will be the yield of production with respect to operation energy demand, as well as information on the relevant phase equilibria, an extraction rate model, and specific process control aspects for the raw materials. The operation energy audit will probably be the most important economic factor in optimizing the operating conditions for the process. In the pilot-scale SFE process, for instance (Fig. 6.3), the major operation loop consists of extractor → pressure reduction valve 1 → separator 1 → pressure reduction valve 2 → separator 2 → compressor → extractor, along with appropriate heat exchange. This continuous operation cycle can be represented on a thermodynamic state diagram of CO_2 solvent to determine the thermodynamic availability

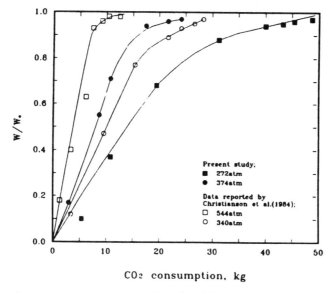

CO₂ consumption, kg

Fig. 6.16. Extraction curves of crude oil from dry-milled soybeans with supercritical CO_2 at a solvent flow rate of 340 mL/min and a temperature of 50°C.

(ΔB, called *exergy* in the European literature) for every stage of the process. The yield can also be evaluated from solubility considerations, allowing the yield to be calculated by the relations

$$\Delta B \text{ (kcal/kgK)} = \sum_{i=1}^{n} (\Delta H - T_0 \Delta S)_i \tag{6.7}$$

$$\Delta x \text{ (wt\%, yield)} = x_3(\text{extractor}) - \sum x_i \text{ (separator } i) \tag{6.8}$$

where x represents solubility in a specific unit and the free energy is defined as the enthalpy, H, and entropy, S, difference for each unit with respect to reference state.

To optimize the yield and the operation energy required to separate the oil from the SC-CO_2, the objective function F, which is to be maximized, is defined as:

$$F = \frac{\text{Extracted oil (wt\%)/cycle}}{\Delta B/\text{cycle}} \tag{6.9}$$

The optimum value of F can be obtained from the state diagram of solubility data for the SC-CO_2–triglyceride system and the Mollier diagram for CO_2 is shown in Fig. 6.17. The thermodynamic information shown in Fig. 6.17 was image-processed into the computer and the given range of process operating conditions. The optimum operating pressures and temperatures in the extractor and separators and the optimum value of the objective function F were then auto-searched, based on the cubic spline

interpolation algorithm, from the stored (graphical) data in the computer. One of the
results is illustrated in Fig. 6.18 for the case where only a single separator is installed.
The same method can be directly extended to any complicated large-scale SFE
process to obtain design criteria that give maximum separation with minimum operation energy demand.

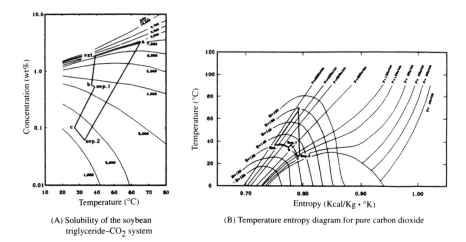

(A) Solubility of the soybean
triglyceride–CO_2 system

(B) Temperature entropy diagram for pure carbon dioxide

Fig. 6.17. Separation yield and corresponding energy demand for the case of dual-
separator soybean-SCF process.

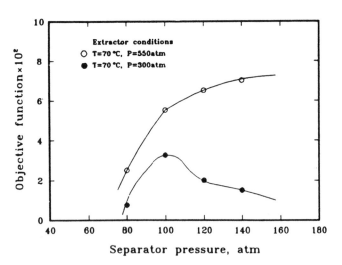

Fig. 6.18. Effect of single-stage separator pressure on the objective function, yield/energy, for the soybean SCF process (separator temperature is fixed at 40°C).

Concluding Remarks

This chapter briefly presented a study of SFE process development for the extraction of edible oil from soybeans with SC-CO_2. Fundamental considerations of supercritical fluid phase equilibria and correlation of mass transfer rates have been presented. A fuzzy expert control system for a large-scale process has been discussed, and a method of optimizing the separation yield with respect to energy demand for the operation of the SFE process has been presented. However, due to the limited space available, the subjects of process control and extraction optimization have only been touched on here. References 30, 32, and 34 are recommended for further reading.

References

1. Johnston, K.E., in *Supercritical Fluid Science and Technology*, edited by K.P. Johnston and J.M.L. Penninger, ACS Symposium Series 406, American Chemical Society, Washington, DC, 1988, pp. 1–13.
2. McHugh, M., and V.J. Krukonis, *Supercritical Fluid Extraction*, Butterworth, MA, 1986, pp. 118–230.
3. Krukonis, V.J., in *Supercritical Fluid Extraction and Chromatography*, edited by B.A. Charpentier and M.R. Sevenants, ACS Symposium Series 366, American Chemical Society, Washington, DC., 1988, pp. 26–43.
4. Brunner, G., in *Supercritical Fluid Technology*, edited by J.M.L. Penninger, M. Radosz, M.A. McHugh, and V.J. Krukonis, Elsevier, Amsterdam, 1985, pp. 245–264.
5. Zosel, K., in *Extraction with Supercritical Gases*, edited by G.M. Schneider, E. Stahl, and G. Wilke, Verlag Chemie, Deerfield Beach, FL, 1980, p. 124.
6. Penninger, J.M.L., in *Chemical Engineering at Supercritical Fluid Conditions*, edited by M.E. Paulaitis, J.M.L. Penninger, R.D. Gray, Jr., and P. Davidson, Ann Arbor Science, Ann Arbor, MI, 1983, pp. 375–376.
7. Brennecke, J.F., and C.A. Eckert, *AIChE J. 35:*1409 (1989).
8. Zosel, K., *Angew. Chem. Intl. Ed. Engl. 17:*702 (1978).
9. Brunner, G., *Ber. Punsenges. Phys. Chem. 88:*887 (1984).
10. Zosel, K., U.S. Patent 3,969,196 (1976).
11. Friedrich, J.P., U.S. Patent 4,466,923 (1984).
12. Friedrich, J.P., and A.C. Eldridge, U.S. Patent 4,493,854 (1985).
13. Christianson, D.D., and J.P. Friedrich, U.S. Patent 4,495,207 (1985).
14. Friedrich, J.P., G.R. List, and A.J. Heakin, *J. Am. Oil Chem. Soc. 59:*288 (1982).
15. Friedrich, J.P., and E.H. Pryde, *J. Am. Oil Chem. Soc. 61:*233 (1984).
16. Christianson, D.D., J.P. Friedrich, G.R. List, K. Warner, E.B. Bagley, A.C. Stringfellow and G.E. Inglett, *J. Food Sci. 49:*229 (1984).
17. List, G.R., J.P. Friedrich and J. Pominski, *J. Am. Oil Chem. Soc. 61:*1847 (1984).
18. Snyder, J.M., J.P. Friedrich, and D.D. Christianson, *J. Am. Oil Chem. Soc. 61:*1851 (1984).
19. Stahl, E., W. Schilz, E. Schutz, and E. Willing, in *Extraction with Supercritical Gases*, edited by G.M. Schneider, E. Stahl, and G. Wilke, Verlag Chemie, Deerfield Beach FL, 1980, pp. 93–114.
20. Stahl, E., K.W. Quirin, A. Glatz, D. Gerard and G. Rau, *Ber. Bunsenges. Phys. Chem. 9:*900 (1984).

21. Krukonis, V.J., paper presented at the *75th Annual Meeting of the American Oil Chemists' Society,* Dallas, TX, April, 1984.
22. Eisenbach, W., *Ber. Bunsenges. Phys. Chem. 88:*882 (1984).
23. Hannigan, K.J., *Chiltons Food Eng. 53:*77 (1981).
24. Robey, R.J., and S. Sunder, paper presented at the *Annual Meeting of the American Institute of Chemical Engineers,* San Francisco, CA, November, 1984.
25. Hong, I.K., S.W. Rho, K.S. Lee, W.H. Lee and K.P. Yoo, *Korean J. Chem. Eng. 7:*40 (1990).
26. Hong, I.K., *Fluid Phase Equilibria of Biomolecules in Supercritical Fluids,* Ph.D. Thesis, Sogang University, Seoul, Korea, 1991, pp. 55–69.
27. Lee, B.S., W.H. Lee, C.S. Lee and K.P. Yoo, in *Proceedings of the Second Korea-Japan Symposium on Separation Technology,* Seoul, Korea, 1990, D9, pp. 306–310.
28. You, S.S., J.W. Kang, B.O. Choi, K.P. Yoo and C.S. Lee, in *Proceedings of the Second Korea-Japan Symposium on Separation Technology,* Seoul, Korea, 1990, D11, pp. 314–318.
29. Lee, M.S., *Extraction Characteristics of Red-Pepper Seed Oil with Supercritical Carbon Dioxide,* M.S. Thesis, Sogang University, Seoul, Korea, 1988, pp. 4–26.
30. Yoo, K.P., W.H. Lee, C.S. Lee and K.S. Lee., *Supercritical Fluid Extraction as an Energy Saving Alternative,* Final Report, Korea Sci. Eng. Foundation 87-0604-03, 1990, pp. 160–211.
31. Kim, W.C., K.S. Lee, K.P. Yoo and W.H. Lee, *J. Korean Inst. Chem. Eng. 28:*131 (1990).
32. Kim, S.Y., K.S. Lee, W.H. Lee, and K.P. Yoo, *J. Korean Inst. Chem. Eng. 28:*93 (1990).
33. Zadeh, L.A., *IEEE Trans. Syst. Man. Cybern., Vol. SMC-1:*1 (1973).
34. Yoo, D.S., Control of Supercritical Fluid Extraction Process Using Fuzzy Logic, MS. Thesis, Sogang University, Seoul, Korea, 1990, pp. 4–35.
35. Reid, R.C., J.M. Prausnitz and B.E. Poling, *The Properties of Gases and Liquids,* 4th edn., McGraw-Hill, New York, 1987, pp. 95–147.
36. Mandelbaum, J.A., and V. Böhm, *Chem. Eng. Sci. 28:*569 (1973).

Chapter 7

Design and Economic Analysis of Supercritical Fluid Extraction Processes[1]

Miriam Cygnarowicz-Provost

Food and Drug Administration, Center for Devices and Radiological Health,
9200 Corporate Blvd., Rockville, MD 20850

Supercritical fluid extraction (SFE) of lipids has been extensively studied on a laboratory scale. Some of the many applications that have been reported include the decaffeination of coffee and tea (1), the extraction of soybean and canola oil (2,3), the extraction of hops (4) and spice resins (5), the fractionation of milk fat (6) and fish oils (7), and the preparation of protein concentrates (8) and cholesterol-reduced eggs (9) and meats (10). Although the list of potential uses of this technology is large, only a few examples of commercial-scale SFE processes exist. These are the production of decaffeinated coffee and tea, the preparation of hops extracts, and the production of spice and flavor concentrates.

Although SFE offers several advantages over conventional processes, such as the use of nontoxic, nonpolluting solvents, easy solvent recovery, and improved fluid transport properties, a serious disadvantage is the large capital costs associated with high-pressure operation. Clearly, SFE applications will not progress beyond the initial development stages unless the economic feasibility of the proposed processes can be estimated. Such an assessment can be obtained through process design studies. Design studies are typically accomplished by constructing mathematical models that simulate the process. Such models use experimental solubility and mass transfer data to estimate parameters for the equilibrium and mass balance equations that describe the process. Once a representation of the process is obtained, operating and equipment costs can be estimated as a function of key process variables (extractor temperature and pressure, solvent flow rates, etc.), and the economics of the design can be assessed. Mathematical optimization procedures can then be used to determine the optimal design for a given economic objective function. Design studies thus explore the effect of critical process parameters on the economics. Areas for improvement can be identified, and a decision can be made whether to continue development of the process.

Of course, the predictions of a design study are only as good as the mathematical models that are used to describe the process. Therefore, they should not be considered a replacement for experimentation. However, by using process simulation in conjunction with basic experimental measurements, the feasibility of a proposed application can be assessed early in its development, thereby guiding the course of future work.

---

[1]Mention of brand or firm names does not constitute an endorsement by the U.S. Department of Agriculture over others of a similar nature not mentioned.

Simulation of SFE processes requires solubility and mass transfer models and a means of estimating the operating and equipment costs. Each of these aspects will be addressed in the sections that follow.

Solubility Models

Equations of state are widely used to represent the solubility of solids in supercritical fluids. An equation of state (EOS) can predict the solubility as a function of the temperature, the pressure, and the solute and solvent physical properties. Equations of state have been widely accepted because they are relatively simple and easy to use. Some of the most commonly used models are the Peng–Robinson (11), the Soave–Redlich–Kwong (12) and the Group-Contribution (13) equations. To represent the solubility of a solid in a supercritical phase, the compressed gas model (14) is typically used. In this approach, the fugacity of species i in the supercritical phase is set equal to the fugacity of the pure solid, giving

$$y_i = \frac{P_i^{\text{sat}} \exp\left[\dfrac{V_i (P - P_i^{\text{sat}})}{RT}\right]}{P \Phi_i} \tag{7.1}$$

where y_i is the mole fraction of species i in the supercritical phase, P is the pressure, Φ_i is the fugacity coefficient of species i in the supercritical phase, P_i^{sat} is the vapor pressure of pure species i, V_i is the molar volume of the solid (assumed to be independent of pressure), R is the gas constant, and T is the temperature. An equation of state may be used to compute Φ_i. For example, the expression for the fugacity coefficient given by the Peng–Robinson equation is

$$\ln (\Phi_i) = \frac{b_i}{b} (Z - 1) - \ln(Z - B) \frac{A}{2\sqrt{2B}} \left(\frac{2 \sum_k x_i a_{ki}}{a} - \frac{b_i}{b} \right) \ln \left(\frac{Z + 2.414B}{Z - 0.414B} \right)$$

$$Z^3 - (1 - B)Z^2 + (A - 3B^2 - 2B)Z - (AB - B^2 - B^3) = 0$$

$$A = \frac{aP}{(RT)^2} \qquad B = \frac{bP}{(RT)}$$

$$a = \sum_i \sum_j x_i x_j a_{ij} \tag{7.2}$$

$$b = \sum_i x_i b_i$$

$$a_{ij} = (1 - \delta_{ij}) \sqrt{a_i} \sqrt{a_j}$$

$$a_i = 0.45724 \ \frac{(RT_{ci})^2}{P_{ci}} \left(1 + \kappa_i \ \sqrt{\frac{T^2}{T_{ci}}} \right) \quad b_i = 0.0778 \ \frac{RT_{ci}}{P_{ci}}$$

$$\kappa_i = 0.37464 + 1.54226\omega_i - 0.26992 \ \omega_i^2 \qquad (7.2)$$

In these equations the δ_{ij}'s are interaction coefficients that are fitted using experimental phase equilibrium data. Note that the pure component properties, such as the critical temperature and pressure (T_{ci}, P_{ci}), the acentric factor (ω_i), and the vapor pressure must be known to solve these equations. For hydrocarbons and other low-molecular-weight species this data is readily available from standard handbooks (15), but obtaining these properties ordinarily represents a problem for most food and lipid mixtures. The critical properties are usually not known and cannot be measured, because these species will decompose before the critical point can be reached experimentally. In addition, very little vapor pressure data exists for the solutes of interest for food and lipid processing. Researchers have attempted to address these difficulties by using group-contribution methods to estimate the critical properties (16,17) and the vapor pressure (18). For example, Schaeffer et al. (19) studied the solubility of a pyrrolizidine alkaloid, monocrotaline, in supercritical CO_2 and CO_2 mixed with 5 to 10% ethanol. The data was modeled using the Peng–Robinson equation of state, and the UNIFAC vapor pressure correlation (18) was used to estimate the vapor pressure. Vapor pressure predictions for other pyrrolizidine alkaloids were compared with experimental data, and the relative errors were found to range from 5 to 58%. Critical properties were estimated using the Ambrose group contribution method. Figure 7.1 shows a comparison of the experimental solubility data and the model predictions. Qualitative agreement is obtained, although the model predictions are particularly poor in the low-pressure regions. This was attributed to inaccuracies inherent in using equations of state, because similar difficulties are also noted for well-characterized systems such as CO_2–naphthalene. Cygnarowicz et al. (20) followed a similar approach in representing the solubility of β-carotene in supercritical CO_2.

Bamberger et al. (21) measured the solubilities of pure trilaurin, trimyristin, and tripalmitin and their corresponding fatty acids (lauric acid, myristic acid, and palmitic acid) in supercritical CO_2 at 313 K and pressures ranging from 80 to 300 bar. The solubility data was modeled by a lattice gas equation of state (22) that was developed to correlate equilibrium data from mixtures of molecules with disparate sizes. The model requires two pure component parameters: a molecular interactive energy (ε_{11}) and a reducing volume (v^*). These were estimated by matching the chemical potentials of the component across the liquid and vapor phases at the vapor pressure of the pure substance. These parameters could be readily determined for the fatty acids, for which vapor pressure data were available. For the triglycerides, however, only the values for trilaurin could be determined, because it has the highest vapor pressure. Therefore, it was assumed that ε_{11} and v^* were the same for all of the triglycerides studied. To represent the mixture data, a mixing rule was used in which a single adjustable parameter was fitted to the experimental data. In contrast to most modeling attempts, the con-

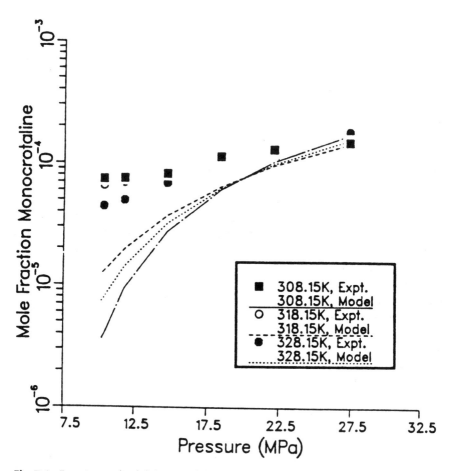

Fig. 7.1. Experimental solubilities and those predicted by the Peng–Robinson equation of state for monocrotaline in supercritical CO_2 (Reprinted with permission from *Fluid Phase Equilibria 43:54*, ©1988, Elsevier Scientific Publishers.)

densed (solid) phases of the mixtures were not considered to be pure. Rather, the lattice gas EOS was used to model both phases. Results of the correlation are shown as solid lines in Fig. 7.2. The predictions are good at higher pressures, but the model cannot accurately describe the results at lower pressures (below 96 atm) because of the proximity to the critical point. Bamberger et al. (21) found that the interaction coefficients (δ_{ij}) that were fitted to the experimental data increased almost linearly with increasing molecular weight, perhaps providing a means to estimate the solubility of other triglycerides for which experimental data may not be available (see Table 7.l).

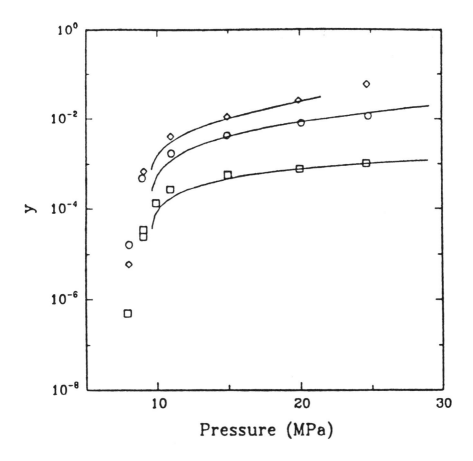

◇ **Lauric Acid**
○ **Myristic Acid**
□ **Palmitic Acid**

Fig. 7.2. Solubility of pure fatty acids in supercritical CO_2. Points represent experimental data and curves are predictions of the lattice gas equation of state. (Reprinted with permission from *Journal of Chemical Engineering Data 33*:331, ©1988 American Chemical Society.)

Representation of solid–supercritical fluid equilibria is often simplified by the assumption that the solid phase is pure (i.e., no supercritical solvent dissolves in it). Such an assumption cannot be made when modeling liquid–supercritical fluid equilibria, and that fact complicates the modeling attempts. Mathias et al. (23) discuss some of the difficulties in modeling phase equilibria for SFE design studies. They studied the extraction of lemon oil and palm oil using supercritical CO_2 and presented equilibrium data

TABLE 7.1 Interaction Coefficients for the Lattice Gas Equation of State used by Bamberger et al. to Correlate Solubilities of Fatty Acids in Supercritical CO_2

Compound	δ_{ij}
Lauric acid	0.06
Myristic acid	0.07
Palmitic acid	0.12
Trilaurin	0.09
Trimyristin	0.105
Tripalmitin	0.13

Reprinted with permission from *J. Chemical Engineering Data* 33:329, ©1988 American Chemical Society.

for both the vapor and the liquid phases. Lemon oil was treated as a mixture of three components: limonene, citral, and sesquiterpene. The valuable component in this mixture is citral, and an objective of an SFE process would be to concentrate this species. Palm oil was assumed to be a mixture of 17 triglycerides, ranging in carbon number from 22 to 54. An SFE process would be designed to concentrate the middle-range triglycerides to obtain a product with desirable melting characteristics. Sufficient vapor pressure data were available for the components of the lemon oil mixture, but only limited, high-temperature data were available for the triglycerides. Mathias and coworkers addressed this limitation by fitting the available data to the Abrams–Massaldi–Prausnitz (AMP) equation (24) and then extrapolating it to the temperatures of interest. The AMP parameters were found to vary linearly with carbon number. Pseudocritical properties were estimated using the AMP equation and the van der Waals volume. The solubility data were correlated to both the Peng–Robinson EOS and an equation of state that uses a theoretically realistic repulsive term (i.e., the Percus–Yevick compressibility equation). One interaction coefficient was fitted for each binary pair. Both models were able to give reasonable representation of the experimental data, although the model based on the Percus–Yevick equation was stated to be slightly better.

Bharath et al. (25) and Ikushima et al. (26) measured vapor-liquid equilibria for mixtures of fatty acid methyl esters and fatty acid ethyl esters in supercritical CO_2. The data were correlated using the Peng–Robinson equation of state, and the pure component parameters were estimated using the Lydersen method. A two-parameter mixing rule, first proposed by de Loos et al. (27) was used to fit the data. As shown in Fig. 7.3, the predicted values agree well with the experimental data, both in the vapor and the liquid phases.

Klein and Schulz (28) also studied the solubility of rapeseed oil, but in addition to applying an EOS, they also considered the use of continuous thermodynamics, in which the composition of a mixture is described through a continuous distribution function with an independent variable that is an appropriate characterizing property, such as the boiling point, molecular weight, or carbon number. This method has been shown to work well for many ill-defined mixtures, including natural gas mixtures and

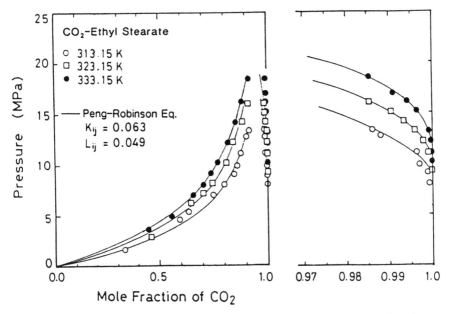

Fig. 7.3. Experimental and calculated vapor-liquid equilibria for the carbon dioxide–ethyl stearate system (Reprinted with permission from *Fluid Phase Equilibria 50:*323, ©1989, Elsevier Scientific Publishers.)

polymer solutions (29). Klein and Schulz used carbon number as the independent variable of the distribution function for the prediction of the solubility of rapeseed oil in supercritical CO_2. Since the composition of rapeseed oil is 90% C52 and C54, these were excluded from the distribution function, and exponential functions were used to estimate the composition of triglycerides with lower and higher carbon numbers. These results were compared to the predictions of the conventional pseudocomponent method. Klein and Schulz found that there was often little difference between the two methods, and in some cases, the predictions of the continuous model were worse than those of the pseudocomponent model. The discrepancies were attributed to uncertainties in the pure component properties for the triglycerides and to inaccuracies in the fitted binary interaction parameters for the equation of state. Klein and Schulz concluded that although it is possible to represent a seed oil mixture using continuous thermodynamics, doing so offers no advantages over the usual approaches.

Clearly, good progress has been made in utilizing equations of state for the prediction of supercritical fluid phase behavior. However, because of the lack of physical property data it is still difficult to apply these models to most systems of interest for SFE. The use of group-contribution methods to predict the critical properties is time-consuming and difficult, especially when the molecule in question contains functional groups for which parameters are not available. These methods, and the other approximations that must be made to utilize equations of state, introduce errors that limit the ranges over which the models are valid.

Other researchers have developed more empirical models. Chrastil (30) and del Valle and Aguilera (31) have fitted coefficients for polynomial equations that estimate the solubility of seed oils in supercritical CO_2 as a function of solvent temperature and density. The equation determined by del Valle and Aguilera is

$$\ln (C) = 40.361 - \frac{18708}{T} + \frac{2186840}{T^2} + 10.724 \ln(\rho) \qquad (7.3)$$

where C is the oil concentration (g/L), T is the solvent temperature (K), and ρ is the solvent density (g/mL). Data from five different seed oil extraction studies were used to determine the coefficients of the equations. As shown by Fig. 7.4, the fit of the equation is quite good, deviating from the experimental values only at very high densities. The coefficients were correlated using experimental solubilities measured from 293 to 353 K and from 150 to 800 bar, and therefore the predictions are valid only for this region. This equation assumes that the oil is composed of a single component, and it cannot predict the composition of the extract should any fractionation occur. However, it is a useful equation because of its simplicity.

Mass Transfer Models

Although many of the experimental studies that are reported are concerned with measuring solubilities in supercritical fluids, it is often the mass transfer rates that are the limiting step in an SFE process. This is especially true when the objective is to extract from a solid material, such as the case with seed oil extractions. A few researchers have attempted to quantify mass transfer rates in supercritical fluids and have developed models that represent the mass transfer as a function of the solvent density and flow rate.

For example, Brunner (32) studied mass transfer rates for the decaffeination of coffee beans, using supercritical nitrous oxide, and the extraction of rapeseed oil, using supercritical CO_2. The extraction process exhibits two distinct periods. Initially the mass transfer rate is constant, but after a certain amount of material is extracted, the rate begins to decline. Brunner attempted to apply correlations for mass transfer coefficients determined from various gas-liquid and liquid-solid fixed-bed processes to model the supercritical fluid extraction process. He presented a correlation, using dimensionless numbers for the mass transfer coefficient, that is valid for a fixed bed of spheres:

$$Sh = C \sqrt[3]{Sc}\, (Re)^{0.6}$$

$$Sh = \frac{\beta_a d}{\rho D_g}$$

$$Sc = \frac{\mu}{\rho D_g} \qquad (7.4)$$

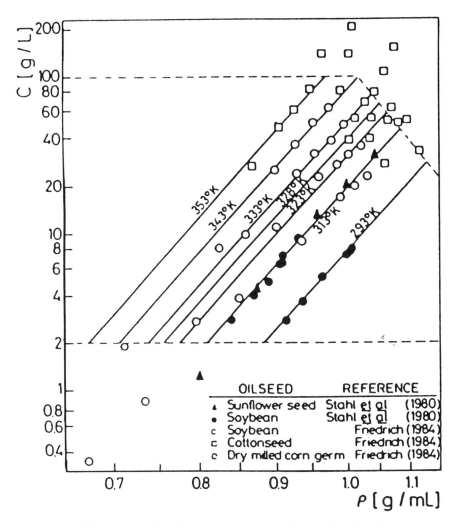

Fig. 7.4. Solubility of vegetable oils (C) in compressed CO_2 as a function of solvent density (ρ). Solid lines represent values predicted by correlation. (Reprinted with permission from *Industrial & Engineering Chemistry Research 27*:1551, ©1988 American Chemical Society.)

$$Re = \frac{\rho u d}{\mu}$$

In these expressions Sc is the Schmidt number, Re is the Reynolds number, Sh is the Sherwood number, β_a is the mass transfer coefficient, d is the particle diameter, D_g is the diffusion coefficient, ρ is the solvent density, u is the solvent superficial velocity,

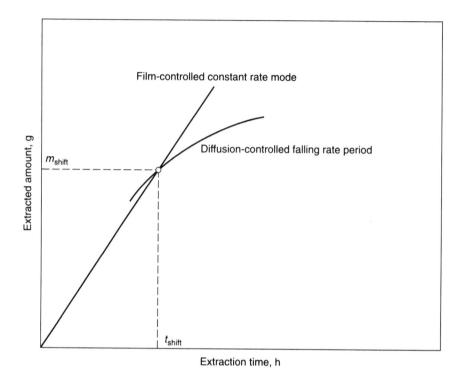

Fig. 7.5. A typical extraction curve for soybean oil production. (Based on Fig. 2 from *Korean Journal of Chemical Engineering* 7(1):40 (1990).)

μ is the solvent viscosity, and C is a constant. During the steady-state period, the quantity of extracted oil is a product of the mass transfer coefficient, the mass transfer area, and a mean concentration difference between the solid and supercritical fluid phase; that is,

$$m = \beta_a A_s V_t \Delta c_m$$

$$\Delta cm = \frac{\Delta c_{in} - \Delta c_{out}}{\ln\left(\frac{\Delta c_{in}}{\Delta c_{out}}\right)} \qquad 7.5$$

$$A_s = \frac{6\,(1-\psi)}{d}$$

In these expressions, m is the mass of extracted oil, A_s is the specific area for mass transfer, V_t is the total volume of the bed, ψ is the volume void fraction of the bed, and Δc_{in} and Δc_{out} are the concentration differences of oil between the mass transfer interface and the bulk at the inlet and the outlet, respectively. To solve these equations, Brunner assumed that the concentration of oil at the mass transfer interface was constant and that the concentration at the outlet of the bed was 90% of the equilibrium concentration. The calculated values were shown to represent the data reasonably well, even though the size distribution of the particles was neglected, and the oil was assumed to be composed of a single component.

Hong et al. (33) followed a similar approach to model the extraction of oil from crushed soybeans They observed that experimental data are well represented by a combined model of steady-state mass transfer at high oil concentrations in the solid matrix, and an unsteady model at low concentrations. The extraction region controlled by a constant rate follows a film-controlling mass transfer, and the region controlled by a decreasing rate follows a diffusion-controlled mass transfer, as shown schematically in Fig. 7.5. The two types of mass transfer calculations were carried out individually, and the results were combined to represent the entire extraction curve. The parameters that define the shift in extraction regime, m_{shift} and t_{shift}, were found from the experimental data. To describe the steady-state regime, Hong et al. used the same expression as Brunner (32), except that the Sherwood number was determined to be

$$Sh\{1 + 1.5(1 - \psi)\} = 0.035 \sqrt{Re} \sqrt[3]{Sc} \qquad (7.6)$$

To describe the mass transfer during the diffusion-controlled regime, Hong et al. used the following approximate expression for the mass of oil extracted as a function of time:

$$\frac{m(t)}{m_0} = \exp(-Kt)$$

$$K = \frac{F_0}{t} \quad F_0 = \frac{D_s t}{d^2} \qquad (7.7)$$

In these equations $m(t)$ is the amount of oil that remains in the solid at any time, t, m_0 is the initial amount of oil in the solid, F_0 is the Fourier number, D_s is the solid diffusivity, and K is a constant that is fitted from experimental data. Hong and coworkers drew parallels between the SFE process and drying. In the first (steady-state) regime the substances to be extracted are removed from the macropores and the particle surface, but in the second regime, mass transfer is controlled by the diffusion of extractants from the micropores of the solid, which reduces the mass transfer rate. As shown in Fig. 7.6, the combined models are able to represent experimental extraction curves well, although this is not surprising, because a significant amount of data fitting is required to use this model. Similar mass transfer studies were published by Kandiah and Spiro (34), who studied the extraction of gingerol; King et al. (35), who studied the extraction of evening primrose oil; and King (36), who studied the extraction of rapeseed oil.

Bulley et al. (3) and Lee et al. (37) took a macroscopic approach to describing the rate of mass transfer in a fixed bed of crushed rapeseeds. They utilized a one-dimensional mathematical model of a fixed-bed extractor, previously presented by

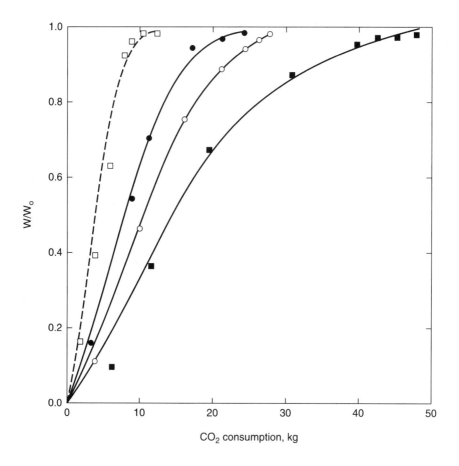

Fig. 7.6. Extraction curves for the production of oil from dry-milled soybeans with supercritical CO_2 at a solvent flow rate of 340 mL/min and a temperature of 50°C. Curves are the predictions of the combined models. (Reprinted from *Korean Journal of Chemical Engineering 7(1)*:40 (1990).)

Bird et al. (38). This model predicts the solvent and seed bed oil concentrations as a function of time and position in the bed. It assumes that plug flow exists in the bed and that axial dispersion is negligible. If the void fraction and the solvent flow rate, temperature, and pressure are assumed to be constant, the material balances are

$$\varepsilon\rho\frac{\partial y}{\partial t} + \rho U\frac{\partial y}{\partial h} = R\{x,y\} \tag{7.8}$$

$$(1-\varepsilon)\rho_s\frac{\partial x}{\partial t} = -R\{x,y\}$$

where h is the distance along the bed, t is the time, y is the weight fraction of oil in the solvent stream (on a solute-free basis), x is the weight fraction of oil in the seed bed (solute-free basis), ρ is the density of the solvent, ρ_s is the density of the seeds, U is the superficial velocity of the solvent, ε is the void volume fraction, and $R\{x,y\}$ represents the rate of mass transfer of the oil from the seed bed to the solvent phase. If the solvent is assumed to be pure at the bed entrance, and all seeds have the same initial oil content, the initial and boundary conditions are

$$y = 0 \text{ at } h = 0 \text{ for t} \geq 0$$
$$x = x_0 \text{ at } t = 0 \text{ for all } h \qquad (7.9)$$

Defining z as

$$z = \frac{\varepsilon h}{U} \qquad (7.10)$$

and substituting into the Eqs. (7.8) gives

$$\frac{\partial y}{\partial t} + \frac{\partial y}{\partial z} = \frac{R\{x,y\}}{\varepsilon\rho}$$

$$\frac{\partial x}{\partial t} = \frac{-R\{x,y\}}{(1-\varepsilon)\rho s} \qquad (7.11)$$

To solve these equations, an expression is needed for the mass transfer rate, $R\{x,y\}$. Lee and coworkers used the simple expression

$$R\{x,y\} = A_p K(y^* - y) \qquad (7.12)$$

where $A_p K$ is an overall volumetric mass transfer coefficient, y^* is the equilibrium concentration of oil in the solvent phase, and y is the actual concentration of oil in the solvent phase. The partial differential equations can be readily solved using the method of characteristics (39,40). To compare the experimental and calculated extraction curves, the oil concentration in the solvent phase at the extractor outlet $y_e\{t\}$ was integrated to compute the total mass of oil extracted (M_e) for a given time period (t_e):

$$M_e = m \int_0^{t_e} y_e\{t\}dt \qquad (7.13)$$

where m denotes the solvent mass flow rate, assumed to be constant. M_e was calculated for various time periods and then plotted for comparison with experimental extraction curves to determine values for $A_p K$. Only experimental values from the constant-rate period were used. The model was able to represent the experimental curves well, as shown in Fig. 7.7 for the solvent phase and Fig. 7.8 for the solid (seed bed) phase. The values of the calculated mass transfer coefficients that gave the best agreement with the experimental results during the constant-rate period are shown in Fig. 7.9. The vertical bars indicate the range of $A_p K$ values that gave good agreement

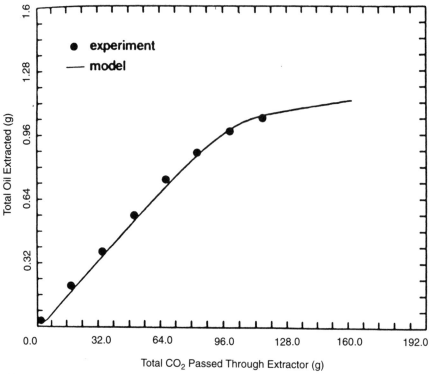

Fig. 7.7. Extraction curve for canola oil. Extraction conditions are 55°C, 360 bar; flow rate is 2.7 g/min. The computed extraction curve was calculated using $A_pK = 33.33$ kg/m³ sec. (Reprinted with permission from *Journal of the American Oil Chemists' Society 63(7):923*, ©1986 American Oil Chemists' Society.)

between the predicted and measured results. Note that the mass transfer coefficient increases with interstitial velocity (i.e., U/ε) and can be represented by the equation

$$A_pK = 32.89(v)^{0.54} \tag{7.14}$$

The units of A_pK and v are kg CO_2/m³ sec and mm/s respectively.

Cygnarowicz-Provost (41) extended this analysis to include a mass transfer coefficient that can represent both the constant-rate and the diffusion-controlled regimes. The following empirical correlation was developed:

$$A_pK = A_pK_0 \exp\left[\ln(C) \frac{(x_0 - x)}{(x_0 - x_{shift})}\right] \tag{7.15}$$

where x_0 is the initial oil concentration in the seeds, x_{shift} is the concentration in the seeds at which the extraction shifts from the mass transfer–controlled to the diffusion-controlled regime, and A_pK_0 is an initial mass transfer coefficient. The above relation-

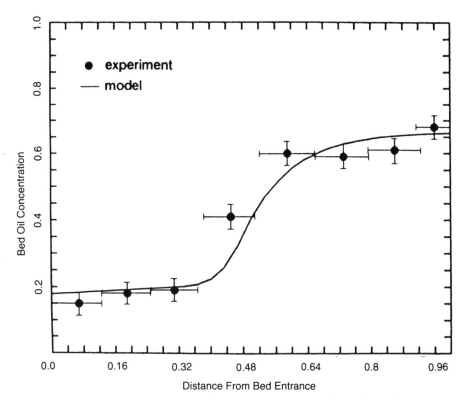

Fig. 7.8. Oil concentration in the canola seeds as a function of normalized distance from the bed entrance after 240 min. Extraction conditions are 55°C and 360 bar; the flow rate is 1.6 g/min. (Reprinted with permission from *Journal of the American Oil Chemists' Society 63(7)*:923, ©1986 American Oil Chemists' Society.)

ship predicts that when $x = x_0$ (i.e., at the beginning of the extraction), $A_p K = A_p K_0$, and that when $x = x_{shift}$, $A_p K = C A_p K_0$. Using this correlation and a C value that is less than one, the mass transfer coefficient will asymptotically approach zero as the seed bed is depleted. The parameter x_{shift} can be determined from experimental data.

Modeling the SFE Process and Estimating Costs

Once descriptions of the thermodynamics and mass transfer have been obtained, an attempt can be made to simulate the SFE process. If the feed to the extractor is a liquid, mass transfer will be fast, and the streams leaving the extractor can be assumed to be in equilibrium. This equilibrium-stage assumption was used in the simulation and analysis of SFE processes to dehydrate ethanol and acetone. For example, Brignole et al. (42) investigated the potential reduction in energy consumption achieved by using supercritical fluids to dehydrate ethanol and propanol. They used the GC equation of

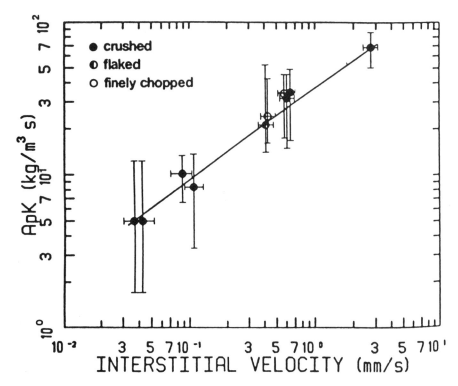

Fig. 7.9. Volumetric mass transfer coefficients (A_pK) as a function of interstitial velocity (v). (Reprinted with permission from *Journal of the American Oil Chemists' Society* *63(7)*:924, ©1986 American Oil Chemists' Society.)

state to model the phase equilibrium and the SEPSIM flowsheet program (43) to simulate the process. Several process configurations were proposed, including cycles with vapor recompression and feed preconcentration. They concluded that SFE processes significantly reduce the energy consumed as compared to azeotropic distillation. Although attempts were made to determine optimal values for important parameters such as the extractor temperature and pressure and the solvent-to-feed ratio, formal optimization methods were not utilized, and thus the interdependence of these parameters was not considered. Furthermore, only the consumption of utilities was minimized; that is, the installed cost of the equipment was not included.

Cygnarowicz and Seider (44) investigated the dehydration of acetone with supercritical CO_2. They developed a model for the process flowsheet (utilizing SEPSIM) and created a nonlinear program to compute designs that minimize either the utility or the annualized cost. They found the variables that had the greatest impact on the process cost and performance were the extractor temperature and pressure, the solvent-to-feed ratio, the separator pressure, and the number of stages in the extractor and separator. In their analysis, the number of stages and the separator pressure were fixed, and the optimization routine was used to locate the best values for the other parameters. When only the cost of utilities was minimized, SFE was shown to be competitive with conventional separation techniques, in agreement with the conclusions of Brignole et al (42). However, Cygnarowicz and Seider also included the installed cost of the equipment in the analysis. These costs were estimated by correlating graphical data of Ulrich (45), which depict the installed cost of basic processing equipment as a function of the unit size and operating conditions. When the annualized cost (which includes a portion of the installed cost of the equipment) was minimized, SFE was no longer competitive with distillation, because of the cost of the high-pressure equipment. Acetone, like ethanol and propanol, is a high-volume, low-cost chemical. These products apparently cannot be separated economically using SFE.

Some work has been done to simulate more promising applications. Moricet (46) simulated the extraction of raw lecithin from soya oil using a supercritical CO_2/propane mixture. The Redlich–Kwong equation of state was used to model the phase equilibria, and a hybrid of the sum rates and the Tomich methods was used to solve the equations for the extractor. Although only the extractor was simulated and no attempts were made to optimize the process, this work is important, because simulated and experimental concentrations were compared. Reasonable agreement was found for the liquid product but not for the vapor product. The errors were attributed to "inaccuracies" in modeling the phase equilibria.

Cygnarowicz and Seider (47) developed a simplified model for a process to isolate β-carotene from fermentation broth using supercritical CO_2. Pure component properties for β-carotene were estimated, and solubility data were fitted to the modified Peng–Robinson equation of state. Then models were prepared to simulate the three-phase (solid, liquid, supercritical fluid) extraction and separation step, assuming equilibrium stages for both processes. Optimal designs were then determined that minimize the annualized cost while producing β-carotene of the desired quality. The results indicate that SFE may be competitive for this application, despite the high operating pressures that were required. β-carotene is a valuable, low-volume product and can support the cost of the high-pressure equipment. In addition, the increases in solubility that were obtained with the addition of cosolvent were shown to have a substantial impact on the economics of the proposed design. An example was given in which the addition of 1 wt% ethanol to the CO_2 solvent transformed an unfavorable design to a competitive one. These results, however, were obtained with a highly simplified model (i.e., the liquid in the fermentation broth was assumed to be pure water, and mass transfer effects were ignored). Additional experimentation, including a study of the mass transfer limitations posed by the extraction from a solid matrix, would be needed to quantify the modeling errors and thus to improve the cost estimates. However, this analysis is useful, because it shows the types of products for which SFE may be most applicable, and it demonstrates the effect of the process parameters on the economics.

Raj and Rizvi (48) developed a process model and estimated the economics of using supercritical CO_2 to fractionate anhydrous milk fat. The objective was to produce fractions that possess desirable physical and chemical properties, including a reduction of cholesterol content. The process that was studied consisted of a packed column extractor, three separators, a recycle compressor, and a heat exchanger. The anhydrous milk fat (AMF) was fractionated by successively lowering the pressure in each of the separators. Since the pressure and temperature determine the composition of product that is precipitated, the quality can thus be specified by setting the process conditions. In designing packed-column extractors, the mass transfer rate is often expressed in terms of the Height Equivalent of a Theoretical Stage (HETS). This quantity, in dimensions of length, expresses the height of apparatus required to accomplish a standard separation. The HETS is given by

$$\text{HETS} = \frac{h}{N} \qquad (7.16)$$

where h is the packed height of the column and N is the number of equilibrium stages. By choosing the appropriate packed length (or number of theoretical stages) and the extractor pressure, temperature, and solvent flow rate, one can extract a fixed quantity of AMF in CO_2. The remainder, or *raffinate*, will consist primarily of high-molecular-weight triglycerides, vitamins, and colorants. In conventional extraction processes the flow rates are usually assumed to be constant throughout the column, because the solute of interest is normally contained in a carrier stream. Raj and Rizvi show that in the AMF fractionation process this assumption is not valid, because a significant amount of the triglycerides are solubilized in the CO_2 as the liquid descends the column, decreasing the flow rate. Therefore, they used the method of Martin (49) to calculate N. HETS values were computed for a spray column and for columns with two different packings (Helipackand knitted mesh). The spray column was found to be much less efficient than the packed columns, and the HETS values decreased (i.e., the columns had greater efficiency) as the solvent-to-feed ratio increased. These trends were consistent with previous studies on SFE of alcohol/water systems (50, 51). Raj and Rizvi also estimated the capital and operating costs for a base-case design and for designs with two and three times the capacity of the base case. Assuming a payout time of four years, the total value added to the product was $1.50/kg AMF. Since this is a significant increase over the market value of unprocessed AMF (i.e., butter), they suggest using process control and automation to reduce the labor costs. A revised estimate gives the value added as $0.34/kg product, which is claimed to be commercially viable.

The design studies discussed thus far have all been for extraction from a liquid. In these extractions the feed material is easily conveyed into and out of the extractor through pumps, and therefore the process can be operated continuously. In extracting from a solid material, however, material transport becomes a major concern. These processes typically must be operated batchwise, which requires time for startup and cleanup and therefore reduces the annual production rates. In addition, mass transfer rates will certainly be the limiting step, and the equilibrium-stage assumption cannot be used.

Eggers and Sievers (52,53) discuss several of these difficulties in their studies on the supercritical fluid extraction of oilseeds. They first considered the mass transfer kinetics and attempted to improve the rate of mass transfer from the seed to the supercritical phase in order to achieve shorter extraction times and thus more economical designs. They investigated the extraction efficiency from several combinations of flaked, decorticated, and cracked rapeseeds. The highest mass transfer rates were found for decorticated and flaked rapeseeds, demonstrating the importance of sample preparation for solid extractions. They also showed the benefits of using a mechanical pressing step, prior to extraction, for high-oil-content seeds such as rapeseed or corn germ. Extraction experiments were performed at 300 bar and 40°C on mechanically deoiled rapeseed press cake (oil content = 22 wt%, bulk density = 561 kg/m^3) and flaked rapeseeds (oil content = 42 wt%, bulk density = 312 kg/m^3). They found that extraction of the press cake required less CO_2; specifically, 26 kg solvent/kg seed were required for the press cake and 70 kg solvent/kg seed were required for extraction of the flaked rapeseed. In addition, the extraction times were shorter, and the extractors were smaller when the press cake was extracted. Eggers and coworkers also investigated the effect of using a mixture of CO_2 and propane as the solvent. They found that rapeseed oil has a much higher solubility in the solvent mixture, and therefore the solvent requirements were reduced as compared with extraction with CO_2 alone. Although the solubility increases with increasing propane content, there is a limit to the amount of propane that should be added, because it presents an explosion and fire hazard. They determined a "safe" upper bound to be 9.5% propane. Even at the lower propane content the solubility was increased such that extraction at 60°C and 350 bar with the solvent mixture was comparable to extraction at 750 bar with CO_2 alone. They also did an analysis of the energy consumption of the SFE process and compared it to a conventional hexane extraction. They found that the SFE process has higher compression and cooling requirements but lower heating requirements, because a significant portion of the heat load can be supplied by the temperature rise that occurs during compression. To minimize the energy requirements, it would be desirable to increase the mass transfer rate as much as possible, because that would decrease the extraction time and increase the annual production rate. As discussed by Eggers and coworkers, this goal could be achieved by conveying the solids continuously into and out of the extractor, countercurrent to the solvent flow. Three different configurations were proposed:

1. The first scheme would allow entry and exit of the solid material through high-pressure airlocks, and the solid would contact the solvent countercurrently via a transport screw. This would involve quick-opening seals that are opened and closed periodically, allowing only quasi-continuous operation.

2. Similarly, the solid could enter and exit through mobile pistons, again permitting only quasi-continuous operation. Such a system requires that the solid be conveyed in a liquid, transforming it to a compressible slurry.

3. The last scheme involves entry and exit of the solid material through a screw press. As shown in Fig. 7.10, the entering seed would be mechanically deoiled initially by a buildup of pressure in the screw press leading to the extractor. Upon reaching the extractor, the solid material can be conveyed countercurrent to the solvent by means of a stirrer. The seeds would exit the system via another screw press in which the pressure would be steadily reduced.

At the present time, however, these ideas are only at the development stage and have not yet been implemented in a processing plant.

Cygnarowicz-Provost (41) also conducted an economic evaluation of a seed oil extraction plant. The extraction model of Lee et al. (37) was used to simulate the extraction process, and the empirical correlation of del Valle and Aguilera (31) was used to estimate the solubility of the oil in the supercritical solvent at various temperatures and pressures. The flowsheet model included estimates for the equipment and operating costs of the extractors, separators, compressors, and heat exchangers. Nonlinear programming was used to determine designs that minimize the annualized cost per kilogram of product. Two modes of operation were considered. A batch process was optimized, in which the total batch time, i.e., the extraction time and the turnaround time, was considered. A process that operates semicontinuously was also studied, in

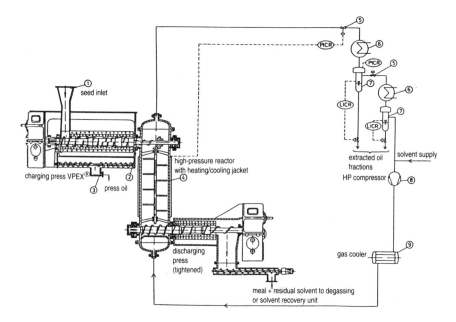

Fig. 7.10. Continuous high-pressure extraction plant for seed oil production. (Reprinted with permission from *Journal of the American Oil Chemists' Society 62(8)*:1229, ©1985 American Oil Chemists' Society.)

which the turnaround time was eliminated by employing multiple vessels and operating them sequentially. A nonlinear program adjusted the extractor temperature and pressure, the flow rate of solvent, the separator pressure, and the extraction time to produce the desired quantity of oil at the lowest cost. The lowest selling price was obtained when the process was operated semicontinuously, at high extractor pressures and low flow rates of solvent. As shown in Table 7.2, the equipment costs for high-pressure operation were actually lower, because of a decrease in the solvent flow rate, which reduced the sizes of the compressors and heat exchangers. In agreement with Eggers et al. (53), it was found that the costs associated with operating the process batchwise are prohibitive. Semicontinuous operation yielded a product cost that was much lower than that obtained for batch operation with a 60 min turnaround time. Although the costs computed for the SFE process are not currently competitive with the hexane-extracted product ($1.40/kg vs. $0.44/kg), the process may be more attractive in the future if environmental costs increase.

Leyers et al. (54) presented a detailed design study of a process to decaffeinate coffee using supercritical CO_2. The base-case design produces 32 metric tons (MT)/day of 97% decaffeinated coffee beans. The design was based on a patented process and proprietary pilot plant data. The extraction section consists of four vessels that operate in semicontinuous mode, with three vessels in extraction service at any one time and the fourth vessel either being filled with fresh beans or discharging extracted beans. The three vessels in the extraction train are cycled such that a countercurrent flow of supercritical solvent is maintained, with the lean solvent contacting the most extracted beans and the freshest beans seeing the most saturated solvent. The extraction conditions given were 140 to 350 bar, 70 to 130°C, and the residence time was 6 to 12 h. The capital costs were estimated using proprietary data or actual vendor quotations. They found a significant increase in the capital and processing costs for a design that was half the base case, and a significant decrease in these costs for a design that was twice the base case. The base-case plant would require $22 million in capital investment, and the processing costs would be $0.41/kg. They also concluded that a truly continuous method of processing the beans would reduce the capital costs and increase the annual production rate.

Dyken et al. (55) describe a prospective design of a supercritical fluid extraction process for the production of monocrotaline. Based on the results of Schaeffer et al. (19,56,57), they proposed a process in which crushed *Crotalaria spectabilis* seeds are extracted with a mixture of CO_2/10% ethanol at 328 K and 274 bar. The extract is passed through an ion exchange column, where the alkaloid, monocrotaline, is retained. The stream next enters a separator, where the lipid, water, and ethanol are precipitated. The CO_2 is then recycled to the compressor. Experimental data was used to size the extractor and separator, and the flowsheet simulator PROCESS was used to estimate the heating, cooling, and compression loads. A selling price for a purified monocrotaline product was estimated to be $17,000/kg, based on a payout time of two years. The high price is a result of the low yield of product (0.3 kg/batch, under the best conditions) and the long cycle time (20 h). The need to remove the ion exchange

TABLE 7.2 Capital and Operating Costs for a Process to Extract Canola Oil with Supercritical CO_2.

	Case 1 $P_e \leq 400$ bar Batch operation	Case 2 $P_e \leq 400$ bar Semicontinuous operation	Case 3 $P_e \leq 880$ bar Semicontinuous operation
Design variables			
Extractor temperature (K)	343	343	343
Extractor pressure (bar)	400	400	880
Separator pressure (bar)	157.6	147.3	137.1
Flow rate of CO_2 (kg/min)	519.6	571.7	52.9
Extraction time (min)	64	59	45
Total capital cost (includes 18% contingency and fees, and working capital) ($)	**14,400,000**	**17,959,000**	**13,037,000**
Raw materials ($/yr)[a]	802,000	3,338,000	4,404,000
Utilities ($/yr)	481,000	361,000	69,000
Labor[b] ($/yr)	354,000	589,000	589,000
Maintenance, repair, supplies, and laboratory charges ($/yr)	1,792,000	2,253,000	1,663,000
Overhead, taxes, and insurance ($/yr)	1,832,000	2,388,000	1,846,000
Depreciation (10% straight-line) ($/yr)	1,440,000	1,796,000	1,304,000
Administrative expenses ($/yr)	314,000	417,000	331,000
Before-tax profit (based on 15% return on investment, 50% income tax rate) ($/yr)	1,440,000	1,796,000	1,304,000
Required revenues ($/yr)	8,455,000	12,938,000	11,510,000
Annual production (kg/yr)	**1,419,000**	**5,973,000**	**8,217,000**
Selling Price ($/kg)	**5.96**	**2.17**	**1.40**

[a]Includes a byproduct credit for canola meal.
[b]Assumes 4-person shifts for Case 1, 6-person shifts for Cases 2–3, 3 shifts per day. Includes supervisory and clerical labor.
P_e refers to the extractor pressure; in Cases 1–2, P_e is bounded by 400 bar, and in Case 3, P_e is bounded by 880 bar. Costs are estimated for a retrofit of an existing plant site. All equipment is assumed to be made of stainless steel. (Reprinted with permission of the author.)

vessel and undertake a three-step wash process was the major contributor to the long cycle times. The authors point out that the process was not optimized and that changing the extraction conditions or increasing the ethanol concentration could improve the recovery of the product. They claim that a selling price of $17/g is not out of line for a purified pharmaceutical product.

Outlook for Supercritical Fluid Technology

Supercritical fluid extraction will continue to be an important technology for food and lipids processing. Feasibility for particular applications will ultimately depend on the economics, i.e., the scale of the process, the cost of the product, and the need for a nontoxic solvent. The production of high-value food additives and pharmaceuticals, such as monocrotaline and β-carotene, appears promising because the high selling price is able to support the capital costs. Fractionation of lipid mixtures, such as fish oil or milk fat, is another promising application, because the use of supercritical solvents allows products of specific composition or physical characteristics to be obtained. Production of lower-valued foods, like soybean or canola oil, is not currently economical. Advances in solids-handling technology, however, may allow continuous operation, which could lower the processing costs of the products.

The outlook for implementation of this technology could change if there is a significant increase in the cost of petroleum products or if environmental concerns become a priority. For example, if the costs of energy were to increase sharply, supercritical fluid extraction could become an attractive alternative for some separations on the basis of energy consumption. In addition, if the cost of disposal of toxic solvents becomes prohibitive, extraction with supercritical fluids may become a preferred method, because no waste is generated. Finally, food products that are prepared without organic solvents (i.e., "naturally") may command a premium price because of health concerns, increasing the incentive to utilize supercritical fluid technology.

References

1. Zosel, K., U.S. Patent 4 247 580 (1981).
2. Friedrich, J.P., G.R. List, A.J. Heakin, *J. Amer. Oil Chem. Soc. 59(7):*288 (1982).
3. Bulley, N.R., M. Fattori, A. Meisen, L. Moyls, *J. Amer. Oil Chem. Soc. 61(8):*1362 (1984).
4. Vollbrecht, R., *Chem. Ind. 19:*397 (1982).
5. Hubert, P., and O.G. Vitzthum, *Angew. Chem. Int. Ed. Eng. 17:*710 (1978).
6. Arul, J., A. Boudreau, J. Makhlouf, R. Tardif, and M.R. Sahasrabudhe, *J. Food Sci. 52(5):*1231 (1987).
7. Rizvi, S.S.H., R.R. Chao, and Y.J. Liaw, in *Supercritical Fluid Extraction and Chromatography:Techniques and Applications,* edited by B.A. Charpentier and M.R. Sevenants, American Chemical Society, Washington, DC, 1988, pp. 89–108.
8. Favati, F., J.W. King, J.P. Friedrich, and K. Eskins, *J. Food Sci. 53(5):*1532 (1988).
9. Froning, G.W., R.L. Wehling, S.L. Cuppett, M.M. Pierce, L. Niemann, and D.K. Siekman, *J. Food Sci. 55(1):*95 (1990).
10. Chao, R.R., S.J. Mulvaney, M.E. Bailey, and L.N. Fernando, *J. Food Sci. 56(1):*183 (1991).
11. Peng, D.-Y., and D.B. Robinson, *Ind. Eng. Chem. Fundam. 15(1):*59 (1976).
12. Soave, G., *Chem. Eng. Sci. 27:*1192 (1972).
13. Skjold-Jorgensen, S., *Ind. Eng. Chem. Res. 27:*110 (1988).
14. Prausnitz, J.M., *Molecular Thermodynamics of Fluid Phase Equilibria,* Prentice-Hall, Englewood Cliffs, NJ, 1969.

15. Reid, R.C., J.M. Prausnitz, and T.K. Sherwood, *The Properties of Gases and Liquids,* 3rd edn., McGraw-Hill, New York, 1977.
16. Ambrose, D., *Correlation and Estimation of Vapour-Liquid Properties I.,* National Physics Lab Report Chem., 92, Middlesex, England, 1979.
17. Lydersen, A.L., *Estimation of Critical Properties of Organic Compounds,* University of Wisconsin, Madison, 1955.
18. Jensen, T., A. Fredenslund, and P. Rasmussen, *Ind. Eng. Chem. Fundam. 20:*239 (1981).
19. Schaeffer, S.T., L.H. Zalkow, and A.S. Teja, *Fluid Phase Equilibria 43:*45 (1988).
20. Cygnarowicz, M.L., R.J. Maxwell, and W.D. Seider, *Fluid Phase Equilibria 59:*57 (1990).
21. Bamberger, T., J.C. Erickson, and C.L. Cooney, *J. Chem. Eng. Data 33:*327 (1988).
22. Kumar, S., U. Suter, R. Reid, *Ind. Eng. Chem. Res. 26:*25 (1987).
23. Mathias, P.M., T.W. Copeman, and J.M. Prausnitz, *Fluid Phase Equilibria 29:*545 (1986).
24. Abrams, D.S., H.A. Massaldi, and J.M. Prausnitz, *Ind Eng. Chem. Fundam. 13:*259 (1974).
25. Bharath, R., H. Inomata, K. Arai, K. Shoji, and Y. Noguchi, *Fluid Phase Equilibria 50:* 315 (1989).
26. Ikushima, Y., N. Saito, and T. Goto, *Ind. Eng. Chem. Res. 28:*1364 (1989).
27. deLoos, Th.W., W. Poot, and R.N. Lichtenthaler, *Ber. Bunsenges. Phys. Chem. 88:*855 (1984).
28. Klein, T., and S. Schulz, *Fluid Phase Equilibria 50:*79 (1989).
29. Cotterman, R.L., and J.M. Prausnitz, *Ind. Eng. Chem. Proc. Des. Dev. 24:*434 (1985).
30. Chrastil, J., *J. Phys. Chem. 86:*3016 (1982).
31. del Valle, J.M., and J.M. Aguilera, *Ind. Eng. Chem. Res. 63(7):*1551 (1988).
32. Brunner, G., *Ber. Bunsenges. Phys. Chem. 88:*887 (1984).
33. Hong, I.K., S.W. Rho, K.S. Lee, W.H. Lee, and K.P. Yoo, *Korean J. Chem. Eng 7(1):*40 (1990).
34. Kandiah, M., and M. Spiro, *Int. J. Food Sci. Tech. 25:*328 (1990).
35. King, J.W., personal communication, 1992.
36. King, M.B., T.R. Bott, M.J. Barr, and R.S. Mahmud, *Sep. Sci. Tech. 22(2,3):*1103 (1987).
37. Lee, A.K.K., N.R. Bulley, M. Fattori, and A. Meisen, *J. Amer. Oil Chem. Soc. 63(7):*921 (1986).
38. Bird, R.B., W.E. Stewart, and E.N. Lightfoot, *Transport Phenomena,* Wiley, New York, 1960, pp. 702–705.
39. Acrivos, A., *Ind Eng. Chem. 48(4):*703 (1956).
40. Dranoff, J.S., and L. Lapidus, *Ind. Eng. Chem. 50(11):*1648 (1958).
41. Cygnarowicz-Provost, M., "Optimal Design of a Process to Extract Seed Oils with Supercritical CO_2," paper IB.4, Conference on Food Engineering, Chicago, IL, March 1991.
42. Brignole, E.A., P.M. Andersen and A. Fredenslund, *Ind Eng. Chem. Res. 26:*254 (1987).
43. Andersen, P.M. and A. Fredenslund, in *Proceedings of the 18th Congress on Chemical Engineering Fundamentals,* Sicily, 1987,
44. Cygnarowicz, M.L. and W.D. Seider, *Ind. Eng. Chem. Res. 28:*1497 (1989).
45. Ulrich, G.D., *A Guide to Chemical Engineering Process Design and Economics,* 1st edn., Wiley, New York, 1984.
46. Moricet, M., in *Proceedings of the First International Conference on Supercritical Fluids,* Nice, France, 1989, edited by M. Perrut, pp. 177–181.
47. Cygnarowicz, M.L., and W.D. Seider, *Biotech. Progress 6(1):*82 (1990).

48. Raj, C.B.C., and S.S.H. Rizvi, *Trans. Inst. Chem. Eng.,* 1993.
49. Martin, J.M., *AIChE J. 9:*646 (1963).
50. Lahiere, L.R., and J.R. Fair, *Ind. Eng Chem. Res. 26:*2086 (1987).
51. Seibert, A.F., and D.G. Moosberg, *Sep. Sci. Tech. 23:*2049 (1988).
52. Eggers, R., U. Sievers, and W. Stein, *J. Amer. Oil Chem. Soc. 62(8):*1222 (1985).
53. Eggers, R., and U. Sievers, *J. Chem. Eng. Japan 22(6):*641 (1989).
54. Leyers, W.E., R.A. Novak, and D.A. Linnig, in *Proceedings of the Second International Symposium on Supercritical Fluids,* Boston, MA, 1991, edited by M.A. McHugh, p. 261.
55. Dyken, J.J., B.L. Knutson, J.F. Morris, and J.T. Sommerfeld, *Process Biochem.* 47 (April 1990).
56. Schaeffer, S.T., L.H. Zalkow, and A.S. Teja, *Biotech. Bioeng. 34:*1357 (1989).
57. Schaeffer, S.T., L.H. Zalkow, and A.S. Teja, in *Supercritical Fluid Science and Techuii;nology,* edited by K.P. Johnston, and J.M.L. Penninger, American Chemical Society, Washington, DC, 1989, p. 416.

Chapter 8

Supercritical Fluid Extraction and Fractionation of Fish Oils

William B. Nilsson

U.S. Department of Commerce, NOAA, National Marine Fisheries Service, NWFSC, Utilization Research Division, 2725 Montlake Boulevard East, Seattle, WA 98112

The processing of oils of marine origin has long been recognized as challenging, in part because they are significantly more complex than seed oils. Fish oils typically contain straight-chain fatty acids that contain from 14 to 22 or more carbons and that are often unsaturated, having from one to six double bonds. Isomers differing only in the position of unsaturation are common, and fish oils are often found to contain more than 60 different fatty acids. Additionally, the high degree of unsaturation of these oils largely limits the application of conventional processing methods, which utilize elevated temperatures. For example, Ackman et al. (1) performed a fractional vacuum distillation of methyl esters derived from herring oil and recovered less than 60% of the highly unsaturated long-chain esters, because of thermal and oxidative decomposition. Specifically, the recovery of all-cis-4,7,10,13,16,19 docosahexaenoic acid (DHA or 22:6ω-3) was only 17%.

The components of fish oils most susceptible to thermal degradation are also those which many biomedical researchers postulate to have nutritional or pharmaceutical value, that is, fatty acids of the ω-3 class. The two major ω-3 fatty acids are DHA and all-cis 5,8,11,14,17 eicosapentaenoic acid (EPA or 20:5ω-3). Research has indicated that ω-3 fatty acids may be useful in the prevention and treatment of cardiovascular diseases (2), some types of cancer (3,4), rheumatoid arthritis (5), and other autoimmune diseases (6). It has also been suggested that DHA may be essential for optimal development of retinal and brain tissues of infants (7,8).

Increasing interest in the biomedical properties of ω-3 fatty acids has driven an expanding demand for well-characterized test materials. Among the materials of interest are purified triglycerides and concentrates of EPA and DHA, usually in the form of ethyl ester derivatives. The most inexpensive and readily available source for these materials is fish oil from a number of species, especially menhaden, but also including walleye pollock, tuna, sardine, and capelin. In this chapter, research efforts to process fish oils and their derivatives to obtain potentially useful test materials by use of supercritical fluids (SFs) will be reviewed.

The solvent properties of SFs have been recognized for over a century (9), although, with few exceptions, their potential for processing and separation

180

applications has not been widely investigated until the last few decades. Many different fluids from SF water (10,11) to SF xenon (12), have been studied. The lower hydrocarbons and supercritical fluid carbon dioxide (SF-CO$_2$) have received most attention, because of their low cost and moderate critical temperatures and pressures. In addition, each has a low normal boiling or sublimation point, thus leading to product recovery without solvent residue. Carbon dioxide is particularly attractive for processing foodstuffs and pharmaceuticals because it is nontoxic and nonflammable. (A recent report (13) of an explosion involving SF nitrous oxide modified with ethanol illustrates the risks of working with fluids other than CO$_2$.) Carbon dioxide has a critical pressure of 73.5 bar and a relatively low critical temperature of 31.1°C, permitting the processing of thermally labile materials. Supercritical fluid CO$_2$ is currently used on an industrial scale to decaffeinate coffee and tea and to produce hops extract (14). Other chapters in this monograph, as well as an excellent book (15), provide further information regarding the properties and other potential applications of supercritical fluids.

This chapter will concentrate on the use of SF-CO$_2$ in the processing of fish oils and other marine products. The ensuing discussion is divided into four separate sections. The first section will describe studies in which supercritical fluids are used to fractionate fish oil derivatives, mainly fatty acid ethyl esters, to concentrate EPA and DHA. The second section discusses the use of supercritical and near-critical fluids to fractionate and purify fish oil triglycerides. The third section reviews applications that best fall into a miscellaneous category, and the final section contains speculation on the future role of SF technology in the processing of fish oils.

Fractionation of Fish Oil Derivatives

The major focus of studies involving the use of SFs for processing fish oil derivatives has been the isolation of EPA and DHA. Nearly all studies thus far utilize SF-CO$_2$. Specifically, most reports have involved the fractionation of fatty acid alkyl esters derived from fish oil triglycerides. Some work on the fractionation of free fatty acids has been reported, but recoveries of EPA and DHA were low. For example, Rizvi et al. (16) reported recovery of only 38.6% of the EPA from free fatty acids derived from herring oil. Inexplicably, these workers indicated recovery of over 90% of the DHA from the same mixture. On the other hand, Krukonis (17) reported a similarly low recovery of EPA and a recovery of only 19% of the DHA from a mixture of fatty acids derived from anchovy oil.

Fish oil alkyl esters are known to be more stable than their fatty acids (18) and appear to be the preferred form for SF fractionation. Successful processes have operated in pressure and temperature regimes in which the SF-ester system consists of two distinct phases: a lighter, fluid-rich phase in which some esters are dissolved and a denser, ester-rich phase containing some dissolved carbon dioxide. Such fractionation processes exploit solubility differences between various components of the mixture in the upper phase. These differences are reflected in values for the *partition* (or *distribution*) *coefficients*. The partition coefficient for any given component A (PC_A), is given by

$$PC_A = Y_A/X_A \qquad (8.1)$$

where

Y_A = the concentration of A in the "extract" or CO_2-rich phase

X_A = the concentration of the same component in the "liquid" or ester-rich phase

Any consistent concentration units may be used. Weight concentration units are used in this chapter.

In the present author's laboratory partition coefficient data is obtained using a dynamic or flow (rather than static) technique (15). The method is briefly described as follows. Esters are loaded into a packed extraction vessel, pressurized, and heated to the desired temperature. Supercritical fluid CO_2, preheated to the same temperature as the extraction vessel, is then passed through the ester charge at a relatively low rate (*ca.* 3 L/min) such that the system remains close to equilibrium. The solute-loaded fluid is then expanded across a heated metering valve, thus trapping extracted material. From gravimetric and gas chromatographic analyses of the extract and measurement of the gas volume in which the extract was dissolved, the value of Y_A in Eq. (8.1) can be determined for each component of interest. With knowledge of the weight and composition of the original charge and all previous fractions, the composition of the liquid phase with respect to any component A (X_A) is deduced from mass balance considerations. Since the lower, ester-rich phase is not directly sampled, X_A is determined on a CO_2-free basis.

Table 8.1 shows partition coefficients for various components of an ethyl ester/CO_2 system determined in this manner (19). The esters were derived from menhaden oil; values for several combinations of pressure and temperature are given. Because of difficulties in ensuring equilibrium that are often associated with flow methods, and because the CO_2 content of the lower phase is unknown, the method introduces errors in the determination of the partition coefficients. These errors are, however, systematic in nature, and it is instructive to examine several fundamental trends displayed by data in Table 8.1. For example, at any given pressure and temperature, the shorter the chain length of the ester, the larger the value of its partition coefficient. Moreover, esters of equal chain length have values that are more similar compared to those with a different number of carbon atoms. Aside from the influence of these structural factors on

TABLE 8.1 Partition Coefficients ($\times 10^3$) for Selected Fatty Acid Ethyl Esters Derived from Menhaden Oil in SF-CO_2 at Three Sets of Pressure and Temperature

P (bar)	T(°C)	14:0	16:0	18:1ω-9	20:1ω-9	20:4ω-6	20:5ω-3	22:6ω-3
138	70	17	9.6	6.0	3.0	3.7	3.6	1.9
151	70	63	32	18	8.1	10	9.9	4.8
151	90	21	11	5.5	2.2	2.9	2.8	1.1

the magnitude of partition coefficients, an increase in the partition coefficients is observed as the pressure is raised isothermally. This behavior is usually attributed to the increasing density and associated solvent power of the fluid and is a general result. Finally, it can be seen that at a given pressure the partition coefficients decrease with increasing temperature. This inverse relation between solubility and temperature is known as *retrograde condensation* (20). Such behavior is commonly observed in other systems in lower pressure regimes, but nonretrograde behavior is often noted at higher pressures (21).

Batch-Continuous Fractionation of Fatty Acid Ethyl Esters

Most studies describing the fractionation of fish oil esters using SF-CO_2 have used "batch-continuous" processes. In this method, a batch of feedstock is loaded into an extraction vessel, and CO_2 is then continuously passed through the charge until the charge is depleted. For process feasibility studies and laboratory-scale production of the desired materials, batch-continuous processing is generally used. On the other hand, continuous-countercurrent operation, which will be described later, would be used for large production-scale processing.

The seminal study on the fractionation of fatty acid alkyl esters using SF-CO_2 in a batch-continuous mode was that of Eisenbach (22). The apparatus used in this work, presented schematically in Fig. 8.1, consisted of an extraction vessel combined with a 4-m packed column of unspecified diameter, into the top of which was inserted a "hot finger." The extractor was charged with ethyl esters derived from cod liver oil (2.35 kg) and pressurized to 150 bar. The extractor, column, and hot finger temperatures were held at 25, 50, and 90°C, respectively. Carbon dioxide was passed through the feed in the extractor at a rate of 25 L/h. A portion of the feed dissolved in the CO_2 and was swept up the column. On reaching the hot finger, the solute-laden fluid was heated, causing retrograde condensation of a portion of the esters. The condensed material fell back down the column, functioning essentially as reflux. The material still dissolved in the fluid was drawn off, the pressure was dropped to condense and recover extract, and the CO_2 was pressurized and recycled. In essence, the described process has much in common with reflux distillation.

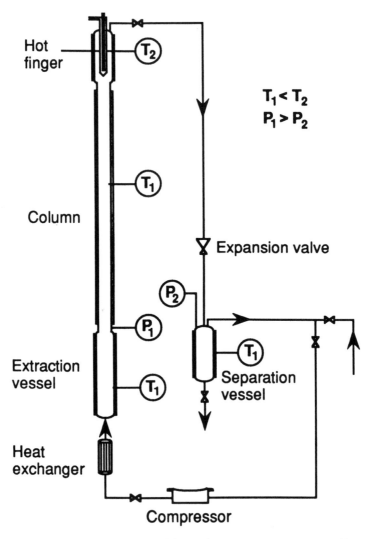

Fig. 8.1. Schematic diagram of the SF-fractionation apparatus used by
Eisenbach. [Adapted from *Journal of Supercritical Fluids* 4:29 (1991).]

By use of this method Eisenbach achieved a rather impressive separation of
the cod liver oil ethyl esters by carbon number. Initially, species of lower carbon number were obtained, a result consistent with the partition coefficient data
of Table 8.1. It was reported that during the course of this fractionation a fraction was obtained, accounting for *ca.* 13% of the feed by weight, containing C_{20}
esters in better than 95% purity. It is important to note that this material was a

mixture of C_{20} esters and *not* high-purity EPA. As is shown in Table 8.1, partition coefficients for esters of equal carbon number are similar when compared to values between esters differing in carbon number. In practice, the smaller the difference in partition coefficients between two species, the greater the difficulty with which they can be separated. Thus, while Eisenbach was quite successful in separating fatty acid esters by carbon number, the method is limited in its ability to separate EPA from other 20-carbon species because of the similarity of their partition coefficients. The cod liver oil ester feed used in this work was 14.5% EPA but also contained about 12% other 20-carbon components, primarily monoenes. Thus, the 96% C_{20} concentrate could not have been over *ca.* 55% EPA.

Had Eisenbach selected a feedstock from a species containing a lower proportion of other C_{20} fatty acids relative to EPA, a higher concentration of 20:5ω-3 could have been obtained. An example of such a feedstock is ethyl esters derived from menhaden oil. A typical fatty acid profile of oil from this species is given in Table 8.2. This table gives the composition not only with respect to specific components but also with respect to components of equal carbon number. It is important to note that this mixture contains 16.5% EPA with a total C_{20} content of 21.8%; thus, EPA accounts for over 75% of the components containing 20 carbons.

TABLE 8.2   Fatty Acid Profile of Ethyl Esters Derived Directly from Menhaden Oil and the Content of the Mixture with Respect to Carbon Number

Component	GC peak area (%)
Major components	
14:0	7.8
16:0	15.6
16:1ω-7	10.9
18:0	3.1
18:1ω-9	7.6
18:1ω-7	3.0
18:4ω-3	2.9
20:1ω-9	1.2
20:4ω-6	1.0
20:4ω-3	1.5
20:5ω-3	16.5
22:5ω-3	2.5
22:6ω-3	10.9
By carbon number	
C_{14}	9.0
C_{16}	32.7
C_{18}	21.3
C_{20}	21.8
C_{22}	14.5

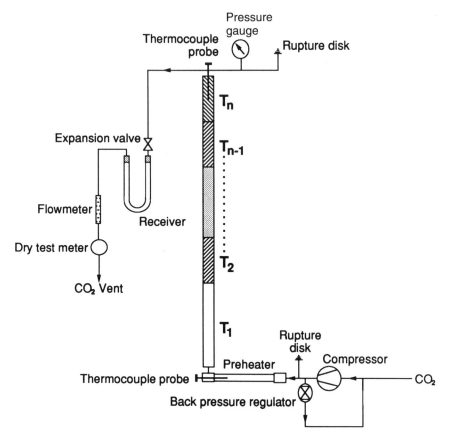

Fig. 8.2. Schematic diagram tor the apparatus used for the fractionation of fatty acid ethyl esters. [Reproduced by permission from *Journal of the American Oil Chemists' Society* 66:1596, © 1989 American Oil Chemists' Society.]

At Nilsson's laboratory the work of Eisenbach was extended (23–27). Reflux within the column was accomplished somewhat differently than by use of a hot finger. A schematic diagram of the apparatus is shown in Fig. 8.2. A 6-ft (1.83 m) packed column with an internal diameter of 0.56 in. (1.42 cm) was wrapped with a number of heating tapes. Each tape was associated with a thermocouple probe, wired to a temperature controller, which in turn supplied current to the tape. A temperature gradient could thus be introduced along the length of the column. The gradients investigated were such that the temperature increased from the bottom to the top of the column. As in the Eisenbach work, processing pressures and temperatures were chosen to correspond to the retrograde region of the fatty acid ester/CO_2 system. In this batch process, the feed

(~20 g) was loaded into the bottom of the column. After establishment of the temperature gradient, SF-CO_2 was passed through the feed carrying a portion of the charge up the column. In each successive temperature zone the solute-laden fluid was heated, causing further retrograde condensation. As a result, in each zone there was a down-flowing reflux enriched in less soluble longer-chain esters. Concomitantly, the fluid leaving each zone was successively enriched in the more soluble components present at a given point in a fractionation.

A fractionation of the feedstock described in Table 8.2 was performed at a pressure of 151 bar (2200 psi) with four temperature zones ($T_1 = 22°C$, $T_2 = 70°C$, $T_3 = 80°C$, and $T_4 = 100°C$, see Fig. 8.2). Figure 8.3 displays fractionation curves for C_{14}, C_{16}, C_{18}, C_{20}, and C_{22} esters as well as the single-component curves for EPA and DHA (26). As in the Eisenbach report, fractions in excess of 95% purity with respect to both C_{20} and C_{22} esters were obtained. C_{20}-rich material was isolated containing EPA in excess of 70% purity. It is to be emphasized that EPA of higher purity than that isolated by Eisenbach was obtained only because the starting oil contained a proportionately lower level of coextracted C_{20} components other than EPA. Again, the limitation of the method in separating components of equal chain length is demonstrated. Results of a study of the influence of structural factors such as degree and position of unsaturation on partition coefficients of fatty acid ethyl esters in SF-CO_2 were published recently (28). Based on the observation that values of partition coefficients for species of equal chain length were found to differ, the data indicate that the separation of esters of equal chain length is theoretically possible. As a practical matter, however, the differences are too small to exploit using batch-continuous processes. Later, a continuous-countercurrent process that potentially could perform such a separation will be discussed.

The maxima of the curves in Fig. 8.3 indicate that if EPA and DHA were the sole C_{20} and C_{22} esters present in the feedstock mixture, these two components could be obtained in better than 95% purity. No known fish oil comes close to filling such a requirement, although some EPA- and DHA-containing algal oils (29) are reported to have fatty acid profiles approaching this ideal. Fortunately, the well-known technique of urea adduction is available for removing virtually all saturates, the bulk of the monoenes and dienes, and even a portion of the trienoic fatty acid esters (30). In this technique esters are mixed with a hot solution of urea dissolved in a solvent such as ethanol. This solution is then cooled to a final temperature chosen to balance product composition and yield considerations. In the presence of long, straight-chain esters, urea crystallizes in a hexagonal structure that forms adducts with molecules that have a diameter of 5 Å or less (31). Less unsaturated fatty acids conform more readily to the crystal structure, whereas the irregularities in the straight chain of the all-*cis* polyunsaturates makes urea adducts of these components thermodynamically less stable. Thus, the more polyunsaturated components primarily remain dissolved in and can be isolated from the solvent phase.

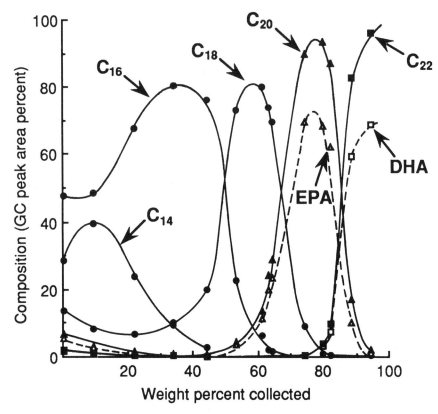

Fig. 8.3.   Curves from the SF-fractionation of fatty acid ethyl esters derived from menhaden oil (see Table 8.2). The fractionation was carried out using CO_2 at 151 bar and zone temperatures of $T_4 = 100°C$, $T_3 = 80°C$, $T_2 = 70°C$, and $T_1 =$ ambient (see Fig. 8.2). [Adapted from *Seafood Biochemistry: Composition and Quality,* Technomic Publishing Co., Lancaster, PA © 1992, pp. 151–168.]

Table 8.3 shows the composition of a urea-adducted menhaden oil ethyl ester mixture. Also shown is the composition of this mixture with respect to esters of equal chain length. Note that EPA and DHA account for the bulk of the C_{20} and C_{22} components, respectively. Therefore, fractionation is expected to yield concentrates of considerably higher purity. Curves for EPA and DHA from the fractionation of 20 g of this mixture at conditions identical to those of Fig. 8.3 are given in Fig. 8.4 (25). EPA of *ca.* 95% purity and DHA of over 90% purity were obtained. The yield of EPA was such that 69% of the EPA originally present in the charge was contained in a 90% pure extract. Similarly, a 90% pure DHA fraction contained 81% of that component present in the original charge.

TABLE 8.3 Composition of Urea-Adducted Ethyl Esters Derived from Menhaden Oil and the Composition of the Mixture by Carbon Number

Ester	GC peak area (%)
Major components	
16:3ω-4	5.2
16:4ω-1	5.6
18:4ω-3	7.4
20:4ω-6	1.4
20:4ω-3	<1.0
20:5ω-3	48.9
21:5ω-3	1.3
22:5ω-3	1.0
22:6ω-3	22.5
By carbon number	
C_{16}	14.1
C_{18}	9.7
C_{20}	50.7
C_{22}	24.2

The amount of SF-CO_2 necessary to carry out the fractionation is indicated by the solvent-to-feed ratio (S/F) expressed as the number of grams of CO_2 necessary to fractionate a gram of feed. In batch processes such as this one, the S/F is an important quantity because at a given flow rate a high value leads to a relatively lengthy time requirement for the fractionation of a given batch. For the test run illustrated in Fig. 8.4, S/F was found to be *ca.* 450, which is a relatively high value. Reduction of S/F is always a desirable goal, and there are two obvious strategies for achieving such a reduction. Since ester solubility increases with pressure, the fractionation could be carried out at a higher pressure. Alternatively, because esters undergo retrograde condensation at 151 bar (2200 psi) and at temperatures below about 100°C, reduction of column temperatures would increase the overall solubility of the ester feedstock, again leading to a reduced S/F. Both strategies are indeed found to reduce the S/F, but unfortunately, they lead to a less successful fractionation because of decreased fluid selectivity at higher pressures or lower temperatures.

By increasing the total number of temperature zones, a fairly effective fractionation of the same urea-adducted ethyl ester mixture can be achieved using somewhat lower temperatures (26). In essence, increasing the number of zones results in more effective internal reflux and better fractionation than could otherwise be achieved at the chosen process conditions. For example, two fractionations were performed at 151 bar (2200 psi) with the top zone at 80°C. In the first, only three zones at 80, 70, and 22°C were used and EPA of just over 80% purity was obtained. In the second, a seven-zone fractionation at the same pres-

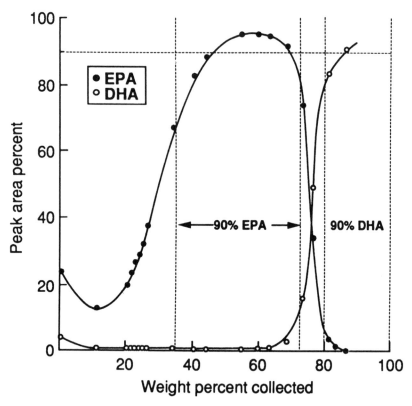

Fig. 8.4. Fractionation curves for EPA and DHA obtained by SF fractionation of urea-adducted menhaden oil ethyl esters (see Table 8.3). The fractionation was carried out using CO_2 at 151 bar and zone temperatures of $T_4 = 100°C$, $T_3 = 80°C$, $T_2 = 70°C$, and T_1 = ambient (see Fig. 8.2). [Reproduced by permission from *Seafood Biochemistry: Composition and Quality*, ©Technomic Publishing Co., Lancaster, PA, 1992, pp. 151–168.]

sure with zone temperatures of 80, 72, 65, 59, 54, 50, and 22°C resulted in isolation of EPA of better than 90% purity. The yield of 90% EPA was 58%. DHA of 90% purity was also obtained in a yield of 77%. While these yield figures are somewhat less than those for the previously discussed fractionation, in which the highest column temperature was 100°C, the overall solvent-to-feed ratio was cut in half (*ca.* 220 vs. 450 to 500). A balance between *S/F* and product yield must be sought as part of the overall optimization procedure.

Introduction of incremental pressure programming (27), also referred to as pressure profiling (32), represents another useful modification to this batch process. In this approach the fractionation is not performed isobarically as in the aforementioned processes, but is increased in increments of, for example, 50 or 100 psi. This is analogous to pressure programming in supercritical fluid chromatography (SFC), except that in SFC the pressure is typically increased in a smooth, often linear fashion (33). The specific programming scheme selected was based on the composition of the feedstock. For example, the urea-adducted esters of Table 8.3 contain *ca.* 14% C_{16} esters, and esters of this chain length were found to be appreciably soluble at 1900 psi and 80°C. Therefore, the initial 14% was "stripped" from the feed at 1900 psi with a maximum gradient temperature of 80°C. The pressure was then raised to strip the next most soluble group of components from the remaining charge. An example of a fractionation, performed using seven zones ranging from 80°C at T_7 to 50°C at T_2 (T_1 was held at ambient temperature; see Fig. 8.2), is given in Fig. 8.5 (27). The plot in Fig. 8.5*a* shows the pressure programming scheme as well as the solvent-to-extract ratio (*S/E*) for selected fractions (defined as the ratio of the weight of CO_2 necessary to collect a given fraction to the weight of the fraction); Fig. 8.5*b* shows four curves from the fractionation of the urea-adducted esters. At each pressure, the *S/E* initially starts out at a lower value and gradually increases as the shortest-chain components present in the charge are depleted. By use of pressure programming, the yield of EPA was over 76%, a considerable improvement over the yield obtained under identical conditions without pressure programming (58%). Again, the improved yield comes at the price of a higher *S/F*. The isobaric test had an *S/F* of 220, whereas the pressure-programmed test has an *S/F* of *ca.* 340 because it was operated at lower pressure during most of the fractionation.

The Charleston, South Carolina, Laboratory of the National Marine Fisheries Service is currently applying supercritical fluid extraction in a batch-continuous mode as one stage in an overall process for production of the ethyl esters of EPA and DHA of high purity. This effort, known as the Fish Oil Biomedical Test Materials Program, was established by a Memorandum of Understanding between the National Institutes of Health, the Alcohol, Drug, Abuse, and Mental Health Administration and the U.S. Department of Commerce. The program was originally devised "to provide a long-term, consistent supply of quality-assured/quality-controlled test materials to researchers in order to facilitate the evaluation of the role that [ω-3] fatty acids play in health and disease" (34). The apparatus in use is a scaled-up version of that shown in Fig. 8.2. The extraction column is 6 ft (1.83 m) in length with an internal diameter of 1.75 in (4.44 cm). The typical charge is about 200 g of ethyl esters that have undergone urea adduction and have a composition similar to that described by Table 8.3. The SF-CO_2 fractionation, a modification of that described by Fig. 8.5, results in two fractions of interest: a mid-fraction, accounting for about 35% of the original charge, containing *ca.* 80% EPA, while a second, smaller fraction is composed primarily of DHA. These two fractions are then used as "seed" for process-scale HPLC, which increases the purity of both the EPA and DHA product to 95 to 99%, using an ethanol-water mobile phase (35). Without prior fractionation of the esters using SF-CO_2,

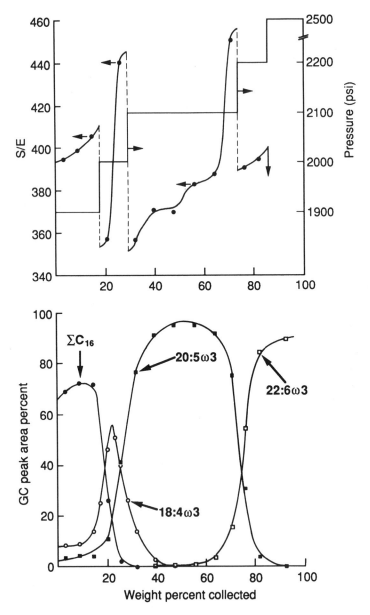

Fig. 8.5. SF fractionation of urea-adducted menhaden oil ethyl esters (see Table 8.3), performed using SF-CO_2 and incremental pressure programming. Column temperatures: $T_7 = 80°C$, $T_6 = 72°C$, $T_5 = 65°C$, $T_4 = 59°C$, $T_3 = 54°C$, $T_2 = 50°C$, and T_1 ambient (see Fig. 8.2). (a) Pressure programming scheme (right ordinate) and curves for the solvent-to-extract ratio (S/E, see text and left ordinate). (b) Fractionation curves. [Reproduced by permission from *Journal of the American Oil Chemists' Society 66*:1596 © 1989, American Oil Chemists' Society.]

purification of EPA and DHA to such high levels in good yields is difficult on a preparative scale because of the presence of shorter-chain components, which, unless removed by the SF fractionation process, would coelute with EPA and DHA.

Continuous-Countercurrent Concentration of EPA and DHA

As mentioned above, batch-continuous processing is appropriate for laboratory scale operation. For large-scale production, however, batch-continuous processing is not normally used, in part because of the high solvent-to-feed ratios that usually characterize such operations and would adversely affect process economics. Krukonis et al. (36) recently published the results of a preliminary design study of a continuous- countercurrent process for large-scale production of the ethyl esters of both EPA and DHA. In continuous-countercurrent operation a stream of the esters is continuously supplied to a column, where it contacts a continuously supplied stream of CO_2 moving in the opposite direction. In this process the columns are vertical, and the specific direction of each stream will depend upon the density of the feed relative to that of the fluid. Sight glass observation of the ester-CO_2 system at the conditions chosen for this design study indicated that the density of the ester feed is greater than that of the SF-CO_2 (Krukonis and Nilsson, unpublished data). Thus, esters introduced into the column will flow downward while the CO_2 will flow in an upward direction.

The process was designed to fractionate the urea-adducted feedstock described in Table 8.3. The results of the previously discussed batch-continuous studies proved the feasibility of obtaining both EPA and DHA of better than 90% purity from this feedstock. For the basis of the design, and for estimating the capital and operating costs of a pilot plant, Krukonis et al. chose to specify an EPA product of 90% purity in a 90% yield at a production rate of 4.55 kg (10 lb)/d. (The original description of the process requires some clarification. As described, the process is actually expected to yield a 90% concentrate of C_{20} esters, not 90% EPA. By inspection of Table 8.3, a 90% C_{20} concentrate would actually be about 87% EPA. As will be discussed later, slight modification of the original design could lead to an EPA product of the originally specified 90% purity. In any case, this oversight has no significant effect on the conclusions reached in the original work.) These yield and purity specifications are arbitrary and could have been made higher, although each refinement would be expected to increase capital and operating costs. The 10 lb/d production figure is similarly arbitrary, and the process, to be described subsequently, could be scaled to virtually any level with the usual economic benefits derived from scaled-up production.

The first step in the design of the continuous-countercurrent process is the acquisition of partition coefficient data for major components of the anticipated feed material at the chosen process pressure and temperature. Based on partition coefficient information available elsewhere (37), Krukonis et al. selected a temperature of 60°C and a pressure of 151 bar (2200 psi). This is likely not "the" optimum combination of conditions, but it probably represents a reasonable approximation of those that would result in the desired product yields and purities with a minimum solvent requirement (i.e., a minimum S/F).

As was discussed previously, the partition coefficients for ester components of equal chain length are found to be quite similar. As a simplifying approximation, in this study all species of equal chain length were treated as a single component and assigned an average partition coefficient. In essence, the feed material of Table 8.3 was treated as a 4-component system containing C_{16}, C_{18}, C_{20}, and C_{22} esters. The average partition coefficients are given in Table 8.4. Also given are values for the *selectivity* (β) of carbon dioxide for the extraction of C_{16}, C_{18}, and C_{20} esters relative to the C_{22} species. The selectivity of CO_2 for component A relative to B is defined as the ratio of the partition coefficient of A to B:

$$\beta = PC_A/PC_B \qquad (8.2)$$

Substitution of Eq. (8.1) and rearrangement give

$$\beta = (Y_A/Y_B)/(X_A/X_B) \qquad (8.3)$$

If the value of β is greater than unity, SFC-CO_2 in equilibrium with the mixture of A and B will extract A preferentially. With reference to Table 8.4, for example, the selectivity of CO_2 for the C_{20} esters relative to the C_{22} esters is 2.3. Substituting in Eq. (8.3)

$$Y_{C20}/Y_{C22} = 2.3 \, (X_{C20}/X_{C22}) \qquad (8.4)$$

TABLE 8.4 Mean of the Partition Coefficients of Components of Equal Carbon Number Present in the Urea-Adducted Menhaden Oil Ethyl Esters at 60°C and 151 Bar and the Values of the Selectively (β) for Extraction of C_{16}, C_{18}, and C_{20} Esters Relative to C_{22} Esters

Components	PC	β
C_{16}	0.090	7.5
C_{18}	0.050	4.2
C_{20}	0.027	2.3
C_{22}	0.012	—

The ratio of C_{20}/C_{22} in the extract is increased by equilibration at a single "stage" by a factor of 2.3 relative to the ratio of C_{20}/C_{22} in the liquid phase. If this extract is then brought to equilibrium with a liquid phase of the same ester content at another stage, vapor leaving that stage will be further enhanced with respect to the C_{20} content by the same factor of 2.3. Equilibration between the liquid and vapor phases at each stage is facilitated by the "internals" of the column, which are mass transfer- and area-enhancing devices such as trays or agitated components. The availability of specific options is well covered in several chemical engineering texts (38,39). This stagewise enhancement forms the basis of continuous-countercurrent operation. Although distribution coefficients for the major components determine the magnitude of the process *S/F*, the selectivity values influence the number of separation "stages" necessary to carry out the desired separation, which in turn leads to an estimate of column height.

Figure 8.6 is a simplified process flow sheet of the continuous-countercurrent plant proposed by Krukonis et al. Ester-containing streams are shown with heavy lines for ease of interpretation. The ester feed is introduced somewhere in the side of the first of two columns (C-1); the point of introduction will be described later. As mentioned above, at 151 bar and 60°C the ester density exceeds that of the SF-CO_2, and esters fall down the column, contacting the CO_2, which is fed in continuously from the bottom of the column. The SF-CO_2

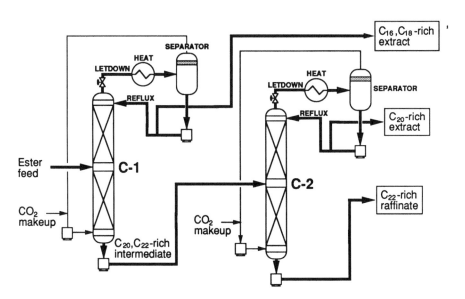

Fig. 8.6. Schematic diagram of a two-column continuous-countercurrent process for the production of high-purity EPA and DHA using SF-CO_2. (Adapted from *Seafood Biochemistry: Composition and Quality,* © Technomic Publishing Co., Lancaster, PA 1992, pp. 169–179.)

preferentially "strips" the shorter-chain length materials present in the feed (i.e., C_{16}, C_{18} esters), resulting in concentration of C_{20} and C_{22} esters in the downward-flowing liquid phase.

At the top of C-1, extract is drawn off, and the esters are condensed. This can be accomplished by reducing the pressure, increasing the temperature, or both, thereby exploiting retrograde condensation. Condensation of the ester stream at the top of the column has several functions. Obviously, it provides a mostly C_{16}- and C_{18} product (which also may be of some value), and it also achieves the necessary cleanup of the CO_2, which is then returned to the bottom of the first column for another extraction cycle. In addition, a portion of the liquid extract is returned to the head of the column to provide a downward-flowing reflux stream. Without such a stream, the ester-laden fluid above the feed stage would have no liquid phase with which to contact in a countercurrent fashion.

At the bottom of C-1, the liquid phase drawn off is concentrated in the higher-molecular-weight C_{20} and C_{22} esters. This intermediate product is then fed into C-2, which operates in a manner analogous to C-1. The SF-CO_2 is forced up from the bottom of C-2, stripping the lighter components (in this case the C_{20} esters and any residual shorter-chain length esters) while concentrating the C_{22} species in the downward-flowing ester stream. Once again, at the top of C-2 the extract is condensed, and part of the product is returned to the column head to function as a downward-flowing stream above the feed stage.

Using distribution coefficient and selectivity data as well as a number of simplifying assumptions that will not be described here, Krukonis et al. computed the composition of the esters at each stage of both columns and, from product purity specifications, were able to determine the number of stages required. The trial-and-error computational procedures used to obtain this information are beyond the scope of this treatment but are available elsewhere (38–40).

Stage composition profiles for both columns, as generated by these procedures, are shown in Figs. 8.7 and 8.8. These curves also indicate the proper locations for introduction of the esters; the feed stage is the stage at which the ester composition is the same as that of the feed introduced, i.e., stage 7 in column 1 and stage 8 in column 2. (Many of the curves exhibit a discontinuity at the feed stage; that is simply an artifact of the computational procedures and has no practical influence on cost estimates, to be discussed subsequently.) The first column is predicted to require 11 stages (the "separator" is not a true stage, because its function is simply to condense esters leaving the top of the column rather than effect further concentration of any ester component). The "bottoms" product, or raffinate, of column 1 contains only about 4% of the shorter-chain C_{16} and C_{18} esters. The second column requires 13 stages to yield the 90% pure C_{20} product containing about 87% EPA. Assuming a column height of 1 ft (30.5 cm) per stage, both columns would be around 15 ft (4.6 m) high.

The overall S/F for this continuous-countercurrent process was estimated to be 30, a figure significantly lower than those of the batch processes described

Composition profiles--column 1

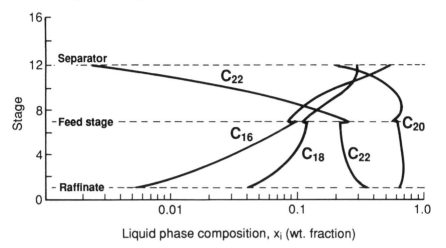

Fig. 8.7.   Computed liquid composition at each stage of column C-1 of the continuous-countercurrent production unit (see Fig. 8.6). (Reproduced from *Seafood Biochemistry: Composition and Quality,* © Technomic Publishing Co., Lancaster, PA 1992, pp. 169–179, courtesy Van Nostrand Reinhold.)

Composition profiles--column 2

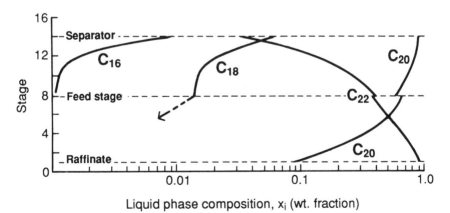

Fig. 8.8.   Computed liquid composition at each stage of column C-2 of the continuous-countercurrent production unit (see Fig. 8.6). (Reproduced from *Seafood Biochemistry: Composition and Quality,* © Technomic Publishing Co., Lancaster, PA 1992, pp. 169–179, courtesy Van Nostrand Reinhold.)

previously. A preliminary analysis of the capital and operating costs of such a process indicates that 87% EPA could be produced at the 4.55 kg/d level at a cost of *ca.* $108 per pound (note that the DHA-rich product is produced "free"). This figure does not include acquisition costs for the urea-adducted feedstock, but nonetheless it is still attractive relative to current prices.

By addition of a few more stages to C-1 and C-2, the EPA content of the light product from C-2 could be increased. Additional stages in the first column would reduce the level of C_{16} and C_{18} components in the raffinate of C-1, whereas more stages in C-2 would lead to a lower C_{22} content in the light product of this column. The result would be further concentration of the light product stream with respect to C_{20} to a level (*ca.* 93%), resulting in a product containing EPA at the 90% level. Such design modifications would entail only a minor increase in production costs, as process controls, not column heights, contribute most heavily to capital costs (private communication, V.J. Krukonis).

Results from an investigation of the partition coefficients for C_{20} esters indicated that with the exception of 20:4ω-6, EPA has a larger partition coefficient in SF-CO_2 than any other 20-carbon species (28), indicating that SF-CO_2 extracts EPA preferentially along with 20:4ω-6. The selectivities of CO_2 for EPA relative to the other C_{20} components (again, with the exception of 20:4ω-6) were on the order of 1.1 or higher. Based upon similar data, Krukonis et al. suggested that it might be possible to design a process for producing high-purity EPA that eliminates the need for urea adduction of fish oil ethyl esters prior to SF-CO_2 fractionation.

One possible configuration of a process using only SF-CO_2 was suggested (36). This process would involve a total of four columns, the first two of which would be identical to those of Fig. 8.6. In the proposed process the light product of C-2, containing *ca.* 90% C_{20}, 6% C_{18}, and 3% C_{22} components (see Fig. 8.8), would be sent to a third column. Selectivity data for C_{20} components, determined at 151 bar and 60°C, indicated that this column could separate EPA and 20:4ω-6 along with the residual C_{18}s from the remaining 20-carbon species and C_{22}s. The light fraction from the third column would then be sent to yet another column for separation of EPA and 20:4ω-6 from the shorter-chain residuals, the former two components being isolated in the raffinate of the fourth column. If the ratio of EPA to 20:4ω-6 is greater than 9:1 in original feedstock, EPA approaching 90% purity could in principle be isolated. Since the selectivities for EPA relative to the other C_{20}s are close to unity, many more separation stages would be necessary to achieve the separation in column 3. This column is thus expected to be considerably taller than columns 1 and 2, perhaps more than 50 ft (15 m). Columns of this height are, however, not unknown in the chemical process or pharmaceutical industries.

Concentration of EPA/DHA Using CO_2 in Combination with Adjuvant Materials

In the present context an "adjuvant" material is one that, when present as part of the SF-CO_2 separation process, results in concentration of EPA, DHA, or both to a higher degree than is possible in its absence. An example of such a material is urea. Arai and Saito showed that urea will form adducts with fatty acid esters in both liquid CO_2 and SF-CO_2 (41), although maximum adduct formation was noted at 40°C. The apparatus used in this work consisted of dual vessels. The methyl ester charge (16 g), derived from sardine oil, was loaded into the first vessel. The second vessel contained *ca.* 100 g of finely powdered urea. Carbon dioxide was pressurized to 100 bar, heated to 40°C, and passed through the ester charge in the first column. Components that dissolved in the SF-CO_2 were then passed through the bed of urea in the second column, where inclusion compounds were formed; as in conventional urea adduction, more stable adducts were formed with saturates and less unsaturated esters. As a result, over the course of the test, both EPA and DHA were enriched in the extract isolated downstream of the second vessel. Specifically, the feed contained 13% EPA by weight, whereas the extract initially contained better than 25% EPA and rose to about 35% as the test proceeded. (The rise in EPA content of the extract was no doubt partly due to fractionation of the charge taking place in the first vessel.) In a similar manner, although the charge contained only 8% DHA, extract in excess of 30% purity with respect to this component was obtained, again in later stages of the experiment. By performing the same experiment in the absence of the urea-packed vessel, Arai and Saito showed that EPA and DHA of such purities were not obtained. Interestingly, it was also shown that inclusion compounds could be decomposed easily by increasing the temperature and pressure, thus leading to recovery of the remaining charge (again, primarily saturates and monoenes). In a more recent report (42) it was shown that modification of the CO_2 with methanol further promotes adduct formation.

In a sense, the stationary phase used in supercritical fluid chromatography (SFC) can be thought of as functioning as an adjuvant to the mobile supercritical fluid phase. Berger et al. (43) demonstrated the feasibility of using certain chromatographic stationary-phase materials as adjuvants in preparative-scale SFC. Specifically, a 6 cm i.d. × 60 cm column was packed with 40-65 μm C18 packing material. The starting materials were methyl ester mixtures derived from an unspecified fish oil and concentrated to 65% or better with respect to DHA in a prior supercritical fluid extraction step. As an example of the degree to which DHA could be further concentrated, a feed containing 73.0% DHA and 14.8% EPA was injected onto the column at a pressure of 135 bar and a temperature of 50°C at a rate of 17.5 g/h, with a mobile phase (SF-CO_2) flow rate of between 20 and 30 L/h and injection cycles of 20 min. Three fractions were obtained, one of which contained DHA of 89.8% produced at a rate of 7.3 g/h. The purity of the final product was found to depend on the concentration of DHA attained in the previous SF extraction step. For example, DHA of 95%+ purity was obtained from a feed containing 81% DHA.

Even better results from the same laboratory have recently been reported (44). In this work, a 6 cm i.d. column was packed with 300 g of 10μ silica. One end of the column was movable, permitting compression of the stationary phase during a chromatographic cycle by use of a hydraulic jack. This technique, known as *dynamic axial compression*, has been in fairly wide use in preparative HPLC processes (45). Application of dynamic axial compression has been shown to increase the efficiency of large-i.d. preparative columns significantly, in part due to minimization of voids in the packing structure that would otherwise adversely affect column performance. The charge in this work contained 56% EPA and 31% DHA (information regarding how this material was processed prior to receipt was unavailable; previous steps probably included, but may not have been restricted to, urea adduction). The mobile phase was SF-CO_2 at 50°C and 142 bar. In the course of each 10-min chromatographic cycle, 3.4 g of EPA of 95% purity and 2.1 g of 85% pure DHA were obtained from each 7.8 g injection. These figures translate to EPA and DHA yields of 78% and 88%, respectively. The authors state that over 3 kg of each product have been produced by use of this method.

Higashidate et al. (46) reported isolation of EPA and DHA in good purities by preparative SFC from sardine oil methyl esters that had undergone no previous concentration steps. The authors used a 125 mm × 10 mm i.d. column packed with silica gel coated with silver nitrate, and SF-CO_2 as the mobile phase. The basis for this chromatographic separation is interaction of silver with double bonds, leading to increased retention times of more highly unsaturated compounds (47). This stationary phase has been found to be effective for separating EPA and DHA from other compounds using liquid-liquid partition (48) and high-performance liquid (49) chromatography. The apparatus included an extraction vessel as well as the separation column. An extraction step prior to SFC was found to be necessary to remove certain impurities present in the feed that were not soluble in SF-CO_2 and would otherwise foul the column. It should be emphasized that the extraction step was not included to achieve any significant concentration of either EPA or DHA prior to preparative chromatography. Just over 350 mg of methyl esters containing 13% EPA and 13% DHA were injected into the column at 40°C with an SF-CO_2 flow rate of 9 g/min. Five fractions were taken over a period of 370 min with pressures increasing from 80 bar initially to 200 bar at the completion of the test. Fraction 3, weighing 12 mg, contained EPA of 93% purity, whereas Fraction 5, weighing 48 mg, contained DHA of 82% purity. These throughputs are obviously much lower than those reported by the French workers (43,44). On the other hand, it is significant that this method requires no concentration of EPA or DHA prior to the preparative SFC step.

Processing of Fish Oil Triglycerides

Triglycerides (or triacylglycerols) are measurably soluble in most commonly studied SFs, although not as soluble as their ester derivatives. As with ethyl esters, the magnitude of triglyceride solubility in a given fluid depends primarily upon the pressure and temperature. It is generally observed that as the pressure is increased, triglyceride solubility also increases. This behavior is usually attributed to the increase in density and solvent power of the fluid with increased pressure. The dependence of triglyceride solubility on temperature is again more complex. At lower pressures an increase in temperature generally causes decreased triglyceride solubility (i.e., retrograde condensation). At higher pressures, the dependence of triglyceride solubility on temperature becomes nonretrograde, i.e., increases with temperature. For the soybean oil/SF-CO_2 system, the crossover to nonretrograde behavior has been found to occur between 200 and 350 bar, depending on the temperature (21). Fish oil triglycerides should behave in a similar manner.

Fractionation of Fish Oil Triglycerides

From a historical perspective, it is interesting to note that a process conceptually identical to the continuous-countercurrent process just described was in use for commercial-scale fractionation of fish oils over 40 years ago (50,51). Although the "Solexol" process did not use supercritical fluids, it did use liquid propane at conditions close to its critical point. (The critical temperature and pressure of propane are 96.7°C and 41.5 bar, respectively.) Near-critical liquids behave in a manner similar to supercritical fluids in that small pressure and temperature adjustments can result in large changes in density and solvent power. Like the continuous-countercurrent process, the Solexol process was operated at conditions at which the process material (in this case, fish oil) and propane form a two-phase system, with the lipid-rich phase being the denser of the two phases.

Passino (50) described the fractionation of sardine oil using three columns operating countercurrently. Again, since the propane's density is less than that of the fish oil, it is fed in from the bottom of the column while the oil is introduced at some point along the side. The overall process is represented by Fig. 8.9. Again, many details are omitted for clarity. The four products are identified and the yields given. Also given are the iodine values (IVs) for each of the products. (The larger the IV of an oil, the higher its content with respect to unsaturation. For purposes of comparison, the ethyl esters of 18:1ω-9, EPA, and DHA have IVs of 82, 384, and 427, respectively.) In the first column, the raffinate product contains the bulk of the color bodies, dissolved in triglycerides containing the longest-chain length fatty acids present in the crude oil. Since by and large these fatty acids are also the ones that are most highly unsaturated, the IV of this material is relatively high. The second column separates the heavier, more unsaturated triglycerides from those of lower molecular weight. Note that the second raffinate product has a relatively high IV of 240. In the third column, the light fraction from the middle tower separates the light triglycerides from vitamins originally present in the crude oil. Again, note the iodine values. In

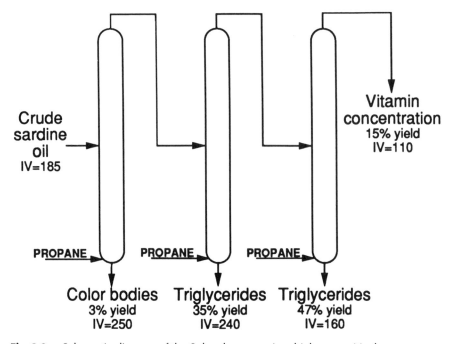

Fig. 8.9. Schematic diagram of the Solexol process, in which near-critical propane was used to fractionate oils. The example given is for the fractionation of crude sardine oil. IV = iodine value. [Adapted from *Industrial and Engineering Chemistry 41*:280 (1949).]

addition to sardine oil, Passino also describes similar processes for fractionating menhaden, cod liver, linseed, soybean, and tall oils.

The first reported attempt to fractionate fish oil using a supercritical fluid was that of Zosel (52). The apparatus used was similar if not identical to that used by Eisenbach in the previously discussed fractionation of cod liver oil esters (see Fig. 8.1). In the Zosel work, SF-ethane at 27°C and at pressures of 100 to 160 bar was passed through the charge, a portion of which was dissolved and swept up the column, which was held at 50°C. On reaching the top of the column, the solute-laden fluid was heated by the hot finger, which was held at 90°C. Since the process pressures chosen for this work were in the region of retrograde solubility for the cod liver oil–ethane system, a portion of the material dissolved in the fluid condensed and fell back down the column. The remaining material was drawn off, and the pressure was dropped to condense and recover the extract. The ethane was then pressurized and recycled.

Zosel isolated 50 fractions and determined the iodine values (IV) of each. The iodine values ranged from about 160 in the earliest fraction to *ca.* 215 in the final fraction. Thus, some fractionation did take place, although it is difficult to quantify its extent without fatty acid profile information. Examination of data from the molecular distillation of menhaden oil from the present author's laboratory (unpublished data, E.J. Gauglitz, Jr.), provides some insight. Five fractions were obtained from oil originally containing 15.2% EPA and 9.2% DHA, with an IV of 176. The first fraction, containing the most volatile, low-molecular-weight triglycerides (i.e., those containing shorter-chain, less unsaturated fatty acids), had an IV of 128. The IV of each successive fraction gradually increased. The final fraction, accounting for 20% of the charge, had an IV of 222. This last fraction contained only about 18.6% EPA and 16.5% DHA. Thus, it can be concluded that Zosel was able to achieve only modest concentration of EPA and DHA in the triglycerides of final fractions. Similarly, Krukonis (17) was able to increase the content of EPA in sardine oil triglycerides by less than 5% using $SF\text{-}CO_2$. Spinelli et al. (53) showed that $SF\text{-}CO_2$ could be used in a three-step process to remove odoriferous volatiles and color bodies to produce a nearly clear, mild-tasting oil, but the EPA and DHA levels were increased only slightly in the final product.

Ender and Steiner recently reported on the phase behavior of soybean oil triglycerides in mixtures of CO_2 and propane (54). Soybean oil is composed of triglycerides containing from 40 to 54 carbons (not counting those of the glycerol backbone). In this work, the oil was treated as a three-component mixture containing a "light" C40/42 component, an "intermediate" C44/46/48 component, and a "heavy" C50/52/54 component. From vapor-liquid equilibrium data the selectivity of the solvent for each of these three "components" relative to one another was estimated. Values of β for C40/C42 relative to C44/46/48 as well as C44/46/48 relative to C50/52/54 were reported to be about 2 at certain conditions. Based upon these data, Ender and Steiner suggested a dual-column continuous-countercurrent process, conceptually identical to that of Fig. 8.6 and also quite similar to the older Solexol process. The light product of the first column of the unit would be concentrated with respect to C40/42 triglycerides, whereas that of the second column would be primarily triglycerides containing 44, 46, and 48 carbons. Finally, the raffinate of column 2 would contain the heavier C50/52/54-containing triglycerides. In principle, this type of process could be adapted for fish oil triglycerides, though it is unclear to what degree any one fatty acid could be concentrated. For example, a triglyceride containing EPA as one of the three fatty acid moieties could contain a total of anywhere from 48 to 64 carbons. Therefore, even a very good separation of triglycerides by carbon number may not lead to a significant concentration of EPA.

The difficulty encountered in attempting to concentrate any one fatty acid moiety from native triglycerides is not surprising. As the number of fatty acid moieties present in the oil increases, the number of possible different triglyceride species increases dramatically. If it is assumed that the different fatty acids that are present are randomly distributed, an oil containing 50 different fatty acids could conceivably contain many thousands of distinguishable

species. In fact, it turns out that the fatty acid moieties are not randomly distributed in fish oils, with the preferential location of the long-chain polyunsaturated
fatty acids such as EPA and DHA at the middle carbon of the glycerol backbone
(55). The oils are therefore somewhat less complex than would be supposed by
the assumption of random distribution. What simplification is derived, however,
actually works against concentration of any given long-chain polyunsaturate
present in triglycerides, because species containing two such moieties are
expected to be relatively minor constituents. For this reason, it is difficult to
effect more than a modest increase in the content in fish oil triglycerides of any
long-chain component such as EPA and DHA by supercritical fluid fractionation
or any other method that does not involve alteration of the original structure
(such as derivatization).

Fractionation of Synthetic Acylglycerol Mixtures

There has been a recent and growing interest in what could be termed "design-
er" or structured triglycerides, i.e., triglycerides containing specific fatty acids
(56). Such materials can be synthesized by either chemical (57) or enzymatic
methods (58). The potential use of such materials remains unclear, but it is conceivable that triglycerides containing high levels of EPA or DHA might be useful as pharmaceuticals. Regardless of the synthetic method, the desired triglyceride product is found to contain the intermediate mono- and diacylglycerides as
well. Fractionation using SF-CO_2 provides a method for purification of such
mixtures.

In the present author's laboratory, SF-CO_2 has been used to purify reaction
mixtures generated by chemical synthesis of triglycerides starting with 90% ethyl
eicosapentaenoate (57). HPLC data indicated that this mixture contained ca. 55%
triglyceride along with ca. 5% diglyceride and 20% unreacted ester, with unchar-
acterized byproducts accounting for most of the balance. In purifying this material, a total of 7 fractions were collected at 60°C and pressures ranging from 138 to
310 bar. The first fraction contained unreacted ethyl esters of very high purity,
whereas the diglycerides and nearly all of the byproducts were collected in fractions 2, 3, and 7. Fractions 4 through 6, accounting for nearly 50% of the 20-g
charge, contained triglycerides of better than 93% purity.

For a better understanding of how SF-CO_2 may be used to remove mono-
and diacylglycerols from synthetic triglyceride mixtures, a recent study of mixtures of mono-, di-, and triolein (MO, DO, and TO, respectively) has been published (59). Partition coefficients for MO/DO/TO in both 10/10/80 and 33/33/33
(w/w/w) mixtures were measured at 60°C and pressures ranging from 172 to
309 bar. At all pressures the order of the partition coefficient values was
MO>DO>TO, thus indicating that MO and DO are extracted preferentially in
accordance with results from the fractionation of the synthetic "triEPA" mixture
just discussed. As expected, the values of the selectivity of SF-CO_2 for extrac-

tion of both MO and DO relative to TO (see Eq. 8.3) were found to decrease with increasing pressure due to the higher fluid density. These selectivity values were also found to depend on the concentration of MO and DO in the acylglycerol mixture. For example, at any given pressure the value of β for MO relative to TO was found to be significantly lower in the mixture containing 33% MO than in that containing 10% MO. Since monoacylglycerols are known to form intermolecular hydrogen bonds (60,61), one possible explanation for the observed behavior is that monooleylglycerol hydrogen-bonds more strongly with itself than with TO. As the TO level increases and the MO level decreases, the intermolecular forces experienced by MO in the lipid-rich phase also decrease, resulting in more facile extraction of MO. Dioleylglycerol exhibits similar behavior, though not quite so strongly. In practical terms, this indicates that the lower the level of mono- and diacylglycerols in a given mixture, the greater the ease with which the mixture can be purified to give the desired triglyceride product.

The Removal of Xenobiotics

Refinement of fish oils is necessary to remove not only impurities that impart objectionable odors and flavors but also fat-soluble pollutants that are present in the marine environment. Such compounds are collectively referred to as *xenobiotics* and include any chemical compound that is foreign to a living organism. Three examples are polychlorinated biphenyls (PCBs), polychlorinated dibenzo-*p*-dioxins (PCDDs), and polychlorinated dibenzofurans (PCDFs). Structures of representative compounds from each class are given in Fig. 8.10. All three classes are quite lipophilic and tend to accumulate in fat-rich tissue (e.g., fish livers). They are generally highly resistant to chemical or thermal breakdown. These three families of halogenated aromatic compounds are all recognized to be toxic above certain levels, and all have been found in fish oils. Conventional methods for their removal from oils include vacuum distillation, but as mentioned previously, the elevated temperatures used can lead to degradation of the oil.

A preliminary study of the behavior of PCBs in the PCB/fish oil/SF-CO_2 system was published recently (62). The test material was cod liver oil containing PCBs at a concentration of 8.7 ppm. Five fractions of the 22-g charge were taken at 70°C at pressures ranging from 168 bar to 438 bar. From these data, partition coefficients for the PCBs (as a class of compounds) were estimated. Values ranged from 0.008 at 168 bar to 0.014 at 438 bar. Also calculated were estimates of partition coefficients for the triglycerides (again, as a class) at 70°C at each of these pressures. From these data, the selectivities of SF-CO_2 for extraction of PCBs from fish oil were found to range from *ca.* 13 at 168 bar (2500 psi) to 1.3 at 438 bar (6500 psi); again the selectivity of SF-CO_2 is seen to be an inverse function of pressure. For purposes of comparison, selectivities for separating fatty acid ethyl esters differing in chain length by two carbons at

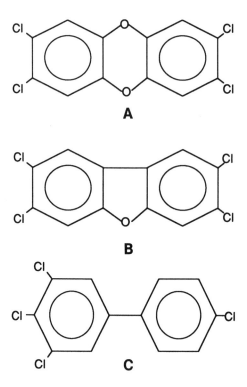

Fig. 8.10. Representative examples of dibenzo-*p*-dioxins, dibenzofurans, and polychlorinated biphenyls. (*a*) 2,3,7,8-tetrachlorobenzo-p-dioxin; (*b*) 2,3,7,8-tetrachlorobenzofuran; (*c*) 3,4,4′,5 tetrachlorobiphenyl.

a given set of conditions are found to be generally on the order of 2 (Table 8.1). A selectivity value of 13 is of practical interest, especially if other techniques of purifying such heat-sensitive compounds present difficulties.

A study of $SF\text{-}CO_2$ extraction of PCDFs and PCDDs from cod liver oil has been published recently (63). Various PCDFs and PCDDs containing from 4 to 6 chlorines were detectable in the oil. Tests using CO_2 were carried out at 150 and 250 bar. Fractions of *ca.* 6 g were taken sequentially until about 40% of the 100-g charge was collected. Analysis of each fraction and the unextracted oil for all compounds of interest indicated preferential extraction of PCDFs and PCDDs as classes from the oil. Not surprisingly, the less chlorinated members of both classes of compounds were seen to be more easily extracted because of their lower molecular weight. Unfortunately, insufficient information is given to provide an estimate of partition coefficients for PCDFs and PCDDs in CO_2 at the studied conditions. PCBs, PCDFs, and PCDDs are but three classes of xenobiotics of importance, and extraction of other toxic xenobiotics appears to be a fruitful area for further research.

Miscellaneous Applications

Extraction of Lipids from Flesh

Several reports appeared in recent years describing the removal of lipids from protein-rich materials using pure $SF-CO_2$ or $SF-CO_2$ modified by addition of an appropriate cosolvent. Generally, these studies were motivated by a desire to obtain a defatted protein-rich product having superior qualities (e.g., better stability) or because the extracted oil was perceived to have some potential value. An example of the latter is work reported by Yamaguchi et al. (64), who studied the extraction of oil from freeze-dried krill as well as krill meal. Both test materials were extracted with $SF-CO_2$ at 250 bar and 80°C. Extracts were found to be composed almost entirely of neutral lipids (acylglycerols, free fatty acids), cholesterol ester, and carotenoids. Polar lipids (i.e., phospholipids) were not extracted—an expected result, as polar lipids are known have a very low solubility in $SF-CO_2$ (65). Oil extracted from the freeze-dried krill was found to contain *ca.* 11% EPA. Of possibly greater value than the acylglycerols are the carotenoids dissolved in the oil. The carotenoids were primarily composed of the mono- and diesters of astaxanthin and free astaxanthin. These compounds may be of value as supplements in the diet of farmed salmon, because the characteristic pink color of wild-caught salmon is due to astaxanthin in the natural diet (66,67). Originally, Yamaguchi et al. noted significantly reduced recovery of all carotenoid components, especially at 80°C. Based upon previous reports (68) that thermal decomposition of all three forms of astaxanthin does not occur at 80°C under atmospheric pressure, the apparent decomposition of astaxanthin was ascribed to the high-pressure CO_2 environment. However, subsequent experiments, in which astaxanthin decomposition was examined at 80°C in CO_2 and N_2 at 250 bar and in air at atmospheric pressure, showed that equal decomposition occurred whether at normal or high pressure, regardless of the gas (69). The conclusion from this work was that astaxanthin decomposition is entirely thermal in nature.

Both liquid and $SF-CO_2$ have also been used to defat meat powder derived from sardines (70). Sardines are an abundant resource, but, in part because of the rather high oil content of the flesh and associated fishy odor, this species has not generally been utilized in making *surimi,* or minced flesh. (Surimi-based products, such as imitation crab meat, are now familiar to the American consumer.) In this work, surimi was made from sardines in the usual manner (71) and dehydrated using high-osmotic-pressure resin, thus forming a meat powder. In the traditional process there are a number of wash steps, but all involve water. The final powdered product thus retains much of the oil originally present in the flesh. Residual oil was extracted at 250 bar using CO_2 at 12°C and a 40°C for 6 to 8 h. Although more oil was extracted at 40°C than at 12°C, protein defatted at the higher temperature was of lower quality, presumably because of more extensive denaturation of the proteins.

Reduction of Cholesterol

Extraction of both lipids and cholesterol from trout muscle using $SF\text{-}CO_2$ has been reported (72). To enhance mass transfer, the flesh was frozen in liquid nitrogen and ground into particles (1 to 2 mm diameter). Extractions were carried out at pressures ranging from 138 to 345 bar at 40 to 50°C using both pure CO_2 and CO_2 modified by addition of 10% ethanol by weight. Extraction of 5 g of muscle with $SF\text{-}CO_2$ for 9 h at 40°C and 269 bar reduced the cholesterol content by a factor of 30 and reduced the total lipid content by *ca.* 80%. Similar treatment with $SF\text{-}CO_2$ modified by the addition of 10% ethanol (w/w) produced essentially cholesterol-free material (<0.001 g/100 g on a dry weight basis) and removed 97% of the lipids. In this test, addition of ethanol facilitated the extraction of phospholipids, leading to the lower total fat content in the extracted muscle. As with the defatted meat powder mentioned above, subsequent analyses of the defatted muscle indicated the occurrence of denaturation.

A study on the extraction of cholesterol from fatty acid ethyl ester mixtures with $SF\text{-}CO_2$ has recently been published (73). The test material in this work was an ethyl ester mixture derived from squid viscera containing cholesterol at a level of 2867 mg/100 g. The most successful test was carried out at 48°C and 103 bar (1500 psi). Cholesterol was depleted early in the extraction, reaching a low level of 14 mg/100 g about midway through the test. Removal of cholesterol was reported to be less effective at higher pressures because of decreasing fluid selectivity.

Future Outlook

The presence of several recent citations in the reference section of this chapter is testament to the ongoing interest in applying supercritical fluids in fish oil processing. An attempt has been made in the foregoing discussion to be inclusive, without editorial comment regarding the practicality of the described applications. As with any proposed use of supercritical fluids, however, enthusiasm for application on anything other than a laboratory scale should be tempered by the recognition that this is a capital-intensive technology. Consequently, supercritical fluid extraction will mainly find commercial application in instances in which the eventual product is of high value or an equivalent product cannot be obtained by more "conventional" means.

The use of $SF\text{-}CO_2$ to isolate EPA and DHA in high purities has been singled out as a potentially profitable application of this technology on an industrial scale (74). What remains unclear, however, is the potential market demand for these materials. The eventual outcome of current and future studies concerning the biomedical properties of EPA and DHA should help define their market. Unfortunately, unambiguous characterization of their physiological activity has been hampered by the lack of reliable and inexpensive commercial sources of these compounds in high purity.

Assuming demand for EPA, DHA, or both on a commercial scale does develop in the future, two processing strategies described earlier seem to have particular potential. The preparative SFC method described by Douget et al. (44) has already been shown capable of producing products on a scale of kg/wk; as a matter of interest, a line of preparative SFC systems was introduced at PITTCON '92 in New Orleans (75). Alternatively, a continuous-countercurrent approach such as that proposed by Krukonis et al. (36) could produce EPA and DHA on virtually any scale. A thorough analysis of the market demands and the capital and operating costs of these two and other competing approaches would be necessary before informed choices could be made.

Acknowledgments

G. Kudo provided valuable translation services. V. Stout also provided translation assistance as well as an especially useful review of this work.

References

1. Ackman, R.G., P.J. Ke, and P.M. Jangaard, *J. Am. Oil Chem. Soc. 50:*1 (1973).
2. Connor, W.E., and S.L. Connor, *Adv. Intern. Med. 35:*139 (1990).
3. Karmali, R.A., J. Marsh, and C. Fuchs, *J. Natl. Cancer Inst. 73:*457 (1984).
4. Reddy, B.S., C. Burill, and J. Rigotty, *Cancer Res. 51:*487 (1991).
5. Kremer, J.M., in *Health Effects of Dietary Fatty Acids,* edited by G.J. Nelson, American Oil Chemists' Society, Champaign, IL, 1991, p. 223.
6. Robinson, D.R., L-L. Xu, W. Olesiak, S. Tateno, C.T. Knoell, and R.B. Colvin, in *Health Effects of Dietary Fatty Acids,* edited by G.J. Nelson, American Oil Chemists' Society, Champaign, IL, 1991, p. 203.
7. Carlson, S.E., P.G. Rhodes, V.S. Rao, and D.E. Goldgar, *Pediatr. Res. 21:*485 (1990).
8. Uauy, R.D., D.G. Birch, E.E. Birch, J.E. Tyson, and D.R. Hoffman, *Pediatr. Res. 28:*485 (1990).
9. Hannay, J.B., and J. Hogarth, *Proc. Roy. Soc. London 29:*324 (1879).
10. Modell, M., U.S. Patent 4,338,199 (1982).
11. Shaw, R.W., T.B. Brill, A.A. Clifford, C.A. Eckert, and E.U. Franck, *Chem. Eng. News 69(51):*26 (1991).
12. Krukonis, V.J., M.A. McHugh, and A.J. Seckner, *J. Phys. Chem. 88:*2687 (1984).
13. Sievers, R.E., and B. Hansen, *Chem. Eng. News 69(29):*2 (1991).
14. Carbonell, E.S., *Cereal Foods World 36:*935 (1991).
15. McHugh, M.A., and V.J. Krukonis, *Supercritical Fluid Extraction: Principles and Practice,* Butterworths, Boston, MA, 1986.
16. Rizvi, S.S.H., R.R. Chao, and V.J. Liaw, in *Supercritical Fluid Extraction and Chromatography,* ACS Symp. Ser. 366, edited by B.A. Charpentier and M.R. Sevenants, American Chemical Society, Washington, DC, 1988, pp. 89–108.

17. Krukonis, V.J., in *Supercritical Fluid Extraction and Chromatography*, ACS Symp. Ser. 366, edited by B.A. Charpentier and M.R. Sevenants, American Chemical Society, Washington, DC, 1988, pp. 26–43.

18. Miyashita, K., and T. Tagaki, *J. Am. Oil Chem. Soc. 63:*1380 (1986).

19. Stout, V.F., W.B. Nilsson, J. Krzynowek, and H. Schlenk, in *Fish Oils in Nutrition,* edited by M.E. Stansby, Van Nostrand Reinhold, New York, 1990, pp. 73–119.

20. Brulé, M.R., and R.W. Corbett, *Hydroc. Process. 63(6):*73 (1984).

21. Friedrich, J.P., U.S. Patent 4,466,923 (1984).

22. Eisenbach, W., *Ber. Bunsenges. Phys. Chem. 88:*882 (1984).

23. Nilsson, W.B., J. Spinelli, V.F. Stout, and J.K. Hudson, *J. Am. Oil Chem. Soc. 63:*470 (1986).

24. Stout, V.F., and J. Spinelli, U.S. Patent 4,675,132 (1987).

25. Nilsson, W.B., E.J. Gauglitz, Jr., J.K. Hudson, V.F. Stout, and J. Spinelli, *J. Am. Oil Chem. Soc. 65:*109 (1988).

26. Nilsson, W.B., E.J. Gauglitz, Jr., J.K. Hudson, and F.M. Teeny, in *Seafood Biochemistry: Composition and Quality,* edited by G. Flick and R.E. Martin, Technomic Publishing Company, Lancaster, PA, 1992, pp. 151–168.

27. Nilsson, W.B., E.J. Gauglitz, Jr., and J.K. Hudson, *J. Am. Oil Chem. Soc. 66:*1596 (1989).

28. Nilsson, W.B., G.T. Seaborn, and J.K. Hudson, *J. Am. Oil Chem. Soc. 69:*305 (1992).

29. Kyle, D.J., *Adv. Appl. Biotech. Ser. 12:*167 (1991).

30. Schlenk, H., *Progr. Chem. Fats Other Lipids 2:*243 (1954).

31. Smith, A.E., *Acta Cryst. 5:*224 (1952).

32. Watkins, J.J., V.J. Krukonis, P.D. Condo, Jr., D. Pradhan, and P. Ehrlich, *J. Supercrit. Fluids 4:*24 (1991).

33. Klesper, E., and F.P. Schmitz, *J. Supercrit. Fluids 1:*45 (1988).

34. Joseph, J.D., *Biomedical Test Materials Program: Production Methods and Safety Manual*, U.S. Dept. of Commerce, Charleston, SC, 1988, NOAA Technical Memorandum NMFS-SEFC-234, p. 1.

35. Krzynowek, J., D.L. Entremont, L.J. Panunzio, and R.S. Maney, *Proc. Ann. Conf. Trop. Subtrop. Fish. Tech. Soc., 12:*74 (1988).

36. Krukonis, V.J., J.E. Vivian, C.J. Bambara, W.B. Nilsson, and R.E. Martin, in *Seafood Biochemistry: Composition and Quality,* edited by G. Flick and R.E. Martin, Technomic Publishing Company, Lancaster, PA, 1992, pp. 169–179.

37. Nilsson, W.B., *Upgrading the Value of Fish Oil,* Final Report, Saltonstall-Kennedy Cooperative Agreement NA85AA-H-SK148, 1987.

38. Treybal, R.E., *Liquid Extraction,* 1st edn., McGraw-Hill, New York, 1951.

39. Treybal, R.E., *Mass Transfer Operations,* 2nd edn., McGraw-Hill, New York, 1968.

40. Robinson, C.S., and E.R. Gilliland, *Elements of Fractional Distillation,* 4th edn., McGraw-Hill, New York, 1950.

41. Arai, K., and S. Saito, *Yukagaku 35:*267 (1986).

42. Suzuki, Y., M. Konno, K. Arai, and S. Saito, *Kagaku Kogaku Ronbunshu 16:*38 (1990).

43. Berger, C., P. Jusforgues, and M. Perrut, in *Proceedings of the International Symposium on Supercritical Fluids, Vol. 1,* Oct. 17–19, Nice, France, edited by M. Perrut, Institute Polytechnique de Lorraine, Lorraine, France, 1988, pp. 397–404.

44. Doguet, L., D. Barth, and P. Jusforgues, in *Proceedings of the International Symposium on Preparative and Industrial Chromatography,* April 6–8, Nice, France, edited by M. Perrut, Institute Polytechnique de Lorraine, Lorraine, France, 1992, pp. 295–300.

45. Colin, H., P. Hilaireau, and J. de Tournemire, *LC•GC 8:*302 (1990).

46. Higashidate, S., Y. Yamauchi, and M. Saito, *J. Chromatogr. 515:*295 (1990).

47. Ozcimder, M., and W.E. Hammers, *J. Chromatogr. 515:*295 (1990).

48. Privett, O.S., and E.C. Nickell, *J. Am. Oil Chem. Soc. 40:*189 (1963).

49. Ghosh, A., M. Hoque, and J. Dutta, *J. Chromatogr. 69:*207 (1972).

48. Passino, H.J., *Ind. Eng. Chem. 41:*280 (1949).

49. Dickinson, N.L., and J.M. Meyers, *J. Am. Oil Chem. Soc. 29:*235 (1952).

50. Passino, H.J., *Ind. Eng. Chem. 41:*280 (1949).

51. Dickinson, N.L., and J.M. Meyers, *J. Am. Oil Chem. Soc. 29:*235 (1952).

52. Zosel, K., Angew. *Chem. Int. Ed. Engl. 17:*702 (1978).

53. Spinelli, J., V.F. Stout, and W.B. Nilsson, U.S. Patent 4,692,280 (1987).

54. Ender, U., and R. Steiner, *Chem. Ing. Techn. 63:*727 (1991).

55. Brockerhoff, H., *Lipids 6:*942 (1971).

56. Kennedy, J.P., *Food Techn. 45(11):*76 (1991).

57. Nilsson, W.B., V.J. Stout, E.J. Gauglitz, Jr., J.K. Hudson, and F.M. Teeny, in *Supercritical Fluid Science and Technology, ACS Symp. Ser. 406,* edited by K.P. Johnston and J.M.L. Penninger, American Chemical Society, Washington, DC, 1989, pp. 434–448.

58. Ergan, F., M. Trani, and G. Andre, *Biotech. Bioeng. 35:*195 (1990).

59. Nilsson, W.B., E.J. Gauglitz, Jr., and J.K. Hudson, *J. Am. Oil Chem. Soc. 68:*87 (1991).

60. Debye, P., and W. Prins, *J. Colloid Sci. 13:*86 (1958).

61. Debye, P., and H. Coll, *J. Colloid Sci. 17:*220 (1962).

62. Krukonis, V.J., *J. Am. Oil Chem. Soc. 66:*818 (1989).

63. Jakobsson, M., B. Sivik, P.-A. Bergqvist, B. Strandberg, M. Hjelt, and C. Rappe, *J. Supercrit. Fluids 4:*118 (1991).

64. Yamaguchi, K., M. Murakami, H. Nakano, S. Konosu, T. Kokura, H. Yamamoto, M. Kosaka, and K. Hata, *J. Agric. Food Chem. 34:*904 (1986).

65. Friedrich, J.P., and E.H. Pryde, *J. Am. Oil Chem. Soc. 61:*223 (1984).

66. Spinelli, J., and C. Mahnken, *Aquaculture 13:*213 (1978).

67. Torrissen, O.J., R.W. Hardy, and K.D. Shearer, *CRC Crit. Rev. Aquat. Sci. 1:*209 (1989).

68. Miki, W., N. Toriu, Y. Kondo, M. Murakami, K. Yamaguchi, S. Konosu, M. Satake, and T. Fujita, *Bull. Japan. Soc. Fish. 49:*1417 (1983).

69. K. Yamaguchi, T. Mori, M. Murakami, S. Konosu, T. Kajiyama, and H. Yamamoto, *Nippon Suisan Gakkaishi 53:*2281 (1987).

70. K. Fujimoto, Y. Endo, S.-Y. Cho, R. Watabe, Y. Suzuki, M. Konno, K. Shoji, K. Arai, and S. Saito, *J. Food Sci. 54:*265 (1989).

71. Lee, C.M., *Food Techn. 38(11):*69 (1984).

72. Hardardottir, I., and J.E. Kinsella, *J. Food Sci. 53:*1656 (1988).

73. Yeh, A.-I., J.H. Liang, and L.S. Hwang, *J. Am. Oil Chem. Soc. 68:*224 (1991).

74. Latta, S., *INFORM 1:*810 (1990).

75. Stevenson, R., *Amer. Lab. 24(8):*28C (1992).

Chapter 9

Supercritical Fluid Extraction of Egg Lipids

N.R. Bulley

Department of Agricultural Engineering, University of Manitoba, Winnipeg, Manitoba R3T 2N2

The past decade has seen the emergence of an increase in consumer concerns related to the quality and safety of foods. One major health concern has been hypercholes-terolemia and its role in coronary heart disease (1). Although there is considerable controversy regarding the risk level involved, there is a general perception, at least in the public view, that high dietary intake of cholesterol should be avoided. Eggs have traditionally been a valued component of the human diet because of their high nutrient content, low caloric value, and ease of digestibility (2). Egg protein is one of the highest-quality proteins for human foods, because it contains all of the essential amino acids. Egg lipids are high in desirable unsaturated fatty acids and contain all the essential vitamins except vitamin C (3). Consequently, there has been considerable interest in lowering the cholesterol content of eggs, either by selective breeding of laying hens, by the manipulation of their diet (4,5), or via a separation process from the actual yolk (6,7,8,9).

Cholesterol is not the only moiety of interest in eggs. The egg white and yolk are very-high-quality protein, whereas the fatty acid composition is highly desirable, for example, in certain foods, such as infant formula (8). Lecithin, a valued component of egg yolk, is used in the manufacture of foods due to its emulsifying and rewetting properties. Lecithin predominantly is composed of phospholipids. One of these phospholipids, phosphatidylcholine (PC), is currently the subject of medical interest. PC has been identified as a significant factor in the treatment of neurological disorders, such as Alzheimer's disease (10). There is a strong demand for high-purity PC, free of triglycerides and phospholipids, such as phosphatidylethanolamine (PE) and sphingomyelin. Because phospholipids tend to be very unstable with regard to oxidation, heat, and light, a benign extraction process with a relatively "inert" solvent would be most desirable. Supercritical extraction (SFE) with CO_2, carried out at moderate temperatures, has such a potential for removing egg yolk lipids without affecting the functional properties of other compounds.

Composition of Egg Yolk

Liquid whole egg consists, on the average, of 64% white and 36% yolk. The white contains approximately 12% solids, which are predominantly protein with small amounts of minerals and sugars and only a trace of fat. Yolk contains about 50% solids, nearly two-thirds of which is fat and one-third protein (11). The constituents of the egg yolk are shown in Table 9.1. Physically, egg yolk can be regarded as a mixture of particulate "granules" and soluble plasma, the latter including low-density "globules" rich in fat.

Lipoproteins are complexes of neutral lipid, phospholipid, and protein having properties that differ from those of their constituents. Lipoproteins are 60% of the yolk (on a dry weight basis) and contain about 12% protein (12). It has been hypothesized that the protein occurs at the surface of the particles, along with charged lipid species, i.e., phospholipids (13). The phospholipids, proteins, and cholesterol interact at the surface of the lipoprotein particles, leaving those lipids that do not contain phosphorus, the neutral lipids, to occupy the center. The neutral lipids coalesce as oily droplets, well protected from the outside by the protein-phospholipid-cholesterol layer. Hence, the lipid in yolk is not free but bound in the lipoprotein particles (12).

The composition of the lipid portion of egg yolk is summarized in Fig. 9.1. Approximately 50% of the yolk is solid matter, consisting of about 32% protein and 64 to 72% lipid. The composition of the yolk lipid is about 65.5% triglyceride, 28.3% phospholipid, and 5.2% cholesterol (19). About 84% of the cholesterol is in the free state, the remainder existing in the esterified form. Polyunsaturated fatty acids make up 18, 18.6, and 28.6% of the triglycerides, phosphatidylcholine and phosphatidylethanolamine, respectively (20). The composition of yolk phospholipid approximates 73% phosphatidylcholine, 15% phosphatidylethanolamine, 5.8% lysophosphatidylcholine, 2.5% sphingomyelin, 2.1% lysophosphatidylethanolamine, 0.9% plasmalogen, and 0.6% inositol phospholipid (21,11). The various triglycerides that make up the yolk lipid contain both saturated and unsaturated fatty acid moieties ranging in chain length from C_{14} to C_{22} (Table 9.2).

TABLE 9.1 Constituents of Egg Yolk

			Percent of egg yolk solids
Proteins	Livetins		4–10
	Phosphoprotein	Vitellin	4–5
		Vitellenin	8–9
		Phosvitin	5–6
Lipoproteins	Lipovitellin		16–18
	Lipovitellenin		12–13
Lipids	Triglycerides		46
	Phospholipids		20
	Sterols		3
Carbohydrates			2
Mineral constituents			2
Vitamins			Traces

[a]Values obtained from (14) and (15,16) for proteins, (17) for lipids, and (18) for other constituents.

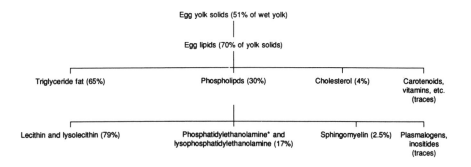

Fig. 9.1. Composition of the lipid portion of egg yolk. [Reproduced from *Journal of Science of Food and Agriculture 17*:746, © 1966.

TABLE 9.2 Fatty Acid Composition of Egg Yolk Lipid[a]

	Total lipids (%)	Saturated acids (%)				Unsaturated acids (%)			
		14:0	16:0	18:0	16:1	18:1	18:2	18:3	20:4
Total lipids	100.02[b]	0.7	27.5	7.3	4.3	44.0	14.0	1.0	0.5
Triglycerides	62.0	0.5	27.5	5.5	6.0	48.0	11.5	0.5	Trace
Free fatty acids	0.4	1.5	25.0	8.5	4.5	46.5	8.5	1.5	4.0
Sterol esters	0.34	4.0	32.5	10.5	6.5	35.0	11.5	Trace	Trace
Phospholipid	(30.5)	0.7	32.0	12.5	2.8	34.5	14.5	1.4	3.8
Lecithin	20.7	1.8	36.5	14.0	4.3	30.0	13.5	—	—
Cephalin	6.7	3.0	22.5	23.0	5.5	34.0	11.0	—	0

[a]Reproduced from *Journal of Science of Food and Agriculture 17*:746, 1966.
[b]Other constituents present were cholesterol (4.1% of total lipid), sphingomyelin (1.2%), lysolecithin (1.9%), and cerebrosides, etc. (2.6%).

Solubility of Egg Lipids in Supercritical Carbon Dioxide

Bulley et al. (22) reported on the effects of temperature and pressure on the solubility of egg yolk triglycerides and cholesterol in SC-CO_2 for pressures from 15 to 36 MPa at 40°C and temperatures from 40 to 75°C at 36 MPa. Lipid solubility was found to strongly depend on the extraction pressure, ranging from 10 mg/g CO_2 at 36 MPa to 0.67 mg/g CO_2 at 15 MPa at 40°C. At 36 MPa, a temperature increase from 55 to 75°C resulted in a solubility decrease from 8.99 to 6.2 mg/g CO_2. Triglycerides and cholesterol constituted the major portion of the extract; cholesterol content as a percent of egg lipid extract was affected by both temperature and pressure. The relative concentration of cholesterol increased with an increase in temperature (5.5 and 6.3% of extracted lipid at 55 and 75°C) and decreased with an increase in pressure (6.5 and 4.0% at 20 and 36 MPa). These relative changes may not be of a sufficient magnitude on which to base an extraction-separation system, because the fatty acid composition of the extracts remained constant throughout the extractions.

Entrainer Effects

Pasin et al. (23) investigated the effects of pressure, temperature, and presence of a methanol/ethanol entrainer on the solubility of cholesterol in supercritical CO_2. The presence of ethanol (5% w/w) increased cholesterol solubility by a factor of 2; the addition of methanol (5% w/w) increases cholesterol solubility by a factor of 10 to 100. Batch extractions of liquid yolk by pure supercritical CO_2 removed up to 18% of the cholesterol content. The use of ethanol (1% w/w) cosolvent in a batch extraction increased the removal of cholesterol to 46%. The authors did not specify the conditions (temperature and pressure) under which the results were obtained.

Conversely Bulley et al. (22) did not find the impact of alcohol entrainers to be as dramatic. They investigated the effects of methanol or ethanol as an entrainer on the overall solubility of egg lipids, triglycerides, and cholesterol in supercritical carbon dioxide. The solubility of egg yolk lipid at 40°C and 36 MPa was 10.46 mg/g CO_2 and for the cholesterol and triglycerides in the extract, 0.42 and 9.32 mg/g CO_2 respectively. When the methanol concentration was increased from 3 to 5%, the oversolubility of the egg lipids increased from 19.7 mg/g CO_2 to 44.4 mg/g CO_2, cholesterol solubility from 0.815 to 1.327 mg/g CO_2, and triglyceride solubility from 17.64 to 38.91 mg/g CO_2. A 3 wt% addition of either methanol or ethanol did not affect the solubilities of egg lipid in the supercritical CO_2 (19.7 compared to 22.1 mg/g CO_2 for the overall solubility of egg lipids, 0.815 to 0.648 mg/g CO_2 for cholesterol, 17.64 to 19.87 mg/g CO_2 for triglyceride). Since methanol and ethanol have almost the same effects on the solubility at the same concentration and ethanol is less toxic than methanol, ethanol was judged to be a better entrainer for the supercritical fluid extraction of egg yolk.

Froning et al. (24) investigated the overall removal of cholesterol and lipid components from dried egg yolk. The goal of their research was to remove the cholesterol from the egg yolk selectively without removing the polar lipids responsible for the functional and sensory properties of the resultant egg product. Extractions were carried out at four different pressure and temperature combinations (163 bar/40°C, 238 bar/45°C, 306 bar/40°C, and 374 bar/55°C). As the temperature and pressure increased, more lipids and cholesterol were recovered. This increase in removal would be expected if a fixed CO_2 mass flow rate was used for a fixed length of time, along with the fact that the solubilities of the triglycerides and cholesterol increased with temperature and pressure as has been reported. If the extractions at the lower temperatures and pressures were carried out to completion, the recorded higher extraction yields could also be due to a change in the composition of the extract or the enhanced solubility of triglycerides and cholesterol at higher temperatures and pressures.

Bulley et al. (22) noted that exhaustive extraction of triglycerides and cholesterol from egg yolks yielded only about 65 to 73% of the total available lipids. This recovery was increased to 82% when 3% ethanol was added as an entrainer and approached 100% when 5% methanol was used. These results support the concept that one or more of the lipid sources in egg yolk are "bound" and are not extractable with supercritical CO_2 for temperatures at 40 to 75°C and only become accessible when an entrainer is added to the SC-CO_2.

If the purpose of an SFE is to remove cholesterol or specific triglycerides selectively, then extraction temperature, pressure, and entrainer combinations must be found that will lead to the desired separation. Froning et al. (24) reported that as the temperature and/or pressure increased (163 bar/40°C to 374 bar/55°C), there was an increase in the ratio of saturated to unsaturated fatty acids in the extracted spray-dried egg yolk. Since the process had extracted the majority of the triglycerides, they speculated that the change in the ratio may reflect the concentration of the phospholipids in the extracted egg yolk rather than any specific selectivity for fatty acid. Bulley et al. (22) reported that the fatty acid composition of the extracts remained constant throughout their extractions. They also reported that the relative concentrations of the triglycerides to cholesterol remained unchanged over the course of the extraction. These results indicate that although the supercritical fluid extraction technology has excellent potential for the production of a low-cholesterol, low-fat-egg product, the preferential removal of cholesterol over other lipids still needs to be better optimized.

Phospholipid Enhancement

To date, SFE research on egg yolk supports the general finding that phospholipids are relatively insoluble in supercritical carbon dioxide. Froning et al. (24) reported that phospholipid concentrations in spray-dried egg yolk had increased significantly after SC-CO_2 extraction, but they did not report on the direct analysis of any of the extracts. For the extracted samples, the phosphatidylcholine (PC) to phosphatidylethanolamine (PE) ratio was about 6.5:1 in the original sample and remained relatively constant in the extracted samples over a range of different extraction temperatures and pressures. It should be noted that the ratio of PC to PE in the extract steadily increased from 6.5:1 to 7.4:1 as the extraction temperatures and pressures were increased, with overall removal of lipids. Bulley et al. (22) reported very low concentrations of phospholipids in extracts collected during their extractions at temperatures from 40°C to 75°C and pressures from 15 to 36 MPa. They also found that the samples collected at the end of an extraction contained higher levels of phospholipid. When 3% methanol or ethanol was present as an entrainer, the concentrations of phospholipid in the extracts collected during the early phase of the extraction remained very low (<0.2%) but increased during the later stages of the extraction (7 to 9%). Even higher phospholipid concentrations were reported when 5% methanol was present, but the authors cautioned that such results should be considered preliminary because of questionable analytical data.

Impact of SFE on the Functional Properties of the Egg Product

In most studies on SC-CO$_2$ extraction of food materials, little attention has been given to the impact of processing on the resulting functional properties of the extract or residue. Froning et al. (24) reported that for dried egg yolk (reduced in weight by 36%), the emulsion stability of mayonnaise was only adversely affected by using material processed at 374 bar/55°C. Although the emulsion stability of the yolk extracted at 306 bar/45°C appeared to be poorer than that of the control, the differences were not significant. These results also showed that sponge cake volume actually improved when it was made with extracted yolk. Since a constant weight of egg yolk was added to each cake batter, the increased protein concentration in the extracted egg yolk likely played a role in the improved functional properties that were observed. At the highest temperature and pressure (374 bar/55°C), Froning and coworkers found a decrease in cake volume, but polyacrylamide gel electrophoresis of the protein fractions did not indicate that any alteration at the extraction temperatures and pressures used in the study. These results are supported by the work of Arntfield et al. (25), which reported that there were no changes in denaturation temperatures or enthalpies of denaturation for SC-CO$_2$-extracted egg yolk. They did find that at an extraction temperature of 75°C there was a reduction in the enthalpy of denaturation for the ovalbumin present in the egg yolk. The use of 3% methanol at 36 MPa and 40°C resulted in a 50% reduction in the denaturation enthalpy of the ovalbumin. Based on these results, high temperatures and the use of entrainers during SC-CO$_2$ extraction can result in significant protein denaturation.

References

1. Gotto, A.M., *J. Amer. Coll. Cardiol. 13:*503–507 (1989).
2. Cook, F., and G.M. Biggs, *Egg Science and Technology,* 3rd edn., AVI Publishing Co., Inc., Westport, CT, 1986, pp. 92–108.
3. Mountney, G.J., *Poultry Products Technology,* 2nd edn., AVI Publishing Co., Inc., Westport, CT, 1976.
4. Hargis, P.S., *World's Poultry Sci. J. 44:*17 (1988).
5. Washburn, K.W., and N.F. Nix, *Poultry Sci. 53:*109 (1974).
6. Blackwelder, J.A., and O.A. Pike, *J. Food Sci. 55:*92 (1990).
7. Warren, M.W., H.G. Brown, and D.R. Davis, *J. Am. Oil Chem. Soc. 65:*1136 (1988).
8. Tokarska, B., and M.T. Clandinin, *Can. Inst. Food Sci. Techn. J. 18:*256 (1985).
9. Larsen, Y.E., and G.W. Froning, *Poultry Sci. 60:*160 (1981).
10. Krishnan, K.R.R., A. Heyman, J.C. Ritchie, C.M. Utley, D.V. Dawson, and H. Rogers, *Biol. Psychiatry 24:*937 (1988).
11. Parkinson, T.L., *J. Sci. Food Agric. 17:*746 (1966).
12. Burley, R.W., R.W. Sleigh, and K. Fretheim, *CSIRO Food Res. 45:*64 (1985).
13. Burley, R.W. and D.J. Kushner, *Can. J. Biochem. Physiol. 41:*409 (1963).
14. Fevold, H.L., *Adv. Protein Chem. 6:*187 (1951).
15. Bernardi, G., and W.H. Cook, *Biochim. Biophys. Acta 44:*86 (1960).

16. Bernardi, G., and W.H. Cook, *Biochim. Biophys. Acta 44:*96 (1960).
17. Lea, C.H., in *Recent Advances in Food Science, Vol. 1,* edited by J.B. Hawthorne, and R. Muil Leitch, Butterworths, London, 1962.
18. Brooks, J., and D.J. Taylor, *G.B., Dept. Sci. Ind. Res.* Food Invest. Board, Spec. Rep. 60, (1955).
19. Privett, O.S., M.L. Bland, and J.A. Schmidt, *J. Food Sci. 27:*463 (1988).
20. Pike, O.A., and I.C. Peng, *Poultry Sci. 64:*1470 (1985).
21. Rhodes, D.N., and C.H. Lea, *Biochem. 54:*526 (1957).
22. Bulley, N.R., L. Labay, and S.D. Arntfield, *J. Supercrit. Fluids 5:*13 (1992).
23. Pasin, G., R.A. Novak, W.J. Reighter, A.J. King, and G. Zeidler, *Proceedings of the 2nd International Symposium on Supercritical Fluids,* Boston, MA, 1991, edited by M.A. McHugh, p. 312, Department of Chemical Engineering, Johns Hopkins University, Baltimore, MD, (1991).
24. Froning, G.W., R.L. Wehling, S.L. Cuppett, M.M. Pierce, L. Niemann, and D.K. Seikman, *J. Food Sci. 55:*95 (1990).
25. Arntfield, S.D., N.R. Bulley, and W.J. Crerar, *J. Am. Oil Chem. Soc. 69(9):*823 (1992).

Chapter 10

Supercritical Fluid Extraction of Cocoa and Cocoa Products

Margherita Rossi

Dipartimento di Scienze e Tecnologie Alimentari e Microbiologiche, via Celoria 2, 20133, Milano, Italy

Three groups of substances give nutritional, technological, and sensory interest to cocoa beans (*Theobroma cacao*). These are lipids, the xanthine stimulants (theobromine and caffeine), and the precursors of the aroma. This last category, according to Martin (1), consists of amino acids; reducing sugars, which react via the Maillard reaction and the Strecker degradation to give many of the aromatic compounds present in cocoa aroma; polyphenols; and some organic acids that contribute as such to the aroma in the fermented bean.

The presence of methylxanthines in the cocoa bean involves toxicological aspects, which are thoroughly discussed by Tarka (2) in an interesting review. These compounds are present in the shell at a level of 0.7 to 1.2% and in the germ at about 2%, while the cocoa powder contains 3.5%. The shells are normally used for the extraction of theobromine for pharmacological uses.

Taking into consideration the whole cocoa bean, lipids are mainly found in the cotyledons (55% fat) and partly in the shell (2 to 3% fat).

Figure 10.1 illustrates a process flowsheet in cocoa technology (3), and a brief description of the main steps of the process is given here.

Roasting is the most important step in the development of chocolate flavor, whose precursors are produced during fermentation (1).

Winnowing is a critical step that is performed to separate the valuable product (i.e., *nibs*) from the by-product (i.e., shell). Variations in the fat content of the shell fraction are caused by nib loss into the shell fraction, which depends on the calibration of the winnowing machine. Thus, the shells may contain a fairly high quantity of fat, including cocoa butter and shell fat itself, which may reach 17% (4).

Milling converts nibs into a fluid paste, known as *liquor*, which is hydraulically pressed for cocoa butter production. Cocoa butter obtained in other ways, such as expressing of beans and shells or solvent extraction, is not considered prime butter (1).

The main products of the process just described are cocoa butter and cocoa powder, while the shells are considered a by-product that are of interest for the possible recovery of lipids or biologically active substances, such as theobromine or antioxidative compounds (5).

The possibility of using supercritical CO_2 in the cocoa industry can extend from the extraction of lipids to the recovery of aroma and xanthines, with the aim of simpli-

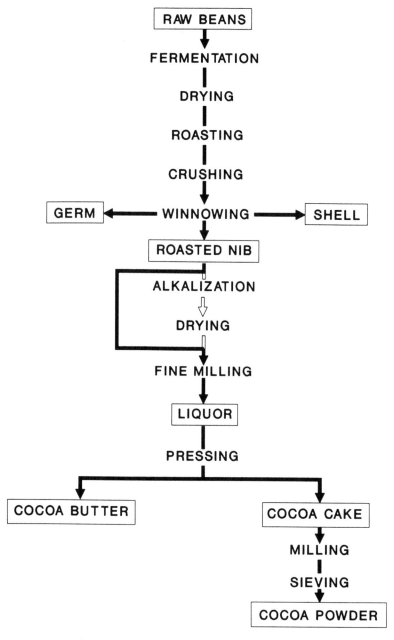

Fig. 10.1. Process flowsheet in cocoa technology. [Reproduced from *Italian Journal of Food Science 1(3):*31 (1989).]

fying traditional extraction methods, of increasing the value of by-products, or in reducing the content of alkaloids in cocoa to modify its dietetic characteristics.

Extraction of Methylxanthines

Theobromine is the main alkaloid in cocoa. It is an N-methylxanthine and has a very similar structure to caffeine (Fig. 10.2). Both can be extracted from cocoa beans with warm water, as described by Magnolato and Isely (6), with a process that involves four phases: an extraction phase; a purifying phase, in which the watery extract is purified on an adsorbent; the recovery of xanthine from the adsorbent; the final addition of the purified concentrated aqueous extract to the cocoa beans in order to reestablish the solids content of the bean.

Margolis et al. (7) described a process in which stimulants are removed from cocoa by supercritical CO_2 extraction of water-swollen nibs at 30 MPa and 90°C. Water's fundamental role in the extraction of theobromine from cocoa must be emphasized. The effect of the water during supercritical extraction of xanthines is described by McHugh and Krukonis (8). The authors report an extensive list of patents, which essentially concern coffee decaffeination, and they hypothesize that caffeine is chemically linked in structures to chlorogenic acid. The water is believed to act as a chemical agent that can free the caffeine linked to the coffee substrate and therefore allow the caffeine to be extracted by the supercritical CO_2. Since caffeine and theobromine are similar in structure, it is thought that water has a similar effect in the supercritical extraction of theobromine from cocoa. In fact, Brunner et al. (9) found that the solubility of theobromine in water-saturated supercritical carbon dioxide is much less than that of caffeine, when xanthines were extracted from moistened cocoabean shells. They used shells containing 1% theobromine and 5% fat and found, in a pressure range between 20 and 40 MPa, that the weight fraction of the theobromine

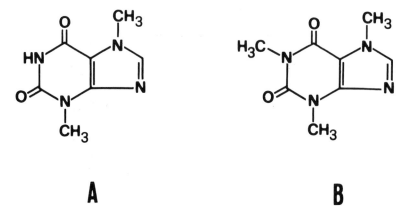

Fig. 10.2.   Chemical structures of theobromine (*a*) and caffeine (*b*).

extracted was dependent on pressure: a doubling of pressure (from 20 to 40 MPa) caused a fivefold increase of the extraction yield (from 20 to almost 100%). An increase in temperature caused an improvement of theobromine solubility in supercritical CO_2 until the saturation level was reached. Their experiments confirm the positive effect of water on the extraction of theobromine, which is extracted extensively when shells moistened to a 50% water level are used. They also advocated the use of water-saturated CO_2. The authors concluded that the results and parameters established for the extraction of caffeine cannot be transferred to the extraction of theobromine.

The extraction conditions just described are well beyond those reported by Sambataro (10) for supercritical CO_2 extraction of cocoa nibs moistened to a 50% level and containing 1.48% theobromine and 57% fat. Figure 10.3 shows the extraction yield obtained for theobromine at 85 and 50°C, in a pressure range between 15 and 45 MPa. Both pressure and temperature have a positive effect on extraction, but the yields obtained are rather low if compared to those found by Brunner et al. (9). We can suggest some explanations to justify this result: particularly that Sambataro (10) used dry supercritical CO_2, but above all, that the extraction was performed on nibs with a high fat content and particle size ranging from 2 to 4 mm. These last two factors negatively influenced the selectivity of the process and the diffusion of the solvent through the cocoa matrix.

Li and Hartland (11) experimentally measured the solubility of pure theobromine and caffeine in supercritical CO_2 at different temperatures (40 to 95°C) and pressures (8 to 30 MPa). They observed that, despite the chemical similarity of the two compounds, the solubility of theobromine was two orders of magnitudes less than that of caffeine. This difference, according to the authors, is due to the difference in the melting points and enthalpies of fusion of the two compounds as well as to the presence of intermolecular hydrogen bonds, which can form between theobromine molecules.

The different solubilities of theobromine and caffeine in supercritical CO_2 therefore explain the results obtained by Brunner et al. (9). Since the concentration of the theobromine in cocoa shell is about one order of magnitude greater than that of caffeine (in the nibs the theobromine-to-caffeine ratio is about 6:1), Brunner et al. (9) observed that during extraction of the shells, because of the remarkable difference in solubility between the two compounds, theobromine saturated the supercritical solvent and that the extraction rate remained constant (about 15 ppm in the solvent phase) until only 40% of the initial theobromine remained unextracted.

Brunner et al. (9) also observed that water was a much more important factor for the extraction of caffeine than for the extraction of theobromine. At a given extraction time, in fact, the extraction yield for theobromine from shells decreased from 67% (obtained at 50% moisture and using water-saturated CO_2) to 16% when dry CO_2 and 8% moisture were used, while that of caffeine, under the same conditions, fell from 100% to values below 5%. Likewise, Lentz et al. (12) demonstrated that in the temperature range of 40–100°C and with a pressure of 15–30 MPa, the solubility of pure caffeine in water-saturated CO_2 was 2.4 to 5.8 times greater than in dry CO_2, while the solubility of theobromine remained the same in both cases.

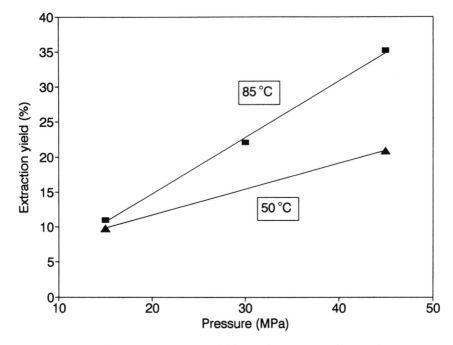

Fig. 10.3.   Theobromine extraction yield from nibs, as reported by Sambataro (10).

Water plays two roles in the extraction of methylxanthines from cocoa. On the one hand, water is necessary to free the xanthines from the matrix; on the other hand, it influences the solubility of xanthines.

Li and Hartland (11) experimented with the use of ethanol as a cosolvent for the supercritical CO_2 extraction of methylxanthines and lipids from cocoa nibs. The characteristics of the nib sample were

Particle size   0.5–2 mm
Theobromine   1.44%
Caffeine   0.232%
Lipids   53.8%
Moisture   6%

When ethanol was used as a cosolvent, the solubility of theobromine increased two orders of magnitude as a function of ethanol concentration in the supercritical solvent. The extraction with an ethanolic cosolvent proved, however, more selective for cocoa butter than for theobromine, contrary to what happens with an aqueous cosolvent (7,10). In fact, the ethanol cosolvent has been successfully tested as an enhancer of the extraction of lipid components in several food stuffs (13,14,15). In the case of the supercritical CO_2 extraction of cocoa, the ethanol enhances not only the extraction of lipids but also that of theobromine. However, the authors (11) measured a selectivity value for cocoa butter relative to total xanthines extracted of approximately 10. In order to extract and separate both lipids and theobromine, they proposed combining traditional separation methods, such as distillation and evaporation, with supercritical extraction.

Extraction of Lipids

Most of the lipids in the cocoa bean are present in the cotyledons, which contain about 55% fat. The products obtained in cocoa technology by grinding of the cotyledons are, in order of particle size, the nibs and the liquor. The latter is hydraulically pressed to produce the so-called cocoa butter. The shells also contain a moderate quantity of fat, which can be extracted with hexane.

The first example of supercritical CO_2 extraction of cocoa butter, from both liquor and nibs, was accomplished at 20 to 40 MPa and 40 to 60°C, as described by Roselius et al. (16), who claimed a complete extraction of fat from liquor and a 74% extraction from nibs. Krukonis (8) verified these results for liquor, but found that less than 5% of the cocoa butter could be extracted from nibs, even when the extraction was carried on for 8 h at about 50 MPa and at 40 to 60°C.

The feasibility of extracting cocoa butter from nibs was also investigated by Rossi et al. (3), who compared the extraction kinetics from liquor, nibs, and shells at 30 to 40 MPa and 50 to 80°C (Table 10.1), and found that the extraction yield depends on the degree of disruption of the lipid-bearing cells. Fig. 10.4 represents the extraction curves that were obtained. The low extraction yields observed for the extraction of nibs (having a particle size ranging from 2 to 4 mm), are caused by the lipid-bearing cells remaining intact in the nib. The curves in Figure 10.4 indicate that the maximum amount of extractable fat will be different for the three cocoa products. The authors concluded that diffusion phenomena had limited influence on the extraction kinetics, and that extraction of cocoa butter appears to be more of a leaching phenomenon.

Cocoa butter solubilities measured by the authors for the supercritical CO_2 extraction of liquor are consistent with the values reported by Li and Hartland (11), who measured cocoa butter solubility in supercritical CO_2 directly on the product prepared by hydraulic expressing of cocoa (Fig. 10.5). This data confirms that supercritical CO_2 extraction of cocoa butter of such a fine-milled product involves a leaching mechanism. This indicates that once the fat has been released from the lipid-bearing cells (through the fine-grinding process), SFE can be readily applied to the extraction of the fat.

M. Rossi

TABLE 10.1   Operative Conditions Maintained During the Extraction Experiments

Cycle identification	Substrate	Pressure (MPa)	Temperature (°C)	Total Time (h)
1	Liquor	40	80	5
2	Liquor	40	50	5
3	Nibs	40	80	5
4	Nibs	40	50	5
5	Shells	30	80	6

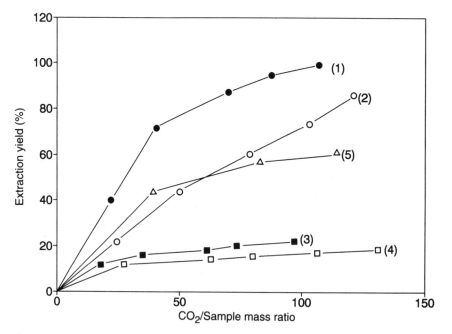

Fig. 10.4.   Fat extraction yield from liquor, nibs, and shells, as reported by Rossi et al. (3). See Table 10.1 for numeric parameters for Cycles 1 through 5.

Rossi et al. (17) also investigated the composition of cocoa butter obtained by supercritical CO_2 extraction of various cocoa products. They observed that the fatty acid and triglyceride composition and unsaponifiable content of cocoa butter, extracted by supercritical CO_2 at 40 MPa and 50 to 80°C, are within the range required for this product. Nonetheless, they observed minor differences in the selectivity (α) (17) of the process for the most important triglycerides of cocoa butter, namely POP (palmitic-oleic-palmitic) (C_{50}), POS (palmitic-oleic-stearic) (C_{52}), and SOS (stearic-

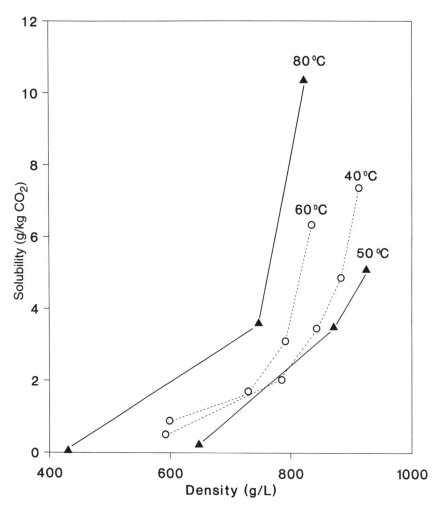

Fig. 10.5. Solubility isotherms of cocoa butter in supercritical CO_2, as a function of CO_2 density. Data from Rossi et al. (3) (filled triangles) and Li and Hartland (11) (open cir-

oleic-stearic) (C_{54}). At 50°C and 40 MPa, the average selectivity of POP relative to POS was 1.05, while that of SOS relative to POS was 0.94. Consequently, Rossi et al. hypothesized that shorter triglycerides are more easily extracted than longer ones.

However, fractionation is negligible at 40 MPa, it may become important when extraction is carried on at lower pressure. On the basis of these results we can hypothesize the possibility of directly fractionating cocoa butter into fractions of different melting points during supercritical fluid extraction of liquor.

Rossi et al. (17) also examined the characteristics of fat extracted from cocoa shell by supercritical fluid extraction, concluding that because of the unsaponifiable matter content and fatty acid composition of the extract, the fat thus extracted can be graded only as shell fat. Nonetheless, its triglyceride composition was quite similar to that of cocoa butter.

Li and Hartland (11) greatly enhanced the solubility of cocoa butter by using ethanol as a cosolvent for the supercritical CO_2 extraction of nibs. They extracted almost 100% fat from nibs at 60°C and 15 MPa using a 30% weight fraction of cosolvent, whereas other authors could not exceed 5% (8) or 20% (3), using pure supercritical CO_2. The effect of ethanol on the composition of the lipid extract should also be investigated in the future.

Extraction of Aroma

Vitzthum et al. (18) studied the aroma obtained from cocoa liquor, using SFE with CO_2. A steam distillation of the extract was characterized by GC/MS, permitting the authors to identify 59 nitrogen-containing compounds not previously reported as constituents of roasted cocoa. Supercritical CO_2 extraction proved to be an attractive and advantageous method compared to the traditional isolation method.

Bundschuh et al. (19) studied the possibility of obtaining highly aromatic extracts from roasted cocoa bean shells by fractional separation of aroma compounds of supercritical fluid–extracted fat. Although they obtained a so-called aroma oil, no quantitative consideration or mass balance was attempted for the SFE.

The effect of supercritical CO_2 extraction on the aromatic fraction of roasted liquor and unroasted nibs were studied by Rossi et al. (17) using high-performance liquid chromatography (HPLC) analysis of the pyrazines. They observed that the pyrazine fraction was almost equally distributed between the solid and the supercritical phase during extraction. Supercritical CO_2 extracted pyrazines along with fat, however, the coextraction did not reduce the aroma of the residue with respect to the pyrazine fraction.

References

1. Martin, R.A. Jr., *Advances in Food Research, Vol. 31,* Academic Press, San Diego, CA, 1987, pp. 211–342.
2. Tarka, S.M. Jr., *CRC Critical Reviews in Toxicology:*275 (1982).
3. Rossi, M., C. Arnoldi, F. Antoniazzi, and A. Schiraldi, *Ital. J. Food Sci. 1(3):*31 (1989).
4. Minifie, B.W., *Chocolate, Cocoa and Confectionery Science and Technology,* 3rd edn., AVI Books, New York, 1989, p. 51.
5. Yamaguchi, N., and S. Naito, *New Food Ind. 26(1):*68 (1984).
6. Magnolato, D., and A. Isely, European Patent 0,070,437 B1 (1985).
7. Margolis, G., J. Chiovini, and F.A. Pagliaro, European Patent Appl. 0,061,017 (1972).
8. McHugh, M.A., and V.J. Krukonis, *Supercritical Fluid Extraction: Principles and Practice,* Butterworths, Stoneham, MA, 1986, pp. 181–188.

9. Brunner, G., F.A. Zwiefelhofer, and A. Simon, in *Abstract Handbook, 2nd International Symposium of High Pressure Chemical Engineering,* September 24–26, edited by G. Vetter, Erlangen, Germany, 1990, pp. 413–417.

10. Sambataro, D. Esperienze di Estrazione della Teobromina da Cacao con Anidride Carbonica Supercritica, Ph.D. Thesis, University of Milan, Italy, 1984, pp. 85–115.

11. Li, S., and S. Hartland, in *Proceedings of 2nd International Symposium on Supercritical Fluids,* May 20–22, Boston, MA, 1991, edited by M.A. McHugh, pp. 13–15.

12. Lentz, H., M. Gehrig, and J. Schulmeyer, *Physica B+C 139–140:*70 (1986).

13. Hardardottir, I., and J.E. Kinsella, *J. Food Sci. 53(6):*1656 (1988).

14. Rossi, M., E. Spedicato, and A. Schiraldi, *Ital. J. Food Sci. 4:*249 (1990).

15. Bulley, N.R., and L. Labay, in *Proceedings of 2nd International Symposium on Supercritical Fluids,* May 20–22, Boston, MA, 1991, edited by M.A. McHugh, pp. 10–12.

16. Roselius, W., O. Vitzthum, and P. Hubert, U.S. Patent 3,923,847 (1987).

17. Rossi, M., C. Arnoldi, G. Salvioni, and A. Schiraldi, *Ital. J. Food Sci. 1(3):*41 (1989).

18. Vitzthum, O.G., P. Werkhoff, and P. Hubert, *J. Food Sci. 40:*911 (1975).

19. Bundschuh, E., M. Tylla, G. Baumann, and K. Gierschner, *Lebensm.-Wiss. u. Techn. 19:*493 (1986).

Chapter 11

Supercritical CO₂ Extraction of Meat Products and Edible Animal Fats for Cholesterol Reduction

Roy R. Chao

Department of Food Science and Human Nutrition, University of Missouri, Columbia, MO 65211

Meat products provide about 36% of energy and 36 to 100% of the major nutrients available from foods. However, they also contribute more than half of the total fat, 75% of the saturated fatty acids, and all of the cholesterol to the diet (1). Despite the total nutritional value of meat consumption, consumers have been advised to lower their dietary fat and cholesterol levels for improving human health. Red meat production rose by only 0.4 million tons in 1991, compared with an average increase of 3.1 million tons per year from 1986 through 1990 (2).

In parallel with the reduction in meat products, the production of rendered animal fats also declines. Of the three animal fats (butter, lard, and tallow), increases in the consumption of beef tallow after 1950 were due to its distinctive flavor and its stability in fryers. In 1985, 460,000 metric tons of edible beef tallow went into baking and frying fats in the United States, but this had declined to 289,000 metric tons in 1990 and is not expected to increase in the near future (2). The decrease in production of beef tallow has also made it unprofitable for renderers to take the extra precautions required for edible tallow.

Meat researchers can help address such concerns by increasing the availability of palatable, easily prepared food products that could help consumers follow the dietary principles. Meat products could thus benefit greatly from a process that selectively removed fat and cholesterol while retaining the positive flavor, textural, and nutritional qualities. A meat product with the minimum fat for palpability and cholesterol for health concerns appears to be in demand. As for edible beef tallow (EBT), the reduced usage of it has prompted a renewed interest in developing fractionated EBT by altering its cholesterol content so that it might be more satisfactorily used as a "tailor-made" edible ingredient, or a fat frying medium, an additive for animal feed, or a replacement for other more expensive imported fats and oils, such as cocoa butter, demanded by confectionery industries.

Supercritical carbon dioxide (SC-CO₂) has many desirable properties as a solvent for biomaterials, including foods. It is widely available, nonflammable, nontoxic, and environmentally sound (3). The technique has overcome many drawbacks imposed by the conventional organic solvent–oriented extraction methods. In addition, it is effective at relatively low temperatures and, in certain cases, does not cause extensive protein damage (4). This is especially important in foods, where proteins contribute extensively to nutrition and functionality. Several reports can be found in the literature that pertain to the use of supercritical CO_2 for fractionation and for extraction of natural products (5–14). The main objective of the present study was to explore the feasibility of using the extraction technique to alter the cholesterol content of ground meat and edible beef tallow using various extraction conditions.

Experimental Design

The Supercritical Fluid Extraction Unit

The extraction apparatus used (Fig. 11.1) was manufactured by the Superpressure Division of Newport Scientific, Inc. (Jessup, MD), modified by adding two more separators. The detailed extraction conditions have been described by Chao et al (15). Essentially, for each test run a sample, around 200 g of frozen and subsequently thawed ground beef, containing about 18% fat, was loaded into the extraction vessel and extracted with supercritical CO_2. The extraction pressures was varied in the range of 100 to 310 bar, and the temperature was in the range of 30 to 50°C. The soluble components were collected at 34.5 bar and 40°C. Edible beef tallow of commercial grade was obtained from Anderson Clayton/Humko Products, Inc., Memphis, TN, and carefully melted, mixed, and subdivided into smaller plastic containers (200 g each) and stored at −27°C until used. Similar procedures to those mentioned for ground beef were followed for the extraction runs of EBT.

Determination of Cholesterol Content

For meat extraction the procedures for determining the cholesterol of extracted lipid sample was described by Chao et al. (15). For edible beef tallow extraction, the preparation of samples for cholesterol content was based on the AOAC method Section 28.110 (16). The prepared sample was then injected to a Supelco (Supelco, Inc., Bellefonte, PA) SPB-1 fused silica capillary column, 30 m × 0.32 mm I.D., on a Varian (Varian Associates, Sugar Land, TX) Model 3700 Gas Chromatograph equipped with dual flame ionization detectors. The initial holdup time was 4 min at 270°C, after which the temperature was programmed to rise to 300°C at a ramp rate of 10°C/min. The helium flow rate and split ratio were set at 1.5 mL/min and 50:1, respectively, and the injector and detector temperatures were set at 310°C.

Determination of Fatty Acid Content

The preparation of samples for total fatty acid analysis was modified from the AOAC method 28.056 (16) by replacing heptane with hexane and employing a Supelco SP-2330 capillary column, 30 m × 0.32 mm i.d. Using the Varian Model 3700 Gas Chromatograph mentioned in the previous section, the prepared sample was injected into a Supelco SPB-1 fused silica capillary column, 30 m × 0.32 mm i.d. at 270°C with an initial hold of 4 min followed by a programmed rise to 300°C at a ramp rate of 10°C/min, helium flow rate 1.5 mL/min and split ratio 50:1. Both the injector and detector temperatures were set at 310°C.

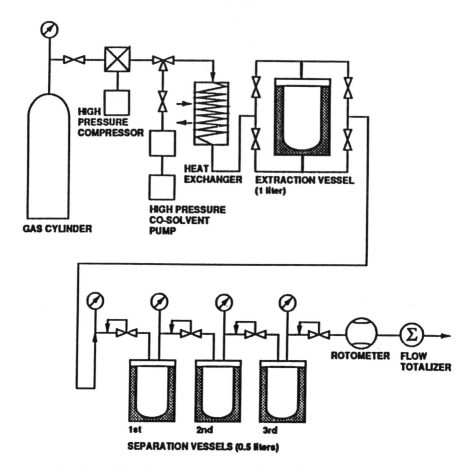

Fig. 11.1.　Flow diagram of the supercritical fluid extraction system (29).

Calibrations

Regular calibrations were performed for the analyses both of cholesterol and fatty acids to maintain the standard deviation of reproducibility and accuracy of gas chromatographic (GC) results to be 1% and 0.5%, respectively.

Results and Discussion

Part A. Supercritical Extraction of Meat: Ground Beef

Table 11.1 shows the results of the total extraction yield of lipids for the runs extracted by SC-CO_2 in a dynamic mode using different pressure and temperature conditions. Comparing the yield of lipids for the 172 bar/35°C run with that of the 310 bar/50°C run indicates that although both runs had a similar density of SC-CO_2, the higher extraction pressure resulted in a significantly higher yield of lipids than the lower pressure. This implies that pressure rather than density was the predominant factor in increasing the extraction yield of lipid from the ground beef samples.

A similar result, indicating that the extraction pressure is the dominant factor for supercritical extraction, was also reported by Zou et al. (17), who studied the solubility of oleic acid, linoleic acid, and their methyl esters in SC-CO_2. They showed that the solubility of a liquid solute in SC-CO_2 increased with pressure despite the temperature difference used. The tested pressure and temperature ranges used by Zou et al. used were from 73 to 284 bar and 40 to 60°C, respectively.

The cumulative yield curves (Fig. 11.2*a*) of total soluble components extracted at 172 bar and 35 to 50°C show a similar slope, but the yield increases with higher temperatures even though different amounts of CO_2 were used for each run. Due to the observed simultaneous extraction of lipid and water (verified via differential scanning calorimetry), each fraction was further separated into the lipid and the water parts via centrifugation. Within the test range, Fig. 11.2*b* shows that the run extracted at 50°C did not cause higher lipid yield than the run made at 35°C. However, the yield of water for the run at 50°C, shown in Fig. 11.2*c*, was significantly higher than those at 35°C and 40°C. The higher yield for water implies that the water solubility of SC-CO_2 was increased as the extraction temperature was increased. Similar runs with the higher extraction pressure of 310 bar and temperatures of 35 ar 50°C were also tried, and the results are shown in Fig. 11.3. In Fig. 11.3*a* the yield curve for lipid at both temperatures showed a sharp increase for the 310 bar/35°C run after 7.7 kg of CO_2 was used, and its accumulated weight rapidly approached the value of the 310 bar/50°C run after 13.9 kg of CO_2 was used. This rapid increase in lipid yield was not observed for water, as shown in Fig. 11.3*b*. The results showed that while the 310 bar/50°C run increased both lipid and water, the 310 bar/35°C run resulted only in higher extraction yield of lipid.

TABLE 11.1   Effect of Extraction Pressure and Temperature on the Extractability of Lipids from Ground Beef by Supercritical CO_2

	Applied Pressure (bar)/Temperature (°C)				
	72/35	172/40	172/50	310/35	310/50
CO_2 density (kg/m³)	876	820	759	992	883
Wt. of CO_2 used (kg)	12.3	7.7	20.1	13.9	15.4
Wt. of extracted lipids (g)	9.8	6.3	14.5	21.8	27.8
Average wt. (g) of extracted lipids per kg of CO_2	0.79	0.82	0.72	1.57	1.80

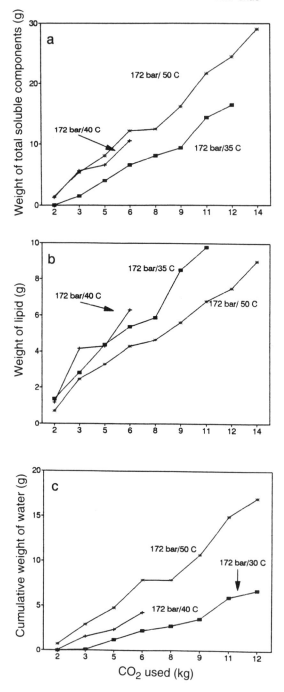

Fig. 11.2. Cumulative yield of (*a*) fractions, (*b*) their lipids, and (*c*) water portions extracted with supercritical CO_2 at 172 bar and different temperature range.

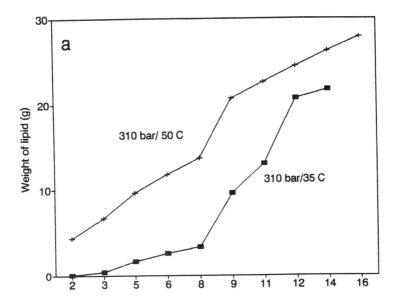

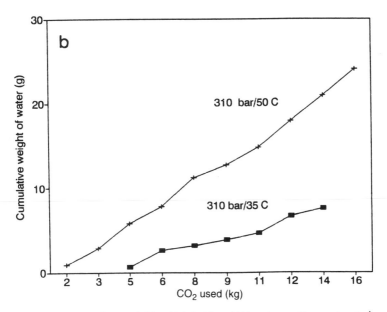

Fig. 11.3.   Cumulative yields of (a) lipid and (b) water portions extracted with supercritical CO_2 at 310 bar and temperatures of 35 and 50°C.

To evaluate the effect of extraction pressure on the yield of lipid and water from ground beef, the three runs with different pressures (i.e. 103, 172, and 310 bar), all at 50°C, were compared. Fig. 11.4a shows that higher pressures increase the extraction yield of lipid, especially for the 310-bar run. The water yield curve for the 172-bar run, however, was similar to that for the 310-bar run, whereas the lipid yield curve of the 172-bar run was closer to that of the 103-bar run, implying that the increase in weight of the fractions collected at 172 bar over those collected at 103 bar was mainly due to increased extraction of water. A similar result with respect to the water yield was also observed for the runs extracted at 35°C (cf. Figs. 11.2c and 11.3b). These results show that although the extractability of both lipid and water increase with higher extraction pressure, at extraction temperatures from 35 to 50°C, the water yield remains about constant despite the increased pressure from 172 to 310 bar. This is consistent with the results of the solubility of water in SC-CO$_2$ as reported by Won (18). Specifically, citing from Stahl et al. (19), the saturation concentration of the water in the SC-CO$_2$ asymptotically approaches a limiting value of 3 mg water/g CO$_2$ at pressures above 200 bar.

Effect of Moisture on Lipid Extractability. Fresh meat usually contains up to 76% water, although it varies with the fat content, protein, and the position from the carcass. King et al. (20) reported that supercritical fluid extraction could achieve maximum results on such meat products as link sausage, luncheon meat, smoked ham, and imported ham when they were finely comminuted and dehydrated, yielding over 96% of the theoretical fat content of the extracted samples. The present author conducted a study to determine the effect of moisture content on the reduction of lipids and cholesterol by extracting freeze-dried ground beef samples prepared with different moisture contents. As a method of food dehydration, freeze-drying is an ideal method for the maintenance of the desirable functional characteristics of the meat. With SC-CO$_2$ at 345 bar and 40°C, Fig. 11.5 shows that when the moisture content of ground beef was reduced to less than 25% prior to extraction, at least 95% of total lipid could be extracted from meat. Further, the percentage of total cholesterol extracted was significantly increased as the moisture content of ground beef was decreased to 5%. The reason for the increased extractability of cholesterol was unknown. It is speculated that the further reduction of moisture content from meat would lead to greater unfolding of the meat muscle matrix. The more unfolded muscle matrix would then provide easy access for SC-CO$_2$ to penetrate into the matrix and extract lipids and cholesterol that are usually embedded in the flesh and hard to reach. A similar study of extracting fish muscle with SC-CO$_2$ mixed with different organic solvents was reported by Hardardottir and Kinsella (11).

Extraction of Cholesterol from Ground Meat. The lipid fractions were also analyzed for their cholesterol contents in order to determine the effect of various extraction conditions on the extraction yield of cholesterol. The average percentages of extracted cholesterol and lipid extracted are shown in Table 11.2 for four selected

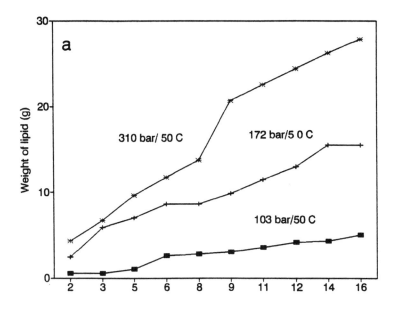

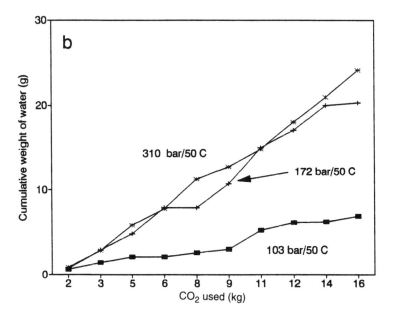

Fig. 11.4. Effect of varying pressure on the accumulated yield of (a) lipid and (b) water extracted with supercritical CO_2 at 50°C.

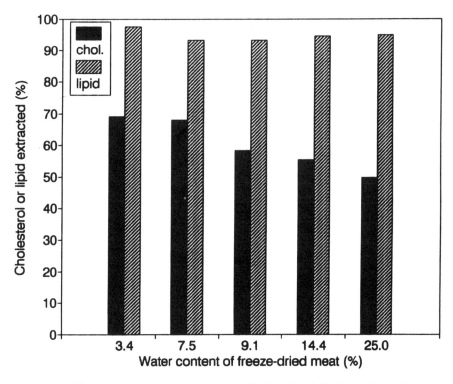

Fig. 11.5. Effect of moisture content on the yield of lipid and cholesterol from freeze-dried ground beef for supercritical CO_2 extraction at 345 bar and 40°C.

extraction conditions. Per kg CO_2 used, both the average percentages of extracted cholesterol and lipid were slightly reduced with increased temperature applied at 172 bar, while the opposite trend was found at 310 bar. Meanwhile, the average percentage of extracted lipid at 310 bar was about twice as much as that found at 172 bar. The ratio of extracted cholesterol to fat was greater than 1.0 at 172 bar, while at 310 bar it was 0.5. The response indicates that while increasing the extraction pressure to 310 bar increases the extraction of lipid, the extractability of cholesterol, however, remains the same. This is consistent with cholesterol solubility data reported for SC-CO_2 extraction of cholesterol from anhydrous butter fat (21), where cholesterol solubility was found to level off at 138 to 172 bar. Chao et al. (15) also discussed other related results concerning the variation of cholesterol content with some extraction conditions.

The chemical and physical characteristics of SC-CO_2-extracted fresh ground pork, as compared with precooked ground pork, was recently described by Clarke et al. (22), who reported that the fresh ground pork extracted with SC-CO_2 at 340 bar and 40°C did not differ in approximate composition from the precooked pork treated at 40°C for 4 hr. Comparison based on compression and penetration measurements between the precooked meat and the prepared precooked meat blended with 50 wt%

TABLE 11.2 Average Percentages of Cholesterol and Lipid of Ground Beef Extracted per kg CO_2 Used with Different Pressure and Temperatures

	Applied pressure (bar)/temperature (°C)			
	172/40	172/50	310/35	310/50
Per kg CO_2 used:				
(1) Average % of cholesterol extracted	2.78	2.09	1.88	2.40
(2) Average % of fat extracted	2.03	1.98	4.03	4.63
Ratio of % of cholesterol/fat extracted (=(1)/(2))	1.37	1.06	0.47	0.52

SC-CO_2-extracted meat showed that the two samples were not different from each other (22). Also, no significant structural differences between the treatments of pre-cooking and SC-CO_2 extraction were revealed in scanning electron micrographs.

Part B. Supercritical CO_2 Extraction of Animal Fat: Edible Beef Tallow

Supercritical CO_2 was also used to reduce the cholesterol content of EBT. For each run, 100 g EBT were extracted, the cumulative yields of the fractions for various operating conditions were plotted, and the results are shown in Fig. 11.6. The SC-CO_2 needed to extract all the EBT charged in the extractor at 345 bar and 241 bar were about 10 kg and 22 kg, respectively. However, as the lower pressure of 138 bar was used, about 22% of total BT was extracted after more than 20 kg CO_2 was used. The results indicated the high dependence of lipid solubility on the applied pressure and temperature conditions. Examining the effect of temperature on the extractability of lipid yield showed that increasing the extraction temperature from 40 to 50°C at 138 bar decreased the lipid yield from 22 to 5%. Such increase in lipid yield was not found as the higher extraction pressure of 241 bar was used. When the extraction was run at 345 bar, the lipid yield was generally higher at 50°C than at 40°C. The results indicated the presence of retrograde extraction pressure around 170 to 175 bar with respect to lipid solubility (19).

Extraction Yield of Cholesterol. The results of cholesterol concentration, expressed in the unit of mg cholesterol/100 g lipid, from the selected fractions for various experimental trials are shown in Fig. 11.7. As expected, the cholesterol concentration of those fractions extracted at 345 bar was the lowest of all, higher for those extracted at 241 bar, and the highest for those extracted at 138 bar. Therefore, extracting EBT with the lower pressure led to higher selectivity in extracting cholesterol than with the higher pressure. Comparing the results of the yields of extracted lipid and cholesterol from Figs. 11.6 and 11.7 shows that within the range of steady extraction rates for the

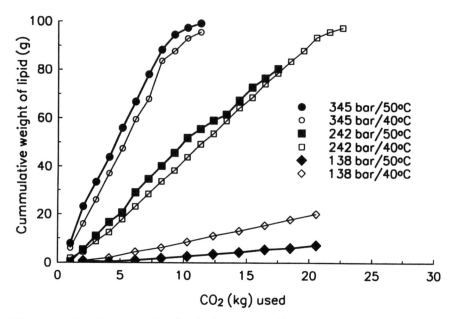

Fig. 11.6. Cumulative weight of beef tallow extracted by CO_2 varying the pressure and temperature from 138 to 345 bar and 40 to 50°C, respectively (29).

runs of 242 and 345 bar at 40°C (i.e., 5.0 and 11.5 g/kg CO2), the average cholesterol yields were 7 to 9 mg and 18 to 20 mg, respectively. The results indicated that the extraction yield of cholesterol increases with the lipid yield.

Effect of Ethanol on the Extractability of Lipid and Cholesterol.   As already mentioned, the lower density of CO_2 formed at lower pressure causes low solvent power and leads to low lipid yield. Studies of the enhancement of the extraction yield of lipid by SC-CO_2 at low extraction pressures by mixing SC-CO_2 with an organic cosolvent prior to extraction have been reported (5,23–24). In our study, pure ethanol (EtOH), listed as a GRAS (generally recognized as safe) food ingredient, was chosen as the cosolvent to exploit its effect on the extraction yields of both lipid and cholesterol of EBT as compared with the condition using SC-CO_2 alone. Figure 11.8 shows that the (CO_2 + EtOH) run mixing 5% (w/w) EtOH as cosolvent with SC-CO_2 at 138 bar/40°C resulted in 3.6 times (from 11 g to 39 g after 12 kg CO_2 was used for extraction) as much cumulative extraction yield of lipid as the run with CO_2 alone. Meanwhile, the cumulative extraction yield of cholesterol was increased to twice (46 g to 102 g) as much as the CO_2 run. Brunner and Peter (25) had similar results, in which the solubility of palm oil in SC-CO_2 with EtOH (10%) as an entrainer was enhanced about 20-fold at 200 bar, 70°C condition. However, Wong and Johnston (26) reported that SFE with EtOH as cosolvent did not induce a large solubility enhancement, due to the formation of a cholesterol-ethanol solid complex.

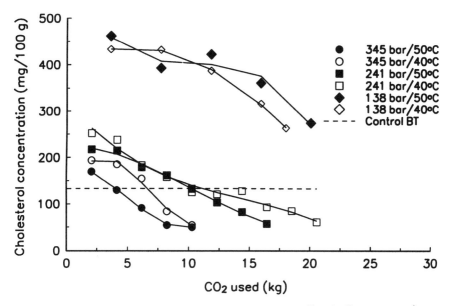

Fig. 11.7. Cholesterol concentration of selected fractions of beef tallow extracted by CO_2 with extraction pressure and temperature varied from 138 to 345 bar, and 40 to 50°C, respectively (29).

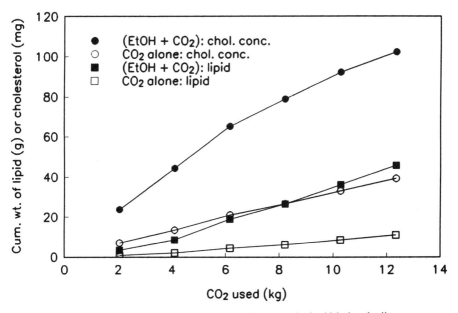

Fig. 11.8. The cumulative weight of lipid and cholesterol of edible beef tallow extracted by CO_2 using SC-CO_2 with and without ethanol at 138 bar and 40°C (30).

Figure 11.9 shows the variation of cholesterol concentration of each fraction for the runs at 138, 241, and 345 bar at 40°C. The rate of decreasing cholesterol concentration was much greater for the (CO_2 + EtOH) run than the run with CO_2 alone at 138 bar. The results indicate that the addition of EtOH to SC-CO_2 caused more extraction of lipid than of cholesterol. Meanwhile, the trend of high and constant cholesterol concentration along the extraction time for the CO_2-only run showed that lower pressure had higher selectivity for cholesterol extracted by SC-CO_2 than higher pressure.

Effect of Adsorbents on Cholesterol Removal from Edible Beef Tallow. Combining supercritical fluid extraction with adsorption affords a further process technique in fractionating solutes dissolved in the supercritical fluid phase. Prior studies utilizing adsorbents in the presence of supercritical fluid media have been reviewed by King (27) and King et al. (28). Shishikura et al. (5) indicated that the polarity of cholesterol is higher than that of triglycerides, whereas the molecular weight of cholesterol is almost equivalent to that of triglycerides with acyl carbon number 24. Following the principle of supercritical fluid chromatography and the similar approach described by Shishikura et al. (5), the tests were run by maintaining the ratio of EBT sample to adsorbents at 1:1 and the extraction pressure and temperature at 345 bar and 40°C. Three absorbents—activated alumina (AA), silica gel (SG), and molecular sieve (MS), a synthetic silico-aluminate (1:1) zeolite—with particle sizes

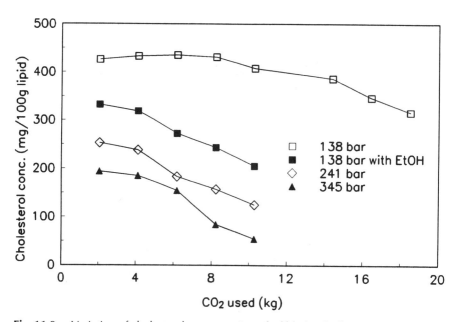

Fig. 11.9. Variation of cholesterol concentration of edible beef tallow extracted with and without ethanol at 138, 241, and 345 bar and 40°C.

ranging between 2.4 and 3.3 mm, were used. Figure 11.10 shows a continuous decrease in cholesterol concentration with extraction time for the fractions of the three runs. The fractions for the AA run had the lowest cholesterol concentration of all, indicating that AA absorbed more cholesterol than the other two adsorbents. Studies of compositional variation in the main fatty acids showed that with the aid of adsorbents the concentration of myristic acid (C14:0) was decreased while that of stearic acid (C18:0) was increased with increased extraction time. Roughly the same concentration levels were maintained for palmitic acid (C16:0) and palmitoleic acid (C16:1), but lower levels were found for oleic acid (C18:1), linoleic acid (C18:2), and linolenic acid (C18:3), as compared with the same fatty acids from the CO_2-alone run and the original EBT sample. Apparently, all the adsorbents used had a strong affinity for triglycerides that contained the fatty acid moieties of unsaturated C18 homologs.

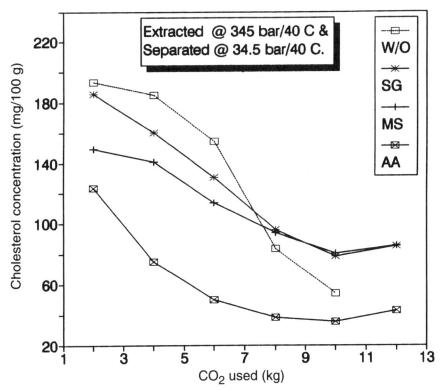

Fig. 11.10. Cholesterol concentration of different fractions of beef tallow extracted without adsorbent treatment (W/O) and using silica gel (SG), molecular sieve (MS), and activated alumina (AA) (30).

Use of Multiple Separators to Enhance Cholesterol Separation. The fractions in the runs just mentioned were collected in one separator. Since the solvent power of CO_2 depends upon the density, a stepwise reduction of separation pressure will alter the density of CO_2 so that the soluble EBT component–loaded CO_2 phase can be further separated into more fractions and collected in different separators. In this study, the separation pressure of the three separators were adjusted at 170 bar/40°C for the first separator (S1); 102 bar/40°C for the second separator (S2); and 34 bar/40°C for the third separator (S3). Table 11.3 shows the yields of lipid and cholesterol of the selected fractions collected from the three separators. In order to accumulate enough sample size, the fraction from S3 was collected for every 6 kg CO_2 used, while the fractions from S1 and S2 were collected at every 2 kg CO_2 used. The table showed that the fractions from S3 contained significantly higher cholesterol concentration than the fractions from S2 and S1. Furthermore, both fractions from S1 and S2 showed that while the extracted lipid weight of fractions was maintained fairly constant as extraction ensued, the cholesterol concentration was actually largely decreased. A material balance for this run showed that the fractions obtained from S3 held 17% and 41%, based on the total accountable yield, of lipid and cholesterol, respectively. Apparently, more cholesterol content was carried over by SC-CO_2 to S3 than to S1 and S2. The application of a multiple separators, therefore, allows one to obtain relatively small quantity of lipid fractions but with markedly high cholesterol concentration. Therefore, a larger separation of cholesterol from EBT with low loss of lipid was possible.

Conclusions

The findings of this study suggest that supercritical CO_2 extraction is feasible in reducing lipid and cholesterol from fresh ground meat. Higher extraction pressure is the

TABLE 11.3   Weights of Lipid and Cholesterol of Selected Beef Tallow Fractions (29)

CO_2 (kg)	Separator	Wt. fraction (g)	Cholesterol concentration (mg/100 g)	Wt. cholesterol (mg)
6	S1	10.2	128	13.0
	S2	4.1	171	6.9
	S3	7.1	433	30.7
12	S1	11.2	92	10.3
	S2	5.1	126	6.4
	S3	10.3	376	38.8
18	S1	11.9	49	5.8
	S2	4.4	75	3.3
	S3	10.7	272	29.2

Charged weight of BT:200 g. Fractions were extracted at 345 bar/40°C and fractionated by the order of 172 bar (S1), 103 bar (S2), and 34.5 bar (S3) at 40°C.

predominant factor for increasing the extractability of lipids. Although higher extraction pressure was the primary factor in increasing the total extraction yield of fat, the lower pressure condition could be used to concentrate cholesterol preferentially. Higher extraction pressures, up to 172 bar, also increased the extraction yield of water from the ground beef. However, when the pressure applied was 172 bar or over, the water yield was essentially independent of pressure, while the lipid yield increased. Decreasing the moisture content of meat prior to extraction can enhance both cholesterol and lipid.

Applying SC-CO₂ to extract cholesterol from EBT with different operating conditions showed that higher pressure increased the extraction yield of lipid than lower pressure. A retrograde behavior of lipid solubility was observed around 170–175 bar. The selectivity of extractable lipid components by SC-CO₂ could be further tuned by the effect of retrograde behavior on the lipid solubility.

High extraction pressure, coupled with multiple separation vessels for gradual reduction of the density of SC-CO₂ during separation, is needed to achieve the full extraction possibility while maintaining shorter fractionation time and smaller quantity of CO₂ used. Research is now being conducted to determine the effect of various adjustments of separation pressures on the separation of cholesterol from those lipid components responsible for the unique flavors of EBT. With the easy manipulation of extraction and separation conditions, small quantities of lipid fractions with markedly high cholesterol content could be fractionated from the original EBT. Therefore, a large reduction in cholesterol content from EBT with low loss of lipid can be accomplished.

Acknowledgment

The author gratefully thanks Dr. A.D. Clarke of the UMC Meat Laboratory and Mr. Y.A. Kwon for their expertise and assistance.

The mention of firm names or trade products does not imply that they are endorsed or recommended by the University of Missouri over other firms or similar products not mentioned.

References

1. Anonymous, in *Designing Foods: Animal Product Options in the Marketplace,* National Academy Press, Washington, DC, 1988, pp. 1–20.
2. Dotson, K., *INFORM 3:*152 (1992).
3. Rizvi, S.S.H., J.A. Daniels, A.L. Benado, and J.A. Zollweg, *Food Techn. 40(7):*57 (1986).
4. Mansoori, G.A., K. Schulz, and E.E. Martinelli, *Bio/Techn. 6(4):*393 (1988).
5. Shishikura, A., K. Fujimoto, T. Kaneda, K. Arai, and S. Saito, *Agric. Biol. Chem. 50(5):*1209 (1986).
6. Temelli, F., C.S. Chen, and R.J. Braddock, *Food Techn. 42(6):*145 (1988).
7. Nilsson, W.B., E.J. Gauglitz, J.K. Hudson, V.F. Stout, and J. Spinelli, *J. Amer. Oil Chem. Soc. 65(1):*109 (1988).
8. Schneider, F., and W. Sirtl, U.S. Patent 4,280,961 (1980).

9. Schwengers, D., U.S. Patent 3,939,281 (1976).

10. Snyder, J.M., J.P. Friedrich, and D.D. Christianson, *J. Amer. Oil Chem. Soc. 61(12):*1851 (1984).

11. Hardardottir, I., and J. Kinsella, *J. Food Sci. 53(6):*1656 (1988).

12. De Fillippi, R.P., and J.M. Moses, *Biotechn. Bioeng. Symp. 12:*205 (1982).

13. Bradley, Jr., R.L., *J. Dairy Sci. 71 [suppl. 1]:*72 (1988).

14. Anonymous, *Food Eng. 61(2):*83 (1988).

15. Chao, R.R., S.J. Mulvaney, M.E. Bailey, and L.N. Fernando, *J. Food Sci. 56(1):*183 (1991).

16. *Official Methods of Analysis,* 14th edn., Association of Official Analytical Chemists, Washington, DC, 1984.

17. Zou, M., Z.R. Yu, P. Kashulines, S.S.H. Rizvi, and J.A. Zollweg, *J. Supercrit. Fluids 3:*23 (1990).

18. Won, K.W., *Fluid Phase Equil. 10:*191 (1983).

19. Stahl, E., K.-W. Quirin, and D. Gerard, *Dense Gases for Extraction and Refining,* Springer-Verlag, Berlin, pp. 1–29 (1988).

20. King, J.W., J.H. Johnson, and J.P. Friedrich, *J. Agric. Food Chem. 37:*951 (1989).

21. Anonymous, *Food Eng. 61(2):*83 (1989).

22. Clarke, A.D., T. Parinyasiri, N.C. Flores, R.R. Chao, and M.E. Bailey, in *Institute of Food Technologists Annual Meeting,* Dallas, TX, June 1–5, 1991, edited by Institute of Food Technologists, Chicago, IL, Abstract No. 439.

23. Panzer, F., S.R.M. Ellis, and T.R. Bott, *Inst. Chem. Eng. Symp. Series No. 54:*165 (1978).

24. Peter, S., and G. Brunner, *Angew. Chem. Ind. Ed. Engl. 17:*746 (1978).

25. Brunner, G., and S. Peter, *Sep. Sci. Tech. 17:*199 (1982).

26. Wong, J.M., and K.P. Johnston, *Biotech. Prog. 2(1):*29 (1986).

27. King, J.W., in *Supercritical Fluids: Chemical and Engineering Principles and Applications,* edited by T.G. Squires and M.E. Paulaitis, ACS Symposium Series No. 329; American Chemical Society, Washington, DC, 1987, pp. 150–171.

28. King, J.W., R.L. Eissler, J.P. Friedrich, in *Supercritical Fluid Extraction and Chromatography: Techniques and Applications,* edited by B.A. Charpenties and M.R. Sevenants, ACS Symposium Series No. 366, American Chemical Society; Washington, DC, 1988, pp. 63–88.

29. Chao, R.R., S.J. Mulveny, and H. Huang, *J. Am. Oil Chem. Soc. 70(2):*139 (1993).

30. Bailey, M.E., R.R. Chao, A.D. Clarke, K.-W. Um, and K.O. Gerhardt, in *Food Flavor and Safety: Molecular Analysis and Design,* edited by A.M. Spanier, H. Okai, and M. Tamura, ACS Symposium Series No. 528; American Chemical Society, Washington, DC, 1993, pp. 117–137.

Chapter 12

Supercritical Fluid Extraction of Algae

M.O. Balaban, S. O'Keefe, and J.T. Polak

Food Science and Human Nutrition Dept., University of Florida, Gainesville, FL 32611

Importance of Algae

Algae are the oldest and most primitive members of the plant kingdom, and are ubiquitous in aquatic or subaerial environments. They account for almost half of the world's organic matter (1). Over 150,000 species are estimated to exist, and between 20 and 30,000 phytoplankton species have been identified (2,3). Algae are the primary producers of organic matter in aquatic environments; they fix carbon dioxide and generate oxygen by photosynthetic activity (4). They form the primary source of food and energy for most of the animal life in aquatic environments. Ackman et al. (5) report that the brine shrimp *Artemia salina* are unable to synthesize linoleic (C18:2ω-6) and linolenic (C18:3ω-3) acids and that the possible source is from dietary intake of microalgae.

Research on commercial utilization of algae has focused on biomass, fuels, hydrocolloids, proteins, and lipids including the nutritionally important fatty acids γ-linolenic (GLA, 18:3ω-6), eicosapentaenoic (EPA, 20:5ω-3) and docosahexaenoic (DHA, 22:6ω-3), as well as other products including β-carotene, xanthophylls, chlorophyllins, ferredoxin, and the phycobiliproteins. Hence, potentially many other specialty chemicals can be extracted from algae.

Classification of Algae

Algae can be separated into two major groups, macroalgae and microalgae, depending on their level of organization. Macroalgae (seaweeds) are more complex and can form colonies with differentiation, resembling higher plants. Microalgae generally exist as single cells, although some species form small colonies. Algae are generally classified using the various pigments found in the cells as the basis of their separation. In 1950 algae were separated into seven divisions, or phyla. The four larger phyla are the Cyanophyta (blue-green algae—photosynthetic procaryotes with many similarities to photosynthetic bacteria, having no nucleus, little intracellular differentiation, and, in some cases, the ability to fix nitrogen), the Chlorophyta (green algae—common in fresh water which can exist in multicellular colonies or single-celled forms), the Phaeophyta (brown algae), and the Rhodophyta (red algae—mainly multicellular marine organisms, e.g., nori and agar). The smaller phyla are the Euglenophyta, the Chrysophyta (including the yellow-green and golden-brown algae and the diatoms; noted for high lipid accumulation), and the Pyrrophyta. This classification is registered with the International Code of Botanical Nomenclature (4).

Algae Culture

Cultivation of macroalgae in Asia is an ancient practice. By the 17th century *Porphyra* was being cultivated in Japan (2). A number of factors contribute to the interest in algae cultivation. High photosynthetic activity and resilience to environmental factors result in versatile growth conditions. Because the chemical composition of algae is affected by growth media and environment, manipulation of culture conditions is a means of modifying algae products to suit specific requirements (6).

Microalgae culture can be divided into four major systems (7):

1. Intensive pond
2. Extensive pond
3. Closed reactor
4. Fermentor

The traditional culture of algae has been in open ponds. The problem of disease limits culture to species with unusual tolerances, such as *Dunaliella* and *Spirulina*, which grow well in high-pH/high-salt environments, and *Chlorella*, which at neutral pH will outgrow many other species. Perhaps a more serious problem is sanitation, especially in developing countries (8). Vertical tubular reactors were described by Miyamoto et al. (9); these solve the problem of oxygen buildup, which if left unchecked, leads to growth inhibition in more common reactor systems.

Dunaliella can be grown in high-salt open ponds in essentially monoculture conditions. This, in concert with its high β-carotene levels, has stimulated commercial enterprises in the United States, Australia, and Israel. The "natural" β-carotene produced from this species has a definite market niche, different from that for the commercial β-carotene produced by synthetic means.

Direct Food Uses of Algae

It is unlikely that algae will be a large source of bulk animal feeds in the foreseeable future, because of the high cost of growth and harvesting compared to traditional feeds (10). However, algae are sometimes used as human foods or nutritional supplements. The estimated annual market in China is over $300 million, whereas macroalgae such as *Porphyra*, which is processed and dried to form *nori* (the edible wrapping of sushi), has an even greater value, estimated at $700 million to $1.5 billion (7,2). The most successful uses of microalgae by humans have been as dietary supplements, especially for protein (11). There is a long history of human use of *Spirulina* in Mexico and Chad as well as of *Nostoc* in South America and China (1). The use of *Spirulina* to produce a nori-type product has been examined and appears very promising. Traditional nori is processed from chopped macroalgae that are dried in sheet form. Sanitation and contamination are a major problem with the traditional product. A *Spirulina*–agar formulation "nori" was preferred over traditional nori. The processing from *Spirulina* in culture to the final product was facilitated by a novel concentration technology. One of the major costs in microalgae utilization is the concentration of the dilute algae. Rakow et al. (12) used a flow concentration as feed directly for nori production.

High-Value Algae Products

Algae can be fractionated into as many products as are economically justifiable. A broad product base eliminates the problems encountered with one-product processing. However, processing strategies must be thought out carefully. Hohlberg et al. (13) have pointed out that the number of commercially interesting products available from *Spirulina* include phycocyanin, β-carotene, astaxanthin, chlorophyll a, GLA, and ferredoxin. Agricultural use and utilization as a source of single-cell protein for human consumption await favorable economic trends.

Hydrocolloids

By far the most important products from macroalgae in Western societies are the polysaccharides algin, carrageenan, and agar. The importance of agar in production of penicillin during World War II stimulated macroalgae research in Britain. Glicksman (14) has pointed out that 30, 40, and 500 million kg of alginates, carrageenan, and agar are produced per year.

Proteins

The protein levels in different algae range from 28 to 71% dry basis (15,8). Protein values derived from Kjeldahl nitrogen determination must be corrected for nonprotein nitrogen and nucleic acids to eliminate overestimation. The quality of algae proteins appears to be quite good, with limiting amounts of sulfur amino acids. The protein quality is affected by processing but generally has been reported to be 80% that of casein. However, nucleic acids, which can be present at 1 to 6% of the dry weight, are a nutritional problem and may limit algae protein to 20 g daily unless they are removed (15).

Spirulina and *Chlorella* have attracted interest because of their high levels of good-quality protein. *Spirulina* has been harvested from Lake Chad in Africa and in Central America for centuries. Although some authors suggest that acceptance of algae protein would not be a problem (11), others have disagreed (15), basing their opinion on acceptability studies in which the green color of chlorophylls in algae was not masked or removed. The traditional usages of algae in Central America and Africa suggest that use as a food supplement is acceptable, especially when "health-giving" properties are implied. In the United States, *Spirulina* has a specialty market as a nutritional supplement, encapsulated as dried powder, directly or mixed, with other components. Its total market for health food supplements in the United States was estimated at 100 tons (1).

Lipids and Pigments

The world production of hydrocarbons by marine algae has been estimated to be from 10 to 49 million tons per year (16). Levels of hydrocarbons in algae are generally low. Ranges of reported data for the different algae families are (dry weight basis)

Blue-green	0.02–0.44
Red	0.003–0.073
Brown	0.0005–0.66
Diatoms	0.004–0.38

The hydrocarbon profiles are often characteristic and may be used to facilitate taxonomic classification. In spite of the enormous hydrocarbon production, the low levels in algae make exploitation for hydrocarbon alone unlikely.

The lipid products found in algae have some potential as sources of diesel, gasoline, or other fuel replacement. The levels of total lipid found in some algae grown in N-limiting conditions can reach 60% of the dry weight. Possible fuels obtained or produced from algae include methane, ethanol, triglycerides, fatty acid esters, and gasoline (3). The direct use of triglycerides as diesel replacers would require modifications to diesel engines because of differences in viscosity. Production of ester fuels via transesterification of the triglycerides would relieve many of the problems. The factor that appears to limit the use of algae as fuel sources is primarily economics. Ben-Amotz et al. (17) reported that the low levels of hydrocarbons produced by microalgae make their use as liquid fuels untenable but indicated that high levels of lipid have economic potential. Photosynthetic activity, harvestability, disease resistance, nutrient tolerance, and culture conditions must be improved before fuels could be economically produced from algae.

Tables 12.1 through 12.3 summarize fatty acid compositions reported for three microalgae families. It can be seen that extremely high levels of ω-3 fatty acids are present in many of the *Cryptophyceae* species. It has been theorized that the ultimate source of the ω-3 fatty acids in marine organisms such as fish is algae (5). There is a sizable market in ω-3-fatty-acid oils and concentrates in many countries. The concentrates are invariably derived from low-cost fish oils, algae-derived ω-3 oils may be able to compete with fish oil concentrates. The potential for specialized triglycerides has not been thoroughly examined. Algae may have potential in production of specific triglycerides, either for cocoa butter replacement or for nutritional uses.

The fatty acids that have attracted interest include the ω-3 fatty acids EPA and DHA as well as the ω-6 GLA. Some algae species have been reported to have extremely high ω-3 fatty acid levels. One of the obvious problems in interpretation arises when examining data reported for *Amphidinium carteri* (Table 12.2). Two research groups have reported data that varied considerably for the same species. DHA was not reported by one of the two literature reports but was 24% in the other. Much of the fatty acid compositional data was derived from older techniques of packed-column gas-liquid chromatography. The resolution obtained on these columns was often poor,

TABLE 12.1 Summarized Percent Fatty Acid Composition of Cryptophyceae (36)

Species	SAT	MONO	18:2ω-6	18:3ω-3	18:4ω-3	20:5ω-3	22:6ω-3	Σω-6	Σω-3
Chilomonas									
paramecium A	37	9	12	27	—	6	3	12	36
paramecium B	26	12	8	40	—	—	—	8	40
Chroomonas sp.	19	7	3	23	23	14	6	5	66
Cryptomonas sp.	10	7	—	7	44	16	10	—	77
ovata	11	6	—	17	34	12	7	7	70
appendicu-	18	25	4	12	13	10	—	4	44
lata									
malculata	20	28	—	6	16	17	—	—	60
Hemiselmis									
viresens	31	12	3	22	16	7	2	5	47
rufescens	21	26	1	7	17	8	—	4	52
brunescens	14	23	0	8	31	14	—	—	72
Rhodomonas									
lens	32	15	2	16	13	13	5	3	47
Unknown	17	4	11	23	16	14	1	12	55

A,B represent different sources of data in Ref. 36.

TABLE 12.2 Summarized Fatty Acid Composition of Dinoflagellates (Dinophyceae) (36)

Species	SAT	MONO	18:2ω-6	18:3ω-3	18:4ω-3	20:5ω-3	22:6ω-3	Σω-6	Σω-3
Amphidinium									
carteri A	16	3	1	3	19	20	24	1	66
carteri B	45	8	—	—	10	7	—	—	17
Exuviella sp.	39	7	9	2	8	4	12	9	26
Gymnodinium									
nelsoni	41	21	1	1	2	5	20	1	28
cohni	40	15	—	—	—	—	25	—	25
Prorocentrum									
micans	24	29	—	6	—	12	—	2	18
Glenodinium sp.	32	29	5	6	23	2	19	5	50
Gonyaulax									
catenella	47	10	3	6	14	1	12	4	33
tamarensis	42	9	4	5	12	3	18	4	38
polyedra	38	4	2	3	14	14	23	2	54
Peridinium									
triquentum	35	7	9	3	13	2	19	9	37
trochoideum A	45	17	3	2	9	1	15	3	27
trochoideum B	53	11	—	3	7	13	—	—	23

A, B represent different sources of data in Ref. 36.

TABLE 12.3 Fatty Acid Composition of Haptophyceae (36)

Species	SAT	MONO	18:2ω-6	18:3ω-3	18:4ω-3	20:5ω-3	22:6ω-3	Σω-6	Σω-3[a]
Dicrateria									
inornata	17	25	5	13	20	8	—	6	45
Emiliania									
huxleyi	24	38	2	1	1	17	—	2	20
carterae A	18	24	3	—	2	20	—	4	26
carterae B	30	50	4	—	—	9	16	4	25[b]
elongata	17	23	2	1	1	28	—	2	32
Imantonia									
rotunda	18	19	1	—	—	31	—	4	36
Isochrysis									
galbana A	17	20	11	14	17	3	—	13	34
galbana B	30	31	8	—	—	3	11	10	14[c]
carterae	24	11	10	8	24	4	9	11	45
Pavlova									
lutheri A	20	26	2	—	1	19	—	5	24
lutheri B	26	27	1	—	4	16	13	2	33
Prymnesium									
pavum	32	37	18	11	2	4	—	18	18

[a]Does not include C_{16}ω-3 fatty acids.
[b]No 18:3ω-3 or 18:4ω-3 reported, yet 22:6ω-3 16%.
[c]No 18:3ω-3 or 18:4ω-3 reported, yet 22:6ω-3 11%.

and extended retention times for the long-chain polyunsaturated fatty acids contributed to inaccuracies. Additional problems of accuracy arose when using stainless steel columns. Identifications were often made only tentatively and, coelution of different fatty acids on less polar columns was often a problem (18). With these considerations in mind, the fatty acid compositions must be viewed with caution. A comprehensive comparison with standardized culturing conditions and harvesting at a defined growth stage has not been done.

The diatoms *Navicula pelliculosa* and *Skeletonema costatum* were grown in normal and stressed conditions in lab scale (19). Stress caused a greater accumulation of lipids, but at a slower rate. Under stress, most of the lipids were triacylglycerols, with EPA and DHA as the major fatty acids. The authors suggested these diatoms as an alternative to fish oils for ω-3 fatty acids.

Cohen et al. (20) have reported GLA levels in several *Spirulina* species. GLA ranged from 8 to 31.7% of the fatty acids, corresponding to 0.3 to 1.4% of the biomass, with averages of 23.5 and 1%, respectively, in 18 species. The temperature of cultivation played a significant role in total lipid and GLA levels with lower temperatures yielding higher GLA but lower total lipid. The net effect was an increase in GLA at lower temperatures.

Hohlberg et al. (13) have reported a simple scheme for fractionation of *Spirulina* into three products: GLA, phycocyanin, and protein concentrate. The total lipid and phycocyanin in the *Spirulina* they obtained from open-pond culture were 3.3 and 6.9% wet basis respectively.

The factors that must be considered before placing emphasis on reported fatty acid compositions include culture conditions such as light intensity, nutrients, culture density, age of culture, and strain and purity of cultures; possible misidentification of fatty acids, especially with older, packed GC columns; method of FAME analysis; etc. The importance of culture age is illustrated in an example reported by Ackman et al. (5) for *Skeletonema costatum* in Fig. 12.1. The levels of eicosapentaenoic and docosahexaenoic acids decreased markedly during the 12 days of culture, with concomitant increases in oleic acid. The authors proposed that light, culture density, or nutrient limitation was responsible for this effect. This time-dependent variability in fatty acid composition in *S. costatum* was also reported by the present authors (21).

GLA is currently produced from borage, blackcurrant, or evening primrose oils. There is considerable medical interest in this fatty acid. There is some evidence that some diseases respond to GLA treatment.

Phycobiliproteins

The phycobiliproteins include phycocyanins, phycoerythrin, and allophycocyanins, depending upon their visible absorption spectra (13). They are deeply colored, fluores-

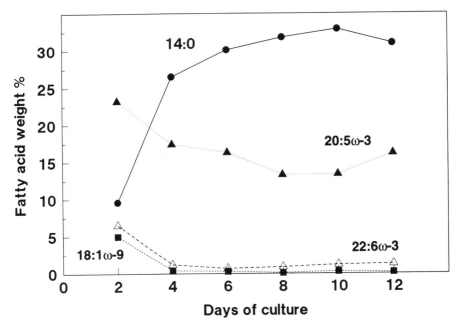

Fig. 12.1. Change in fatty acid composition of with culture time for *Skeletonema costatum.* Source: Ackman R.G., et al., *J. Fish. Res. Bd. Can. 21*:747 (1964).

cent, complex molecules containing linear tetrapyrroles, which function in light reception. They occur as large-molecular-weight aggregates called *phycobilisomes*, which are attached to photosynthetic thylakoid membranes.

Phycocyanin is a blue, water-soluble protein-pigment complex. It is used both as a natural food colorant in Japan and in biochemical tracing experiments. Phycobiliproteins can be used in fluorescence microscopy, in flow cytometry, and as fluorescent labels in tracing experiments. The estimated price of reagent-grade phycocyanin has been reported as $1000 to 5000/kg (13). The phycocyanin isolation scheme proposed by Hohlberg et al. (13) included aqueous salt extraction, centrifugation, ammonium sulfate precipitation, dialysis, gel filtration chromatography, and finally freeze-drying. The high value of the product justifies this complicated scheme, but in improved extraction would be of great interest.

Carotenoids

Most *Dunaliella* species have been shown to contain large amounts of β-carotene (22). *Dunaliella* is salt-tolerant and can grow at salt concentrations ranging from 0.5 to 5.5M salt while producing large amounts of β-carotene. Since high salt concentrations deter growth of other microorganisms, and since vast surfaces of salt beds are available and unused in some areas of the world, *Dunaliella* has attracted a great deal of attention for open-pond culture. The species that has attracted the most attention, due to its very high β-carotene production, is *Dunaliella salina*. Up to 8% of the dry weight of this species can be β-carotene (23), but much higher values may be obtained by species selection and mutagenesis (1). The β-carotene content of *Dunaliella salina* can be 30 times as great as in carrots on a dry weight basis. Levels of β-carotene in a continuous culture have been reported to be sustainable at 3 mg β-carotene/L water (24) and represent up to 20% of the dry weight in some cultures (1).

The most pressing difficulty in the use of *Dunaliella* in β-carotene production are technical limitations that are due to the low levels (2–5 mg/L) that are generally found. Recovery is difficult and expensive, but new processing technology appears promising (10). Cellular adsorption onto siliconized glass beads can reach 80% of cellular mass as the ionic strength is increased.

β-carotene has been reported to be valued between $500–1000/kg (7). The production of β-carotene from *Dunaliella* with current economics is not competitive with chemical synthesis. However, the market interest in "natural" compounds and in the antineoplastic activity of β-carotene suggest that the natural product will be successful in its smaller and higher-valued "natural products" market.

Pharmaceuticals

The high value of pharmaceuticals and the difficulty in synthesis of complex stereo-chemicals suggest that this will be one of the most exciting areas for algae utilization in the coming years.

Antineoplastic agents. Hydroquinone, which has anticancer and antineoplastic activity, has been extracted from *Bifurcaria* (2). *Bonnemaisoniaceae* species have been found to contain halogenated ketones. A complex sulfonated polysaccharide has been found in *Dumontiaceae* that has been reported to have antiviral activity against *Herpes simplex.* Spatol, an anticancer drug, has been recovered from *Spatoglossum* species, whereas another anticancer drug, strypoldinone, has been extracted from *Stypopodium.*

Antimicrobial extracts (Fungicides). The antibiotic activity of lipid-soluble extracts from a number of different Caribbean marine algae has been examined by Ballantine et al. (25). Of the 103 species studied, the extract of only one species, *Asparagopsis taxiformis,* inhibited all five test microorganisms: *Bacillus subtilis, Pseudomonas aeruginosa, Escherichia coli, Staphylococcus aureus,* and *Candida albicans.* Species activity variability was found and attributed to differences in reproductive state, locale, or season.

Mayer et al. (26) have reported that fucoidan is the likely compound responsible for antiviral activity in *Macrocystis pyrifera.*

Supercritical Fluid Applications

Cell Disruption

An important unit operation in the recovery of valuable chemicals from microalgae and other unicellular sources is cell disruption (27). Cell walls can be disrupted and chemicals recovered by supercritical (SC) or near-critical gases such as nitrous oxide, argon, carbon dioxide, or nitrogen by explosive decompression and permeability enhancement. This is possible because of the large specific volume changes with relatively small changes in pressure near the critical point. Continuous operation at 35 MPa is possible. In experiments with *E. coli, S. cerevisiae,* and *B. subtilis,* nitrous oxide was the best fluid, followed by CO_2 and nitrogen (28). High critical fluid densities and low polarities were postulated to cause solubilization of lipids and hydrophobic compounds in cell walls and cytoplasmic membranes.

Extraction

Lipids. Oils present in microalgae are of specific interest because of the rich content of polyunsaturated fatty acids. SC-CO_2 extraction of lipids and pigments from *Scenedesmus obliquus,* with and without ethanol as entrainer, was tested by Choi, et al. (6). The algae were grown in densities of 0.55 mg algae dry wt/mL solution at 32°C and pH = 7. KNO_3 was used as the N source, and 2% atmospheric CO_2 as the C source. Extraction at 40°C and 37.8 MPa was carried out at 3 L gas (at STP)/min. The extract had a light green color. The authors mention that chlorophyll a was extracted in their studies, but they point out results of other studies in which chlorophyll was insoluble in SC-CO_2 (29,30). It is possible that entrained algae particles may have caused this discrepancy. Fig. 12.2 represents the authors' results of analyses of SC-

CO_2 extracted lipids from the protein concentrate. As expected, there were trace amounts of phospholipids. When ethanol was added to CO_2 as an entrainer at a 15% level (Fig. 12.3), there was a significant increase in the amount of phospholipids in the extract, and the color was dark green. The authors give yields (based on freeze-dried protein concentrate) of 42.5% for total extracted lipids and pigments.

SC-CO_2 was also used in the extraction of secondary metabolites from the Mediterranean brown algae *Dilophus liqulatus* by Subra et al. (31). About 2.7 g algae was mixed with 50-micron glass balls and placed in the extraction cell. Experimental temperature and pressure were achieved, and the system was isolated and equilibrated for 30 min. The CO_2 was then recirculated for 30 min. Fresh CO_2 was then introduced to flush the "saturated" gas from the cell. The extracts were trapped in hexane after compression and then analyzed and used in bioassays. Pressures and temperatures that were utilized were 8, 10, 15, 20, and 25 MPa and 35, 45, and 55°C. The extracts for CO_2 densities below 0.74 g/cm³ were colorless. At 15 MPa and 45°C the color of the extract was orange. For densities up to 0.8 g/cm³ the extracts were yellow; for densities higher than that, the color was green. The authors attributed the latter to chlorophyllian pigments. Extraction at 28 MPa and 45°C gave the highest yield (0.89), and were the most effective in inhibiting the growth of the fungus *Aspergillus fumigatus* in a disc-diffusion analysis. HPLC and TLC profiles of various extract fractions were

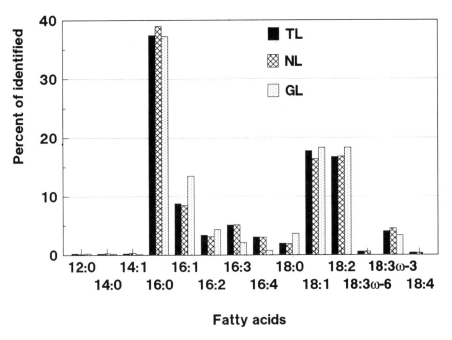

Fig. 12.2. Fatty acid composition of total lipids (TL), neutral lipids (NL), and glycolipids (GL) in SC-CO_2 extracts of *Scenedesmus obliquus* at 40°C and 37.8 MPa. *Source:* Choi, K.J., et al., *Biotechn. 1(2):263* (1987).

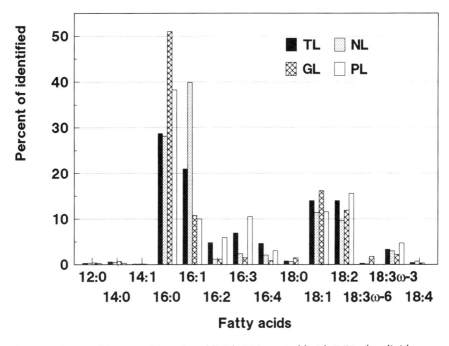

Fig. 12.3. Fatty acid composition of total lipids (TL), neutral lipids (NL), glycolipids (GL), and phospholipids (PL) in SC-CO$_2$+ethanol extracts of *Scenedesmus obliquus* at 40°C and 37.8 MPa. *Source:* Choi, K.J., et al., *Biotechn. 1(2):263* (1987).

given. The authors also tried ethyl acetate as an entrainer added to CO$_2$ but failed to notice any change in extraction.

There is also interest in the extraction of lipids from fungi. Sakiki et al. (32) grew the fungus *Mortierella ramanniana* under controlled conditions, harvested it, and treated it at 120°C to inactivate lipases. The cells were crushed in a ball mill for 3 hrs with twofold ethanol and were dried to a 6% water level. The authors gave the lipid composition of the cells by saturation and by fatty acids. The extraction fluids used were CO$_2$, CO, N$_2$O, CHF$_3$, and SF$_6$. A 20-g sample was placed in the extractor, and the temperature and pressure were maintained at 60°C and 24.5 MPa while equilibration took place for 1 hr. Then the system was allowed to operate under flow conditions, and 1-g fractions were collected. Lipid analysis was performed by thin-layer chromatography (TLC) and GC. The solubilities of fungal oil in the extraction fluids (% wt) were given as

N$_2$O   2.3
CO$_2$ 0.48
CHF$_3$ 0.0099
SF$_6$ 0.0012

TABLE 12.4 Fatty Acid Compositions of *Skeletonema costatum* and of *Ochromonas danica*

| | Skeletonema costatum | | | | | | | | Ochromonas danica | |
| | Batch 1 | | Batch 2 | | Batch 3 | | Batch 4 | | Literature | Exp. |
	(mg/g)[a]	%	(mg/g)[a]	%	(mg/g)[a]	%	(mg/g)[a]	%	%[b]	%[a]
C14:0	4.53	12.9	10.57	21.7	3.72	15.4	7.27	16.8	15	18.06
C14:1	0.10	0.3	0.32	0.7	0.25	1.0	0.16	0.4		0.31
C16:0	2.67	7.6	4.10	8.4	1.24	5.1	2.28	5.3	10	10.99
C16:1	4.97	14.1	9.78	20.1	3.48	14.4	7.24	16.8		1.50
C16:2ω-7	0.70	2.0	0.96	2.0	0.51	2.1	0.68	1.6		
C16:2ω-4	2.16	6.1	3.47	7.1	1.25	5.1	2.31	5.6		
C16:3	3.87	11.0	3.81	7.8	3.38	14.0	5.90	13.7		
C16:4	2.64	7.5	4.06	8.3	3.50	14.5	5.68	13.1		
C18:0	0.34	1.0	0.27	0.6	0.14	0.1	0.04	0.1		
C18:1ω-9	1.62	4.6	0.38	0.8	0.17	0.7	0.16	0.4	3	3.23
C18:1ω-7	0.78	2.2	1.28	2.6	0.45	1.9	1.14	2.6	8	6.42
C18:2	1.31	3.7	1.06	2.2	0.71	2.9	0.72	1.7		
C18:3ω-6	1.50	4.3	1.90	3.9	1.15	4.7	0.52	1.2	16	17.55
C18:3ω-4	0.67	1.9	0.87	1.8	0.57	2.4	0.96	2.2	2	
C18:4	0.24	0.7	0.31	0.6	0.19	0.8	1.04	2.4	10	14.77
C20:0									1	1.15
C20:2										0.25
C20:3										4.83
C20:4	0.03	0.1	0.35	0.7	0.05	0.2	0.05	0.1	5	11.50
C20:5ω-3	5.50	15.6	3.55	7.3	2.44	10.1	5.49	12.7	11	3.83
C22:0										0.95
C22:1										4.64
C22:5									2-4	
C22:6ω-3	1.53	4.4	1.62	3.3	1.00	4.1	1.54	3.6		
	35.16	100.0	48.66	100.0	24.20	100.0	43.18	100.0	85	100.0
Original lipid (mg/g)	75.50		115.00		56.30		103.00			

[a] Freeze-dried Skeletonema costatum.
[b] Haines et al. (37).

The authors also studied the effects of temperature and pressure on oil solubility in CO_2 and N_2O. The logarithm of oil solubility vs. fluid density was linearly increasing below a density of 0.8 g/cm^3, whereas at higher density the slope of the curves decreased. The authors observed that the components of low molecular weight were extracted more easily, and they were removed early in the extraction, compared to triglycerides.

In another study by Cygnarowicz-Provost et al. (33), the filamentous fungi *Saprolegnia parasitica* was extracted with SC-CO_2, CO_2-ethanol (10%), and CO_2-methanol (10%), between 40 and 80°C and 31 to 38 MPa. Ninety-two percent lipid recovery was achieved with 10% ethanol at 60°C and 31 MPa. There was a very significant increase in both oil and EPA yield when an entrainer was added. However, lower selectivity towards EPA was observed when cosolvents were used, despite the higher overall oil yield. Some fractionation was also observed with pure CO_2: triglycerides rich in shorter chain fatty acids were extracted first (33).

Skeletonema costatum, a salt-water diatom of the class Bacillariophyceae, has been reported to produce relatively large amounts of lipids of interest, especially EPA. *Ochromonas danica*, a freshwater phytoflagellate, was also studied because it has a high lipid content. Table 12.4 displays the fatty acid content of different batches of

TABLE 12.5 Supercritical CO_2 Extraction of Freeze-Dried Skeletonema costatum at 17 MPa

Fatty acid	Original mg/g[a]	%	SC extract mg/g[a]	%	Residue mg/g[a]	%
C14:1	0.10	0.28	0.05	1.66	0.11	0.53
C16:0	2.66	7.57	0.40	13.29	1.43	6.86
C16:1	4.95	14.08	0.72	23.92	3.02	14.49
C16:2ω-7	0.69	1.96	0.03	1.00	0.37	1.78
C16:2ω-4	2.17	6.17	0.08	2.66	1.32	6.33
C16:3	3.85	10.95	0.03	1.00	2.04	9.79
C16:4	2.64	7.51	0.01	0.33	1.21	5.81
C18:0	0.34	0.97	0.02	0.66	0.15	0.72
C18:1ω-9	1.63	4.64	0.13	4.32	0.33	1.58
C18:1ω-7	0.78	2.22	0.04	1.33	0.44	2.11
C18:2	1.31	3.73	0.03	1.00	0.62	2.98
C18:3ω-6	1.51	4.29	0.36	11.96	1.65	7.92
C18:3ω-4	0.69	1.96	0.00	0.00	0.38	1.82
C18:4	0.26	0.74	0.01	0.33	0.12	0.58
C20:4	0.03	0.09	0.07	2.33	0.10	0.48
C20:5	5.50	15.64	0.05	1.66	3.81	18.28
C22:6	1.53	4.35	0.00	0.00	0.96	4.61
Total	35.16	100.00	3.01	100.00	20.84	100.00

[a]Freeze-dried algae.

Skeletonema and *Ochromonas*. These parameters necessary to evaluate the technical feasibility of SC-CO_2 extraction of microalgal lipids were studied (34). Different extraction pressures (17, 24, and 31 MPa) were used at 40°C. Supercritically extracted oils and remaining lipids were analyzed by gas and thin layer chromatography. Effects of phospholipids present in *Skeletonema* on the extraction yield, and the difference in fatty acid profiles from 2 and 3-wk-old cultures were also studied. Solubility data for *Skeletonema* indicated no difference in extraction yields between extractions at 24 and at 31 MPa. Tables 12.5 (page 259) through 12.7 show fatty acid composition of original, extracted, and remaining lipids in *Skeletonema* at 17, 24 and 31 MPa, respectively. Lower pressures preferentially extracted fatty acids of shorter chain length. At higher pressures, long-chain fatty acids such as EPA were extracted much better. Enzymatic treatment of the samples with phospholipase C increased the yield of oils extracted with supercritical CO_2 by reducing the phospholipid levels and increasing neutral fractions. Three-wk-old cultures exhibited higher amounts of total lipids, but 2-wk-old cultures contained a greater variety of polyunsaturated fatty acids (34). *Ochromonas* had a higher oil level than *Skeletonema,* but the relative fraction of polyunsaturated fatty acids such as EPA was lower. The oil yield from *Ochromonas* increased with

TABLE 12.6 Supercritical CO_2 Extraction of Freeze-Dried *Skeletonema costatum* at 24 MPa

Fatty acid	Original mg/g[a]	%	SC extract mg/g[a]	%	Residue mg/g[a]	%
C14:0	10.57	21.72	3.91	24.64	3.95	19.06
C14:1	0.32	0.66	0.08	0.50	0.17	0.82
C16:0	4.10	8.43	1.21	7.62	2.00	9.65
C16:1	9.78	20.10	3.19	20.10	4.12	19.88
C16:2ω-7	0.96	1.97	0.33	2.08	0.40	1.93
C16:2ω-4	3.47	7.13	1.10	6.93	1.44	6.95
C16:3	3.81	7.83	1.31	8.25	1.48	7.14
C16:4	4.06	8.34	1.44	9.07	1.44	6.95
C18:0	0.27	0.55	0.04	0.25	0.21	1.01
C18:1ω-9	0.38	0.78	0.12	0.76	0.37	1.79
C18:1ω-7	1.28	2.63	0.19	1.20	0.91	4.39
C18:2	1.06	2.18	0.24	1.51	0.87	4.20
C18:3ω-6	1.90	3.90	0.47	2.96	0.39	1.88
C18:3ω-4	0.87	1.79	0.24	1.51	0.39	1.88
C18:4	0.31	0.64	0.43	2.71	0.14	0.68
C20:4	0.35	0.72	0.10	0.63	0.12	0.58
C20:5	3.55	7.30	1.07	6.74	1.49	7.19
C22:6	1.62	3.33	0.40	2.52	0.83	4.01
Total	48.66	100.00	15.87	100.00	20.72	100.00

[a]Freeze dried algae.

TABLE 12.7 Supercritical CO_2 Extraction of Freeze-Dried *Skeletonema costatum* at 31 MPa

Fatty acid	Original mg/g[a]	%	SC extract mg/g[a]	%	Residue mg/g[a]	%
C14:0	4.52	12.86	1.29	14.97	3.96	17.04
C14:1	0.10	0.28	0.03	0.35	0.09	0.39
C16:0	2.66	7.57	0.69	8.00	1.72	7.40
C16:1	4.95	14.08	1.10	12.76	3.75	16.14
C16:2ω-7	0.69	1.96	0.13	1.51	0.45	1.94
C16:2ω-4	2.17	6.17	0.50	5.80	1.69	7.27
C16:3	3.85	10.95	0.99	11.48	1.65	7.10
C16:4	2.64	7.51	0.75	8.70	1.65	7.10
C18:0	0.34	0.97	0.06	0.70	0.30	1.29
C18:1ω-9	1.63	4.64	0.72	8.35	0.46	1.98
C18:1ω-7	0.78	2.22	0.13	1.51	0.45	1.94
C18:2	1.31	3.73	0.34	3.94	0.88	3.79
C18:3ω-6	1.51	4.29	0.51	5.92	0.52	2.24
C18:3ω-4	0.69	1.96	0.14	1.62	0.46	1.98
C18:4	0.26	0.74	0.05	0.58	0.15	0.65
C20:4	0.03	0.09	0.01	0.12	0.02	0.09
C20:5	5.50	15.64	1.01	11.72	3.91	16.82
C22:6	1.53	4.35	0.17	1.97	1.13	4.86
Total	35.16	100.00	8.62	100.00	23.24	100.00

[a]Freeze dried algae.

pressure at all levels. Tables 12.8 through 12.10 show fatty acid contents of original, extracted, and remaining lipids at 17, 24, and 31 MPa, respectively. Chlorophylls in both species of algae were not extracted by SC-CO_2.

The seaweed *Palmaria palmata,* which contains 45% EPA (dry basis), was extracted with SC-CO_2 and with a mixture of CO_2 and 10% ethanol (35). Pressures of 20.7, 41.5, and 62 MPa and temperatures of 35, 45, and 55°C were used. Solvent rates were maintained at 300 mL/min. The extracted and remaining lipids were analyzed by TLC and GC. The authors observed no significant changes in the solubility of oils under the extraction conditions (Fig. 12.4). Observed solubilities were 2.88 to 3.63 mg oil/L CO2 at STP. The extract at 20.7 MPa contained the highest concentration of EPA. Addition of ethanol increased solubilities threefold, slightly improved the yield of EPA, and changed the composition of the extracted oil, as shown in Table 12.11.

Carotenoids. When *Dunaliella* is grown under high light intensity, high salinity, nitrate and phosphorus deficiency, and extreme temperatures, it can accumulate as much as 8% dry weight β-carotene (23). A continuous flow-through extractor was used to measure the solubility of pure β-carotene in experimental conditions (24 MPa, 40°C).

TABLE 12.8 Supercritical CO_2 Extraction of Freeze-Dried *Ochromonas danica* at 17 MPa

	Initial		SC extract		Residue	
	mg/g	%	mg/g	%	mg/g	%
C14:0	26.39	18.06	3.80	24.73	16.32	18.45
C14:1	0.46	0.32	0.24	1.59	0.00	0.00
C16:0	16.07	10.99	1.70	11.02	9.30	10.52
C16:1	2.18	1.49	0.25	1.61	1.47	1.66
C18:0	4.73	3.23	0.38	2.49	2.38	2.69
C18:1	9.38	6.42	0.89	5.76	5.56	6.29
C18:2	25.65	17.55	2.35	15.29	16.93	19.14
C18:3	21.59	14.77	2.25	14.65	13.81	15.62
C20:0	1.68	1.15	0.14	0.94	1.04	1.17
C20:2	0.37	0.25	0.06	0.37	0.32	0.36
C20:3	7.06	4.83	0.59	3.85	5.06	5.72
C20:4	16.81	11.50	1.72	11.18	11.01	12.44
C20:5	5.60	3.84	0.73	4.76	3.31	3.75
C22:0	1.39	0.95	0.10	0.63	0.92	1.04
C22:1	6.79	4.64	0.08	0.52	0.55	0.63
Total	146.14	100.00	15.37	100.00	88.44	100.00

TABLE 12.9 Supercritical CO_2 Extraction of Freeze-Dried *Ochromonas danica* at 24 MPa

	Initial		SC extract		Residue	
	mg/g	%	mg/g	%	mg/g	%
C14:0	26.39	18.06	6.63	24.39	17.70	19.27
C14:1	0.46	0.32	0.43	1.44	0.03	0.03
C16:0	16.07	10.99	2.90	10.89	9.72	10.59
C16:1	2.18	1.49	0.37	1.44	1.37	1.50
C18:0	4.73	3.23	0.68	2.58	2.40	2.61
C18:1	9.38	6.42	1.53	5.73	5.80	6.32
C18:2	25.65	17.55	4.11	15.29	17.70	19.29
C18:3	21.59	14.77	3.98	14.79	14.73	16.06
C20:0	1.68	1.15	0.26	0.98	0.99	1.08
C20:1	0.00	0.00	0.05	0.13	0.00	0.00
C20:2	0.37	0.25	0.11	0.39	0.29	0.32
C20:3	7.06	4.83	1.08	3.99	4.94	5.39
C20:4	16.81	11.50	3.08	11.35	11.25	12.26
C20:5	5.60	3.84	1.31	4.89	3.25	3.55
C22:0	1.39	0.95	0.19	0.72	0.90	0.98
C22:1	6.79	4.64	0.11	0.41	0.24	0.27
C22:6	0.00	0.00	0.02	0.31	0.44	0.48
C24:0	0.00	0.00	0.09	0.25	0.00	0.00
Total	146.14	100.00	26.93	100.00	91.78	100.00

TABLE 12.10 Supercritical CO_2 Extraction of Freeze-Dried *Ochromonas danica* at 31 MPa

	Initial		SC extract		Residue	
	mg/g	%	mg/g	%	mg/g	%
C14:0	26.39	18.06	9.87	23.40	13.78	17.08
C14:1	0.46	0.32	0.51	1.21	0.00	0.00
C16:0	16.07	10.99	4.60	10.91	7.81	9.68
C16:1	2.18	1.49	0.57	1.34	1.20	1.48
C18:0	4.73	3.23	1.10	2.60	2.05	2.54
C18:1	9.38	6.42	2.47	5.85	5.11	6.33
C18:2	25.65	17.55	6.49	15.38	15.83	19.61
C18:3	21.59	14.77	6.24	14.79	13.06	16.18
C20:0	1.68	1.15	0.40	0.94	1.13	1.40
C20:1	0.00	0.00	0.23	0.55	0.00	0.00
C20:2	0.37	0.25	0.14	0.33	0.34	0.42
C20:3	7.06	4.83	1.67	3.97	4.51	5.58
C20:4	16.81	11.50	4.71	11.16	10.72	13.28
C20:5	5.60	3.84	2.23	5.29	3.11	3.85
C22:0	1.39	0.95	0.34	0.82	0.90	1.12
C22:1	6.79	4.64	0.59	1.41	0.38	0.47
C22:6	0.00	0.00	0.01	0.03	0.49	0.61
C24:0	0.00	0.00	0.02	0.04	0.29	0.36
Total	146.14	100.00	42.19	100.00	80.72	100.00

Since the concentration of β-carotene in the CO_2 effluent was dependent on the original charge and on gas flow rate, it was concluded that contact time between SC-CO_2 and β-carotene was the limiting factor. Increasing contact time from 0.3 to 84 sec increased the mole fraction of β-carotene in the CO_2 from 0.2×10^{-6} to 6×10^{-6}. The authors also cautioned against oxygen contact of the final extract after depressurization, since β-carotene can oxidize rapidly. Extraction of *Dunaliella* with SC-CO_2 showed carotenoid and xantophyll pigments as extract. There was a larger amount of 9-*cis* and 13-*cis* isomers in the extract relative to the total percent of β-carotene. The authors attributed this to their possibly higher solubility in SC-CO_2.

Conclusion

Algae contain many potentially harmful biogenic compounds. The presence and solubility of these compounds in SC fluids needs to be determined before algae extracts can be used for food/feed purposes. However, SC fluids appear to be promising as extraction and/or fractionation tools for obtaining high-valued chemicals from micro or macroalgae.

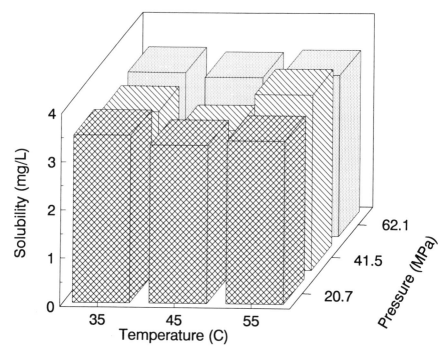

Fig. 12.4. SC-CO$_2$ solubility of lipids from *Palmaria palmata* at different pressures and temperatures. Solvent rate = 300 mL/min. *Source:* Mishra, V.K., et al., in *Proceedings of the 8th World Congress on Food Science and Technology,* 1992.

TABLE 12.11 Effect of Ethanol on the Fatty Acid Profile of the Lipids Extracted by SC-CO$_2$ at 20.7 MPa and 45°C (data taken from Ref. 35)

Fatty acid	Without ethanol (%)	With ethanol(%)
14:0	3.3	4.2
16:0	15.7	18.3
16:1 (ω-5, ω-7)	4.5	4.9
18:0	1.8	3.1
18:1 (ω-9,ω-7)	5.0	9.2
18:2 (ω-6)	1.3	1.3
18:4 (ω-3)	1.4	Trace
20:4 (ω-6)	2.8	3.0
20:5 (ω-3)	64.1	56.0

References

1. Klausner, A., *Bio/Techn. 4:*947 (1986).
2. Harvey, W., *Bio/Techn. 6:*488 (1988).
3. Neenan, B., D. Feinberg, A. Hill, R. McIntosh, and K. Terry, *Fuels from Microalgae Technology, Status, Potential, and Research Requirements,* Solar Energy Research Institute, Golden, CO, (1986).
4. Bold, H.C., and M.J. Wynne, *Introduction to the Algae,* 2nd edn., Prentice-Hall, Englewood Cliffs, NJ, (1985).
5. Ackman, R.G., P.M. Jangaard, R.J. Hoyle, and H. Brockerhoff, *J. Fish. Res. Bd. Can. 21:*747 (1964).
6. Choi, K.J., Z. Nakhost, V.J. Krukonis, and M. Karel, *Biotechn. 1(2):*263 (1987).
7. O'Sullivan, D. *FINS. 21(2):*23 (1988).
8. Kyle, D., *J. Amer. Oil Chem. Soc. 66(5):*648 (1989).
9. Miyamoto, K., O. Wable, and J.R. Benemann, *Biotechn. Lett. 10(10):*703 (1988).
10. Curtain, C.C., S.M. West, and L. Schlipalius, *Austral. J. Biotechn. 1(3):*51 (1987).
11. Fox, K.D., *Hydrobiol. 151/152:*95 (1987).
12. Rakow, A.L., M.L. Chapell, and R.L. Long, *Biotechn. Prog. 5(3):*105 (1989).
13. Hohlberg, A., J.M. Aguillera, and A. Herrera, in *Engineering and Food. Advanced Processes, Vol. 3,* edited by W.E.L. Spiess and H. Schubert, Elsevier Applied Science, New York, 1990, pp. 79–88.
14. Glicksman, M., *Hydrobiol. 151/152:*31 (1987).
15. Becker, E.W., in *Micro-Algal Biotechnology,* edited by M.A. Borowitza and L.J. Borowitza, Cambridge University Press, Cambridge, UK, 1988.
16. Nevenzel, J.C. in *Marine Biogenic Lipids, Fats and Oils, Vol. I,* edited by R.G. Ackman, CRC Press, Boca Raton, FL, 1989, pp. 3–71.
17. Ben-Amotz, A., T.G. Tornabene, and W.H. Thomas, *J. Phycol. 21:*72 (1985).
18. Ackman, R.G. in Analysis of Fats, Oils and Lipoproteins, edited by E.G. Perkins, American Oil Chemists' Society, Champaign, IL, 1991.
19. Drewicz, G.A., and A.P. Handel, *Institute of Food Technologies Annual Meeting,* New Orleans, LA, June 19–22, 1988, p. 207.
20. Cohen, Z., A. Vonshak, and A. Richmond, *Phytochem. 26(8):*2255 (1987).
21. Balaban, M.O., J. Polak, A. Peplow, and E. Philips, in *Engineering and Food. Advanced Processes, Vol. 3,* edited by W.E.L. Spiess and H. Schubert, Elsevier Applied Science, London, 1990, pp. 166–182.
22. Moulton, T.P., L.J. Borowitzka, and D.J. Vincent, *Hydrobiol. 151/152:*99 (1987).
23. Lorenzo, T.V., S.J. Schwartz, and P.K. Kilpatrick, in *Proceedings of the Second International Symposium on Supercritical Fluids,* Boston, May 20–22, 1991, edited by M.A. McHugh, Johns Hopkins University, Baltimore, MD, pp. 297–299.
24. Anonymous, *Austral. J. Biotech. 1(2):*38 (1987).
25. Ballantine, D., W.H. Gerwick, S.M. Velez, E. Alexander, and P. Guevara, *Hydrobiol. 151/152:*463 (1987).
26. Mayer, A.M.S., A. Diaz, A. Pesce, M. Criscuolo, J.F. Groisman, and R.M. Lederkremer, *Hydrobiol. 151/152:*497 (1987).
27. Kula, M.R., and H. Schutte, *Biotechnol. Progr. 3:*31 (1987).
28. Castor, T.P., and G.T. Hong, in *Proceedings of the Second International Symposium on Supercritical Fluids,* Boston, May 20–22, 1991, edited by M.A. McHugh, Johns Hopkins University, Baltimore, MD, pp. 139–142.

29. Hyatt, J.A., J. Org. Chem. 49:5097 (1984).

30. Dandge, D.K., P. Heller, and V. Wilson, Ind. Eng. Chem. Prod. Res. Dev. 24:162 (1985).

31. Subra, P., R. Tufeu, and Y. Garrabos, in Proceedings of the Second International Symposium on Supercritical Fluids, Boston, May 20–22, 1991, edited by M.A. McHugh, Johns Hopkins University, Baltimore, MD, pp. 304–307.

32. Sakaki, K., T. Yokochi, O. Suzuki, and T. Hakuta, JAOCS 67(9):553 (1990).

33. Cygnarowicz-Provost, M., D.J. O'Brien, R.J. Maxwell, and J.W. Hampson, in Proceedings of the Second International Symposium on Supercritical Fluids, May 20–22, 1991, edited by M.A. McHugh, Johns Hopkins University, Baltimore, MD, pp. 17–20.

34. Polak, J.T., Supercritical Carbon Dioxide Extraction of Lipids from Skeletonema costatum, M.Sc. Thesis, University of Florida, 1988.

35. Mishra, V.K., F. Temelli, and B. Ooraikul, in Supercritical Fluid Processing of Food and Biomaterials, edited by S.S.H. Rizvi, Blackie Academic and Professional, New York, 1994, pp. 214–221.

36. Kayama, M., S. Araki, and S. Sato, in Marine Biogenic Lipids, Fats and Oils, Vol. II, edited by G.R. Ackman, CRC Press, Boca Raton, FL, 1989, pp. 3–48.

37. Haines, T.H., S. Aaronson, J.L. Gellerman, and H. Schlenk, Nature 194:1282 (1962).

Chapter 13

Effect of Supercritical Fluid Extraction on Residual Meals and Protein Functionality

Fabio Favati[a], Jerry W. King[b], and Gary R. List[a]

[a]Dipartimento di Biologia DBAF, Università della Basilicata, Via N. Sauro 85, 85100 Italy;
[b]Food Quality Safety Research Unit, National Center for Agriculture Utilization Research, ARS, USDA, 1815 North University Street, Peoria, IL 61604 USA.

Introduction

Over 80 years ago Bridgeman (1) reported that pressure (700 MPa) promotes the coagulation of egg albumin, an effect that Grant (2) later showed to result from cleavage of sulfhydryl groups in the protein. Early studies also showed that tobacco mosaic virus is denatured by pressure (3); several other papers published in the 1940s and 1950s noted the general effects of pressure on protein and enzymes (4, 5). However, aside from these early reports, the effect of pressure combined with a supercritical fluid did not receive much attention until the last decade, when several groups showed that supercritical carbon dioxide (SC-CO$_2$) could be used to recover lipids and oils from a variety of substrates (6–12). Other commodities processed by supercritical fluid extraction (SFE) during the 1970s and 1980s included coffee, hops, cocoa, and tea (13–15).

It is probably fair to say that the primary goal in many of the cited applications of supercritical fluid media is to produce an extract that is superior or equivalent to that obtained from a different processing or extraction procedure. Generally the effect on the material remaining in the vessel, after the extraction ends, is of secondary importance. Investigations to examine the extracted residuals have been motivated by curiosity as well as economic considerations. Certainly the presence of a proteinaceous meal that is solvent-free presents some interesting opportunities for incorporation into the animal and human food chain.

This chapter is primarily concerned with the effect of supercritical fluid processing on residual protein-based meals and their functionality in an anticipated end use. Unlike most of the chapters in this monograph, we have tried to focus on what is left over after the extraction rather than on the extract. Of course, extraction conditions and sample preparation have been noted, particularly when lipid removal is crucial to the production of the desired end product.

We have specifically avoided any coverage on enzymatic reactions in supercritical fluid media, because that is covered in Chapter 15. Likewise, the relative new area of ultrahigh-pressure food processing is outside the scope of this review. Fundamental studies on the effect of pressure on proteins have been noted where relevant, but that subject is normally in the realm of biophysical chemistry. What follows here, then, is a highly focused review describing what is known about natural-product residuals remaining after supercritical fluid extraction, their proximate properties, and applicability to dietary requirements in man and animal, and their organoleptic properties and morphological alteration, which could play a role in food compounding.

Functional Properties of Proteins

Nowadays consumers are more discriminating about the quality of food items found on the market, and the success of several food products is strictly related to the satisfaction of the consumers' expectancies about the perceived quality, represented by attributes such as flavor, color, odor, and texture. The food industry is therefore increasingly looking for ingredients that have versatile but consistent functional properties and are also compatible with product formulation.

Besides their nutritive value, proteins constitute one of the most important classes of ingredients, because they can be incorporated in a variety of food items and have wide ranges of functional properties. Some of the most important functional properties of proteins are solubility, emulsifying properties, foaming properties, fat absorption and water absorption.

Solubility is of extreme importance for use of proteins in most food systems. Furthermore, several other functional properties (e.g. gelation, emulsifying and foaming properties) are strictly related to an initial protein solubilization step. Protein solubility is influenced by many factors, of which the most important are pH and ionic strength value. Plots of protein solubility data as a function of these two variables can give some information on the extent of the denaturation processes that could have occurred during previous technological treatments or processes, and that information can also be used to predict the kinds of food or beverage into which the proteinaceous material could be incorporated.

The emulsifying properties of proteins are related to their capacity to form and stabilize emulsions in food systems. Emulsion capacity represents the amount of oil that a protein solution or suspension can emulsify before the emulsion collapses. Emulsion stability is a measure of the stability of the emulsion in presence of stress factors such as heat, blending, and centrifugation (16,17).

Foaming properties are important for the use of proteins in cakes, whipped toppings, and frozen desserts. In this case the parameters that are taken into account are foam volume and foam stability (measured as the amount of foam that remains after a definite period of time), usually as a function of pH and temperature (16,17).

Juiciness, texture, binding of structure, and appearance are some of the qualities of food items that depend on the fat and water absorption properties associated with proteins. Interactions between proteins and water are important to such characteristics as hydration, swelling, viscosity, and gelation; entrapment of fats in meat emulsions, flavor absorption, and dough preparation are a few of the aspects related to the interactions between proteins and lipids. Fat absorption and water absorption are generally evaluated by measuring the amount of oil or water retained by the proteins under defined conditions.

Effects of SFE on Proteins

Pressure alone or extraction with $SC\text{-}CO_2$ may cause denaturation of protein, destruction of individual amino acids, or loss of enzymatic activity. Proteins, peptides, and amino acids are polar materials showing little if any solubility in $SC\text{-}CO_2$ and after extraction they are found in the residuals (18). Ideally, if the residue is to be consumed as a food, no alteration should occur during extraction with the supercritical fluid, or at least functional and nutritional properties should be preserved. Unfortunately, many factors influence the effect of supercritical fluids on proteins and enzymes, including moisture, temperature, pressure, and residence time in the extractor as well as the material to be extracted or treated itself.

In two different studies Weder reported that, in the presence of water, proteins treated with carbon dioxide or nitrogen at 30 MPa and 80°C undergo unfolding, partial oligomerization, and some fragmentation (19–20). However, these alterations are not related to the kind of gas utilized but to the concurrent presence of heat and water.

Weder studied also the influence of $SC\text{-}CO_2$ on some amino acids with isoelectric points ranging from 3.2 to 9.7 (19). The experimental data show that these components do not react with carbon dioxide at supercritical conditions. Of the amino acids tested, only L-glutamine showed some alteration, being partially converted to 2-pyrrolidinone-5-carboxylic acid. However this conversion can occur even in the absence of carbon dioxide; it is caused by heat alone. On the basis of these results, Weder concluded that deterioration to proteins and amino acids should not occur under the conditions likely to be employed in SFE of foods.

Results of other studies conducted on some pure proteins (cholesterol oxidase, α-chymotrypsin, lysozyme, penicillin amidase, ribonuclease, and trypsin) treated with carbon dioxide at mild temperatures and pressures (35 to 80°C, 10 to 30 MPa) suggest that under static conditions (2 to 24 h) these substances suffer minimal damage as long as water is absent from the system (18). Cholesterol oxidase showed little or no denaturation at 35°C and 10 MPa, as measured by electron paramagnetic resonance spectroscopy. After treatment with humid $SC\text{-}CO_2$ at 37°C and 10 MPa, fluorescence emission spectroscopy showed a partial unfolding of α-chymotrypsin, penicillin amidase, and trypsin, although gel electrophoresis failed to show any oligomerization or fragmentation. Treatment of lysozyme and ribonuclease with humid $SC\text{-}CO_2$ at 80°C and 30 MPa caused some denaturation of these proteins, as evidenced by increasing tryptic digestibility; furthermore, some oligomerization and fragmentation were detected by gel electrophoresis. However, available lysine content and amino acid composition were unaltered.

Soybean proteins

Eldridge et al. (21) extensively studied the extraction of soy flakes with $SC\text{-}CO_2$ under conditions where triglycerides show complete miscibility in the supercritical fluid. The effects of $SC\text{-}CO_2$ extraction conditions on nitrogen solubility index (NSI), flavor score, and enzyme activity of soy flakes in this study are shown in Table 13.1. Where high NSI and flavor scores are desired, 81.6 MPa and 85°C give the best results at moisture levels of 10.5 to 11.5%. NSI and flavor scores approached 70 and 7 respectively. Lipoxygenase activity was usually destroyed at moisture levels above about 6.5%, but urease activity remained unaffected in all 20 samples, since a consistent pH change of 2.0 to 2.1 was observed throughout. The usual grassy-beany and bitter flavors of hexane-defatted soybean flakes were only minimally detectable in the optimally SC-CO_2 extracted materials. Bland defatted soybean meal prepared by SC-CO_2 extraction was further processed into high-quality protein concentrates and isolates that were shown to be stable when stored under adverse conditions.

TABLE 13.1   Effect of SC-CO_2 Extraction Conditions on NSI, Flavor Score, and Enzyme Activity of Soy Flakes (21)

| | Extraction conditions[a] | | | | | | |
Sample	Pressure (MPa)	Temperature (°C)	H_2O (%)	NSI[b]	FS[c]	LU[d]	TI[e]
1	78.2	100	9.0	48	6.9	0	17.5
2	78.2	90	5.0	82	4.8	589	31.3
3	78.2	90	9.0	67	6.3	3	21.6
4	78.2	90	9.0	68	6.6	0	19.4
5	78.2	90	9.0	57	6.7	8	19.9
6	78.2	90	9.0	65	7.2	0	27.3
7	78.2	90	9.0	69	6.2	0	27.5
8	78.2	90	9.0	69	6.0	ND	27.7
9	78.2	90	12.4	62	6.4	16	29.5
10	78.2	80	9.0	80	5.8	779	31.3
11	74.8	84	6.5	80	4.2	688	ND
12	74.8	84	11.4	63	6.1	57	27.3
13	74.8	96	6.5	67	6.8	0	22.8
14	74.8	96	11.4	33	7.2	0	10.0
15	81.6	84	6.5	81	5.6	625	21.4
16	81.6	84	11.4	62	7.2	3	24.8
17	81.6	96	6.5	72	6.0	0	28.2
18	81.6	96	11.4	34	7.3	0	ND
19	72.1	90	9.0	66	6.3	10	27.7
20	84.4	90	9.0	69	6.6	80	27.4

[a]In all extractions, time and CO_2 flow remained constant.
[b]Nitrogen solubility index (NSI), a measure of protein solubility.
[c]Flavor score: 1 to 10 where 1 is strong and 10 is bland.
[d]Lipoxygenase units: $\mu M\ O_2$ consumed/min/mg protein.
[e]Soybean trypsin inhibitor activity, mg/g.
ND = not determined.

Various investigators have studied the effects of SC-CO_2 extraction conditions on the activity of the trypsin inhibitor present in soybeans (21–24) the results are summarized by Weder (18). At moisture levels of 9.8% and extraction conditions of 40 to 50°C and 13.5 to 68.9 MPa, no trypsin inhibitor inactivation was detected after extended reaction times. Pancreatic hypertrophy, changes in digestive enzymes, and reduced growth rate were observed in chicks fed the extracted meals. However, by careful control of the moisture content of the flakes, temperature, and pressure during extraction, the trypsin inhibitors could be completely inactivated (Table 13.1). Judicious control of moisture, temperature, and pressures during SFE of soy flakes could also inactivate lipoxygenase, which can catalyze the formation of lipid hydroperoxides known to be precursors of undesirable flavors. However, SFE conditions that destroy both trypsin inhibitors and lipoxygenase resulted in considerable protein denaturation (Table 13.1, run 18), as shown by the low NSI value.

Treatment with SC-CO_2 does not prove efficacious in some instances. Sessa et al. (24) contrasted SC-CO_2 with hexane defatting of whole and cracked soybeans and found no apparent differences in trypsin inhibitor activity after steeping the beans in 0.1 M Na_2SO_4. The SC-CO_2 extraction conditions of 85 MPa and 84°C for 15 min were not sufficient to deactivate the enzyme. These investigators also claimed that SC-CO_2 treatment did not reduce the residual sulfite levels in the steeped beans.

Corn Germ Proteins

Dry milled corn germ (CG) was extracted with SC-CO_2 at 50°C and pressures ranging from 34 to 54.4 MPa (25). Flavor evaluation of the resulting meals, against hexane-extracted controls showed that SC-CO_2–extracted meals were superior; after tempering to 8% moisture followed by an additional extraction, further improvement in flavor was achieved. Two factors are responsible for these observations.

1. SC-CO_2 is much more effective in removing triglycerides and bound lipids than hexane is.
2. Residual triglycerides left in extracted meals can be oxidized either by air or enzymes, giving rise to grassy, beany, bitter flavors.

The bound lipid content of hexane-extracted meal was nearly four times that of the SC-CO_2–extracted flour. The improvement in flavor afforded by SC-CO_2 extraction can be attributed to peroxidase inactivation as well as to the lowered lipid content. Peroxidase activity is normally difficult to destroy by the normal toasting process; however, Christianson et al. (25) reported that treatment under supercritical fluid extraction conditions resulted in a sevenfold reduction in peroxidase activity. Some protein denaturation was observed, as shown by reduction in nitrogen solubility index of the SC-CO_2–extracted flours. The amino acid profile of protein in both SC-CO_2–extracted and hexane-extracted CG flours had an excellent balance of essential amino acids and compared favorably with published FAO/ALTO standards for highly nutritious proteins.

SFE-defatted corn germ proteins can exhibit quite different fat-binding capacity and water retention properties, from the proteins prepared by hexane defatting. Lin and Zayas (26) found the SFE germ to be whiter—less red and less yellow—than corresponding germ defatted with hexane. Although temperature-dependent, the fat-binding capacity and water retention of the SC-CO$_2$–produced CG was significantly better than those of the solvent-extracted germ. Fat-binding capacity could be made equivalent for both germs by the application of heat. Both germs exhibited higher protein solubility at higher pH. Some of these observed differences may be attributed to the lower lipid level in the starting SC-CO$_2$–treated germ (almost 10% lower than in hexane-processed germ) or to morphological differences in germ structure.

Tömösközi et al prepared defatted CG proteins using hexane and SC-CO$_2$ (26 MPa/60°C); to better characterize the influence of the SFE treatment, these authors also studied the properties of alkaline-extracted protein isolates prepared from the defatted CG meals. Compared to meal prepared with hexane extraction, SC-CO$_2$–defatted meal showed a slightly higher *in vitro* nutritional value, but this difference disappeared in the isolates. The emulsion activity of meals and isolates was similar for both extraction procedures, but the emulsion stability of the meals was very poor compared to that of the isolates. The supercritical extraction process did not significantly affect the solubility profile of the treated proteins, but it negatively influenced their foaming properties. In contrast with the results obtained by Lin and Zayas (26), water and oil absorption values did not differ significantly between CG proteins defatted with SC-CO$_2$ and with hexane.

Other similar corn-based substrates treated with SC-CO$_2$ also exhibited improved properties critical to their incorporation into food systems. Wu et al. (28) evaluated corn distillers' dried grains (CDG) after extraction with SC-CO$_2$ and showed that the extracted CDG had lower fat and higher neutral detergent fiber content relative to untreated CDG. Perhaps more important were the acceptable flavor scores given the CDG substrate after SC-CO$_2$ treatment, as summarized in Table 13.2. Although individual flavor descriptions (cereal/grain, fermented, astringent) were higher than those of a standard substrate (untreated wheat flour) and in some cases those of untreated CDG, the overall flavor scores for the SC-CO$_2$ CDGs were significantly blander than those for untreated CDG. These results were achieved by using only small amounts of GRAS (Generally Recognized as Safe) cosolvents (water or ethanol) with the SC-CO$_2$.

Wu et al. (29) similarly characterized corn gluten meal that had been extracted with SC-CO$_2$ versus hexane/ethanol-treated meal. Extraction pressures ranged from 13 to 68 MPa, while extraction temperature was varied from 40 to 80°C. The resultant properties of the extracted meals were found to depend somewhat on the meal particle size, which influenced the degree of lipid removal, particularly since oil levels are extremely low in gluten meal (approx. 2%). Both SC-CO$_2$ treatment and hexane/ethanol-extraction of the meal were found to decrease the intensity of the fermented off-flavor significantly.

TABLE 13.2 Flavor Scores and Descriptions for Corn Distillers' Grains (CDG) Defatted with Supercritical Carbon Dioxide at 81.6 MPa (28)

		Flavor description[b]		
Treatment	Flavor score[a]	Cereal grain	Fermented	Astringent
Wheat flour				
Untreated	8.0 D	0.8	0	0
CDG				
Untreated	5.0 G	0.5	2.0	0.6
86°C, 1 mL ethanol	5.9 F	0.7	1.4	0.4
86°C, 1 mL H_2O	6.7 E	0.8	1.0	0.5
86°C, 1 mL ethanol, 1 mL H_2O	6.4 E	0.8	1.2	0.3
101°C, 1 mL ethanol	5.8 F	0.7	1.3	0.6
101°C, 1 mL H_2O	6.3 E	0.7	1.2	0.2
101°C, 1 mL ethanol, 1 mL H_2O	6.4 E	0.7	1.2	0.4

[a]Based on a 1–10 scale with 10 = bland, 1 = strong flavor. Flavor scores with no common letter (D–G) are significantly different (95% confidence level, $P < 0.05$). Least significant difference = 0.6.

Sunflower Proteins

In recent years several papers have been published on the extraction of sunflower oil with SC-CO_2 but only a few have reported the effects of the extraction process on the residual protein (27, 30–34). Studies conducted over a wide range of pressures (20 to 70 MPa) and temperatures (40 to 80°C) showed that the highest solubility of sunflower oil in the supercritical fluid was reached at 80°C and 70 MPa, similar to the SFE of other seed oils (31). At these conditions it was possible to remove more than 90% of the available oil contained in 280 g of seeds in only 6 min.

In subsequent research, Favati et al. (34) studied the effects on the proteinaceous substrate, conducting the SFE process at two sets of pressure and temperature conditions (20 MPa/40°C *vs.* 70 MPa/80°C) chosen to have a low and high impact on the matrix. Hexane extraction was also performed on the sunflower seeds for comparison purposes. Compared to the raw sample, both SFE and hexane extraction procedures reduced the sunflower protein solubility over the pH range 1.5 to 8.0, while only the sample extracted at 20 MPa and 40°C showed a slightly higher solubility at higher pH values (Fig. 13.1*a*). For protein solubility as a function of the ionic strength of the solution in which the sunflower proteins were solubilized, the hexane-treated meal and the unextracted sample showed similar profiles, whereas the meals extracted with SC-CO_2 had slightly higher solubilities above $\mu = L$ (Fig. 13.1*b*). Water absorption data were similar for both the meals treated with hexane and those treated with SC-CO_2, while fat absorption levels were somewhat inversely related to the amount of lipids left in the matrix after the extraction. The meals treated at 20 MPa and 40°C had a residual oil content of about 32% and a fat absorption level lower than that of samples extracted with hexane or with SC-CO_2 at 70 MPa and 80°C.

Similar results were reported by Tömösközi et al. (27) for sunflower proteins extracted with CO_2 at 30 MPa and 40°C. The protein solubility was not affected by the SFE treatment; fat absorption was lower in comparison to the hexane-extracted counterparts, because of partial removal of the native oil. The emulsion activity and emulsion stability of the SC-CO_2 extracted meals were slightly higher, but foam activity and foam stability were adversely affected by the supercritical extraction.

Peanuts and Peanut Butter

Santerre et al. (35) extracted roasted peanuts with SC-CO_2 at 41.36 MPa and temperatures of 50 to 60°C for up to 4 h. The extracted peanuts were evaluated for shatter, color, shear compression force, moisture, lipid content, and sensory attributes. The extracted peanuts were also processed into peanut butter and evaluated for color, relative torque resistance force ratio, and sensory attributes. After 4 h extraction SFE reduced the lipid content from 51.6 to 40.6% and the moisture from 1.4 to 1.2%. Peanut shatter was highest after 4 h extraction, but no trend was observed between shatter and extraction time. The color of peanuts was affected by SFE extraction time. This apparent bleaching effect was observed on the outside of partially extracted peanuts and was associated with lipid removal from the kernels. The color of the peanut butter was consequently affected by SFE but these changes were small. The shear compression force of peanuts increased with extraction time; other workers had demonstrated a corresponding inverse correlation with sensory crispness (36). The peanut butter also showed increased relative torque resistance with extraction time, attributed to reduction of oil content in the peanut butter. Sensory evaluations made on SFE-extracted peanuts showed a reduction in roasted-peanut aroma, fracturability, and moistness with increasing extraction time. Similar evaluations made on peanut butter showed increased adhesiveness but no trends in aroma intensity with extraction time. The authors concluded that because SC-CO_2 does not extract phospholipids, SFE of oil from roasted peanuts could potentially result in extended storage stability of the residual peanuts.

Miscellaneous vegetable proteins

Stahl et al. (37) extracted soy flakes, lupine seed, cottonseed, and jojoba meals with SC-CO_2 at 40°C and 35 MPa. Control extractions were made with hexane. The proteins were isolated and characterized by gel electrophoresis. The results showed no significant difference in the relative proportions of the subunits or reduced polypeptides recovered from CO_2-treated meals compared to the respective hexane-extracted control. In addition, there was no evidence of protein crosslinking. Amino acid analysis also showed no significant differences between the control and the CO_2-extracted samples.

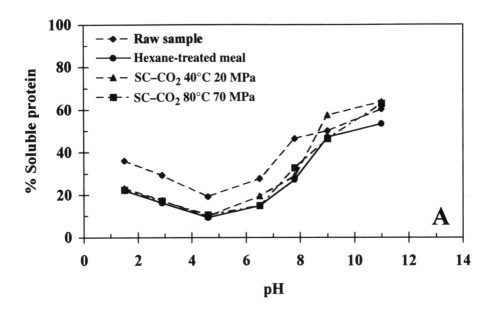

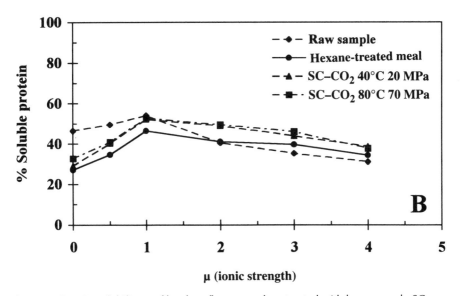

Fig. 13.1. Protein solubility profile of sunflower meals extracted with hexane and SC-CO$_2$ as a function of pH (a) and ionic strength (b). (Source: Favati, F., et al., in *Fluid: Supercritici e le laro Applicazioni,* edited by I. Kikic and P. Alessi, Centro Stampa dell' Università degli Studi di Trieste, Trieste, Italy, 1995, pp. 121–128.)

Tömösközi et al. (27) prepared protein isolates from SC-CO$_2$–extracted wheat germ and pumpkin seed meals and compared them to hexane-extracted counterparts. The *in vitro* nutritional values of the SC-CO$_2$– and hexane-treated samples were similar, while the content of cysteine and lysine in the whole-protein meals was higher in the SFE materials; however, after conversion to isolate this difference decreased. The two different oil extraction procedures did not cause any appreciable effect on the protein solubility profiles of meals and isolates; only pumpkin seed isolates treated with the supercritical fluid showed a better profile. Analytical data for emulsifying activity, emulsifying stability, water absorption, and fat absorption did not allow definite conclusions to be drawn. Foam activity of all the samples was low but acceptable, but foam stability was adversely affected by the extraction with SC-CO$_2$.

Special Applications

Previous sections of this review have been primarily concerned with the characterization and use of proteins and matrices that represent commodity agricultural products treated with SC-CO$_2$. This section focuses on a variety of substrates that illustrate the potential of SFE for producing a desired effect on the product remaining after the extraction. The coverage is probably not exhaustive but will allow the reader an appreciation of how widespread the application of supercritical fluids as "treatment" agents in processing foodstuffs can be. Of course, many of the cited studies are at best bench-scale in scope, and no claims can be made for the practicality or economic viability of their application.

Protein Hydrolyzates

Froschl et al. (38) have reported on lipid removal from protein hydrolyzates using SC-CO$_2$ as well as SC-CO$_2$/ethanol mixtures in a pilot plant whose upper pressure limit was 32.5 MPa. Both a protein hydrolyzate of high viscosity and a powder were extracted; however, a specific description of the matrices was lacking because of their proprietary nature. The target value was 97% removal with minimal loss in nitrogen content; the traditional process, using methylene chloride as the extraction solvent, resulted in a loss of 20% of the nitrogen content.

Despite attempts to increase the fat solubility in the SC-CO$_2$ by increasing both the temperature and pressure (to 60°C and 30 MPa, respectively), the targeted removal value could not be obtained. In addition, foaming problems occurred in the extractor; to eliminate foaming a demister was installed in the extraction vessel. Introduction of approximately 6 wt % ethanol into the SC-CO$_2$, however, resulted in over 90% of the lipids being removed; a 96.3% decrease was achieved at 30 MPa and 60°C. However, this gain was offset by a reduction in nitrogen value, even when using SC-CO$_2$ alone. High levels of amino acids were found in the ethanol phase of the extract after SFE, a result in stark contrast with earlier reported studies. Subsequent experiments with ethanol-laden SC-CO$_2$ showed that amino acids were not solubilized to any appreciable extent in this mixture; hence, the investigators theorized that the amino acids reached the collection vessel via entrainment with lipoproteins or water present in the sample. This nonspecificity in the extraction forced the investigators to abandon the SFE process, but their results suggest that a SC-CO_2 approach may have applicability in dissolving low-molecular-weight proteins or peptides from natural products. Such dissolutions have been demonstrated, particularly in supercritical fluid chromatography (39) or using ultrahigh-pressure SC-CO_2 above 70 MPa (40).

Meat

Extraction of fat from meat has been covered in Chapter 11. Many of these studies have characterized both extract and residual substrate in terms of fat or moisture content, but few note the effect of SFE on the proximate values, such as protein and ash. King et al. (41) extracted preformed beef patties with SC-CO_2 varying the extraction conditions, such as the moisture level of the meat patty, pressure, temperature, precooking, and spatial orientation of the patties in the extractor vessel. The results are reported in Table 13.3 on a dry matter basis for both control and extracted beef patties.

Higher extraction pressures tended to increase the protein content of the patties as fat and moisture content was reduced. Ash content, on the other hand, was not appreciably affected. Contrast analysis of the results given in Table 13.3 yields the trends noted in Table 13.4. Precooking, for example, increased the protein and ash content, as might be expected, while decreasing the cholesterol of the patties. The freeze-drying process, crucial to the removal of fat in meat products with SC-CO_2 (42), had no effect on fat, protein, or ash content.

Contrast analysis of control versus extracted patties suggested that protein and ash were significant altered at a P value of less than 0.05. The contrast of extraction pressure was significant (17.0 vs. 54.4 MPa) for sample fat, protein, and ash content. This suggests that a patty high in protein may be produced by SFE with CO_2.

Eggs

Dried eggs find a wide use in the food industry, and of the utmost importance is the retention of their functional properties, which are mainly related to the characteristics of the native proteins and to the presence of phospholipids. In the last few years treatment of dried egg yolk with SC-CO_2 for removal of cholesterol and lipids has been studied by several authors (43–46) and recently a low-cholesterol egg formulation obtained by using SC-CO_2 has become available on the market.

Froning et al. (43) utilized pure CO_2 to extract cholesterol and lipids from dried egg yolk at four different levels of pressure and temperature (16.3 MPa/40°C; 23.8 MPa/45°C; 30.6 MPa/45°C; 37.4 MPa/55°C). The effects of the extraction process on the substrate were assessed, particularly some functional properties of the matrix; emulsion stability (mayonnaise test) and sponge cake volume.

The proximate composition of the treated samples showed that the extraction of the lipids was influenced by the operating conditions; at 37.4 MPa and 55°C more than one-third of the lipid fraction could be removed. Protein content increased

TABLE 13.3 Protein, Ash, Fat, and Cholesterol Content of Control and Supercritical CO_2–Extracted Beef Patties (41)

Treatment[a]	Extraction pressure (MPa)	Protein (%)	Ash (%)	Fat (%)	Cholesterol (mg/100 g)
Raw control		49.19	2.38	48.42	177.09
Raw static	17.0	56.10	2.11	41.79	193.80
Raw dynamic	17.0	55.24	2.24	42.52	196.56
Raw dynamic	54.4	56.82	2.30	40.88	194.38
Raw FD[b] control		44.78	2.03	53.19	167.48
Raw FD static	170	48.00	2.22	49.79	154.07
Raw FD dynamic	17.0	52.19	2.35	45.45	145.79
Raw FD dynamic	54.4	70.49	3.16	26.34	102.97
Cooked control		59.78	1.98	38.25	189.33
Cooked static	17.0	71.20	2.56	26.213	186.55
Cooked dynamic	17.0	72.89	2.62	24.49	182.95
Cooked dynamic	54.4	77.17	2.77	20.06	176.24
Cooked FD control		62.59	2.20	35.22	194.34
Cooked FD static	17.0	62.91	2.33	34.76	163.12
Cooked FD dynamic	17.0	69.13	2.58	28.30	134.98
Cooked FD dynamic	54.4	83.00	3.15	13.85	100.54

[a]Extraction method: "static" = CO_2 is not replenished; "dynamic" = CO_2 is replenished.

because of the decrease in lipid concentration; the moisture level changed with no logical correlation with the extraction parameters. The cholesterol content of the dried egg yolk was reduced by at least 16% when the extraction was conducted in the mildest conditions, and approximately 66% of the total cholesterol was removed when operating at 30.6 MPa/45°C and 37.4 MPa/55°C. The extraction with SC-CO_2 apparently did not cause any reduction in the total phospholipid content—a desirable effect, considering the importance of these components for the emulsifying properties of eggs. The emulsion stability of mayonnaise prepared with the eggs treated with SC-CO_2 did not show significant differences ($P<0.05$) for the different treatments when compared to a control sample. Only for the eggs extracted at 37.4 MPa/55°C was a definite ($P<0.05$) reduction measured in the emulsion stability. Supercritical fluid extraction significantly ($P<0.05$) improved the sponge cake volume values, except for the extractions conducted at 37.4 MPa and 55°C. The authors attributed the decrease in sponge cake volume to partial denaturation of the egg proteins under the extraction conditions. However, electrophoretic analysis of the samples did not show any appreciable protein fraction modification that could be ascribed to the extraction conditions.

Rossi et al. (44) studied the extraction of lipids and cholesterol from egg yolk powder at 15 MPa and 50°C with pure SC-CO_2 and with the addition of ethanol at 1, 3, and 7% (w/w) level. About 50% of the available cholesterol was removed when the

TABLE 13.4   Contrasts of Extraction Parameters for SC-CO$_2$–Extracted Beef Patties (41)

	Contrast comparisons			
Variable	Raw *vs.* cooked	FDa *vs.* NFD^b	Control *vs.* extracted	17 MPa *vs.* 54.4 MPa
Dry matter basis:				
Δ Fat^c	0.13	0.86	***	***
Δ Cholesterol^d	***h	***	0.29	0.12
Cholesterol	0.46	***	0.28	0.11
Fat	***	0.43	**g	***
Protein	***	0.35	0.08	***
Ash	***	0.09	**	***
Moisture-retained basis				
Δ Fat^e	0.12	***	**	***
Δ Cholesterol^f	***	***	0.23	0.09

^{a}FD = Freeze-dried.
^{b}NFD = not freeze-dried.
^{c}Difference in percentage fat from control on a dry matter basis.
^{d}Difference in percentage cholesterol from control on a dry matter basis.
^{e}Difference in percentage fat from control on a moisture-retained basis.
^{f}Difference in percentage cholesterol from control on a moisture-retained basis.
^{g}** = significant difference ($P < 0.05$).
^{h}*** = significant difference ($P < 0.01$).

ethanol concentration reached 7%. However, the presence of ethanol lowered the selectivity between cholesterol and some phospholipids, thus causing extraction of lecithin. Furthermore, the addition of ethanol led to a slight extraction of proteins.

The presence of an alcohol as entrainer can also denaturate of the proteins contained in the egg yolk, as demonstrated by Arntfield et al. (45), who extracted freeze-dried egg yolk with SC-CO$_2$ at 36 MPa and temperatures of 40, 55, 65, and 75°C and used differential scanning calorimetry to evaluate protein conformation changes that could occur as a result of the SFE process. The observed denaturation indicated that over the range of temperatures studied, SFE with pure CO$_2$ did not affect the thermal stability of the proteins. However, the enthalpy of denaturation (ΔH) for ovalbumin was significantly lower for the extraction conducted at 75°C. This effect on the protein conformational changes was then shown to result from the combination of pressure and temperature: Heating the egg yolk at 75°C in presence of air or CO$_2$, but not under pressure, did not cause any change in the ΔH value.

The same authors then studied the stability of the egg yolk proteins in the presence of an entrainer by adding methanol to the CO$_2$. These experiments were conducted at 40°C and 36 MPa with a 3% methanol concentration in CO$_2$. Differential scanning calorimetry analysis revealed that the presence of methanol caused significant conformational changes in the egg protein. The resulting ΔH value for ovalbumin was 50% lower than that of a similar sample treated with pure CO$_2$.

Rice

Defatting has also been practiced on several other substrates to enhance their end use. Taniguchi et al. (47) extracted rice and rice koji (used for brewing applications) with SC-CO$_2$ at 30 MPa and 40°C so as to produce a substrate similar in lipid content to polished rice. As in the case of the protein hydrolyzate, addition of ethanol removed more lipid. The removal of lipids from rice koji did not affect the overall enzyme activity of the rice, but acid carboxypeptidase in particular was found sensitive to the SC-CO$_2$/cosolvent treatment. However, unlike the treatment of rice with neat CO$_2$, the SFE processing of rice koji did not improve the quality of the resultant sake. (See Chapter 14.)

Mustard seeds

Oils can contribute to the deterioration of spices such as mustard seed by oxidation. Any defatting of mustard seeds must be accomplished without lowering the activity of myrosinase, an enzyme essential to the generation of the acrid component of mustard (allylisothiocyanate). As with the case for rice, SFE with CO$_2$ at 30 MPa and 40°C (48) permitted up to 90% of the oil to be extracted, without decreasing myrosinase activity or the concentration of its substrate (synigrin). Pretreatment of the mustard seeds was found to be critical for achieving a successful deoiling; fully pressed seeds yielded the optimum result. (See Chapter 14.)

Leaf Protein Concentrates

Removal of lipids and lipophilic components by SFE has been studied by Favati et al. (49,50) for the production of alfalfa leaf protein concentrates (LPC) for human consumption. LPC have a high protein content and good amino acid composition, but there are major obstacles to their inclusion in the human diet, particularly their color and strong grassy flavor. The former is due to the presence of chlorophylls and carotenoids, the latter to the lipid fraction and to the products of oxidation of the unsaturated fatty acids.

The same authors (50) studied the effects of SFE treatment on some functional properties and organoleptic characteristics of the LPC in comparison with those of LPC treated with acetone. The extractions were conducted at four different pressure levels (10, 30, 50, and 70 MPa) and at the temperature of 40°C. Acetone extraction proved to be more effective for lipid removal; the maximum extraction of fat with SC-CO$_2$ was attained at the pressure of 70 MPa (Table 13.5). Lipid extraction resulted in correspondingly increased protein content, whereas the ash level was only slightly affected by the SFE process. Water absorption data were not significantly different for the untreated LPC and the samples extracted with the two solvents. Conversely, the treatment with SC-CO$_2$ caused a definite reduction in the amount of oil absorbed. The residual meals were further characterized in reference to the solubility of the proteins over the pH range 3 to 11. In comparison with the untreated LPC, acetone extraction significantly ($P < 0.05$) reduced the protein solubility, whereas SFE did not appreciably affect this functional property. Sensory evaluation of the treated samples indicated that the grassy flavor was definitely reduced in both SC-CO_2– and acetone-extracted meals. Despite the almost complete removal of the carotenoid pigments (Table 13.5), SC-CO_2–treated LPC did not show a significant reduction in the color, which was evaluated as "deep green," whereas acetone extraction resulted in a "white-gray" residue. This difference can be attributed to the presence in the LPC of chlorophylls and pheophytins, which at the tested SFE conditions exhibited only minimal solubility in the supercritical fluid (49).

Fish

SFE has also been investigated for the removal of lipids from the muscle of several species of fish, with the main goal of obtaining a proteinaceous residue having chemical, physical, and organoleptic characteristics suitable for its use for human consumption (51–53).

Hardardottir and Kinsella have reported on the extraction of lipids and cholesterol from rainbow trout (*Salmo gairdneri*) muscle using neat and ethanol-enriched (10% by weight) SC-CO_2 (51). The experiments were conducted at 40 and 50°C over the pressure range 13.8 to 34.5 MPa. The addition of ethanol to the stream of CO_2 did not cause a significant increase in the removal of cholesterol; conversely, in the presence

TABLE 13.5   Composition and Properties of Untreated LPC and Residual LPC after Extraction with Acetone and SC-CO_2 (49, 50)

		Extracted LPC[a]			
	Untreated LPC		Extraction pressure (MPa)		
		Acetone	10	30	50	70
Proximate composition (dry basis):						
Crude protein (N × 6.25) (%)	46.2	56.1	48.0	50.6	51.1	51.2
Crude fate (%)	13.0	0.2	7.5	6.5	6.2	5.4
Ash (%)	12.8	15.6	13.2	14.1	14.2	13.9
Functional properties:						
Water absorption (g H_2O/g LPC)	2.90	3.04	2.95	2.93	2.95	2.82
Fat absorption (g oil/g LPC)	2.33	2.31	1.81	1.84	1.89	1.87
Carotenoid recovery:						
Carotene extracted (%)	–	–	9.19	95.98	96.47	98.48
Lutein extracted (%)	–	–	0.52	29.52	61.23	70.20

[a]Extraction temperature 40°C.

of the entrainer the amount of lipids recovered rose from 78 to 97%, with extraction phospholipids as well. SFE caused some denaturation of the proteins, as indicated by the reduction of the solubility of muscle protein; electrophoretic analysis revealed that the extraction process affected myosin especially. As a consequence, the residual meals showed poor emulsifying properties and did not form gels when heated.

Fujimoto et al. (52) attempted to use liquid and supercritical CO_2 to defat sardine (*Sardinops melanosticta*) meat powder as an alternative source of surimi (minced meat) which is utilized in Japan for the production of Kamaboko. (The problems related to the use of sardines for surimi include the high oil content, the strong characteristic odor, and the low thermal stability of the myofibrillar proteins of this fish.) The extractions were conducted at 25 MPa and at temperatures of 12 and 40°C. Both treatments successfully reducing the lipid content, but the quality of Kamaboko produced with sardine defatted with SC-CO_2 was absolutely inferior. This result was ascribed to the negative influence of the higher extraction temperature on the proteins. The authors then investigated the effect of the addition of a modifier (10% ethyl acetate) to the liquid CO_2. The degree of denaturation of the proteins was evaluated by measuring the soluble nitrogen and the ATPase activity of the defatted minced sardine meat. Extraction with neat liquid CO_2 reduced the nitrogen solubility value, and a further decrease was caused by the use of the modifier. ATPase activity was even more significantly affected by the presence of ethyl acetate. The sardine meat powder defatted with liquid CO_2 allowed the production of Kamaboko of good quality, having a faint sardine odor and good storage stability, while the addition of the organic solvent resulted in a Kamaboko of much lower quality and having a dry and heated sardine odor.

Temelli et al. (53) applied SC-CO_2 to remove the oil contained in the muscle of Atlantic mackerel (*Scomber scombrus*) (53). In this research the use of SC-CO_2 had two objectives: recovery of a lipid fraction rich in ω-3 fatty acids, and production of a proteinaceous residue having potential uses for food applications. The experiments were conducted at three different levels of pressure (20.7, 27.6, and 34.5 MPa) and temperature (35, 45, and 55°C), and as a reference the fish muscle was also defatted with hexane. The effects of the extraction conditions on the water-binding potential (WBP) of the residual proteins were then evaluated. Both SFE and hexane extraction caused an increase of the WBP values; however, the WBP of the SC-CO_2–extracted proteins was always lower than that of hexane-extracted proteins. WBP appeared to be related to the amount of lipids removed from the substrate and to changes in protein functionality. The authors attributed this behavior to an increase in the available binding surface area of the protein due to the lipid removal, as well as to changes in the protein conformation.

Phospholipids Concentration and Isolation

Often treatment of a substrate with SC-CO_2 is designed to remove or enrich a specific component, such as phospholipids in eggs or oilseeds. Bulley (Chapter 9) has covered the application of SFE to eggs, primarily to remove cholesterol from the substrate and to increase its protein and phospholipid content. Here we shall focus on attempts to remove phospholipids from matrices after initial defatting so as to recover a value-added product that would otherwise be wasted.

In a previous chapter it was noted that a highly viscous medium such as lecithin can be deoiled by applying a special technique known as "jet extraction" (54). Here the surface area–to-volume ratio of the substrate to be deoiled is maximized by extrusion through a narrow orifice. The droplets or "strings" of lecithin are then immediately contacted by SC-CO_2, which can effectively deoil the lecithin if the contact time is sufficient. A deoiled powder can result by using this approach if all parameters are optimized. Eggers and Wagner (55) have recently studied the original process (54) in detail and offer an alternative method to achieve the same result.

Perhaps the first to specifically address the removal of phospholipids after oil extraction from seed matrices was Temelli (56). She demonstrated what previous investigators had shown, that addition of a cosolvent (entrainer) such as ethanol to SC-CO_2 enhanced the total recovery of lipid matter on a mass base while increasing the phospholipid content of the extract. Both canola flakes and press cake were extracted typically at 62 MPa and 70°C using ethanol mixed with the canola substrates at 5 to 15 wt.% addition. Ethanol was also added to the canola substrates after initial deoiling of the flakes with SC-CO_2, since equipment for continuous addition of ethanol to the SC-CO_2 was not available in this study. This principle has also been practiced in the decaffeination of coffee, where water is added to the beans to enhance extraction of caffeine specifically from the bean matrix.

Dunford and Temelli (57) expanded on these original studies to better characterize the phospholipid recovery from deoiled canola. Again, enhanced recovery was experienced by soaking the canola meal in ethanol after initial deoiling of the canola substrate. A phospholipid recovery level of only 21% was reported for extraction at 55 MPa and 70°C using SC-CO_2 with 8 mol%, provided continuously by a second cosolvent pump. Presoaking of the canola meal prior to SC-CO_2/ethanol extraction yielded 30% recovery of available phospholipids, according to these researchers. SC-CO_2/ ethanol extraction of an acetone-insoluble (AI) fraction yielded a 50% phospholipid-containing extract utilizing ethanol presoaking of the AI fraction. In all of the above cases, phosphatidylcholine (PC) was the major phospholipid recovered. Sample mass balances with accompanying phospholipid yields are presented in Fig. 13.2. This flow schematic could serve as the basis of a future process, designed to accommodate GRAS-approved processing agents (CO_2, ethanol) in several of the described stages. A similar approach has been described by Manohar et al. (58) using response surface methodology to optimize the fractionation of SC-CO_2–deoiled soybean lecithin, except that these researchers used ethanol rather than SC-CO_2/ethanol for PC enrichment.

More recently, Montanari et al. (59) have reported on the recovery of the phospholipid fraction left over from the exhaustive deoiling of soybean flakes. Their approach has been to deoil soybean flakes at 70 MPa and 80°C, which results in nearly complete removal of the soybean oil and retention of the phospholipids by the soybean meal. These investigators then utilized SC-CO_2 with ethanol as a cosolvent to

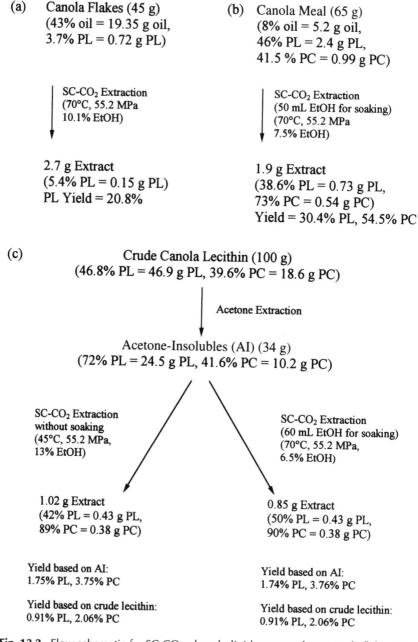

(a) Canola Flakes (45 g)
     (43% oil = 19.35 g oil,
     3.7% PL = 0.72 g PL)

(b) Canola Meal (65 g)
     (8% oil = 5.2 g oil,
     46% PL = 2.4 g PL,
     41.5 % PC = 0.99 g PC)

SC-CO₂ Extraction
(70°C, 55.2 MPa
10.1% EtOH)

SC-CO₂ Extraction
(50 mL EtOH for soaking)
(70°C, 55.2 MPa
7.5% EtOH)

2.7 g Extract
(5.4% PL = 0.15 g PL)
PL Yield = 20.8%

1.9 g Extract
(38.6% PL = 0.73 g PL,
73% PC = 0.54 g PC)
Yield = 30.4% PL, 54.5% PC

(c)   Crude Canola Lecithin (100 g)
     (46.8% PL = 46.9 g PL, 39.6% PC = 18.6 g PC)

Acetone Extraction

Acetone-Insolubles (AI) (34 g)
(72% PL = 24.5 g PL, 41.6% PC = 10.2 g PC)

SC-CO₂ Extraction
without soaking
(45°C, 55.2 MPa,
13% EtOH)

SC-CO₂ Extraction
(60 mL EtOH for soaking)
(70°C, 55.2 MPa,
6.5% EtOH)

1.02 g Extract
(42% PL = 0.43 g PL,
89% PC = 0.38 g PC)

0.85 g Extract
(50% PL = 0.43 g PL,
90% PC = 0.38 g PC)

Yield based on AI:
1.75% PL, 3.75% PC

Yield based on AI:
1.74% PL, 3.76% PC

Yield based on crude lecithin:
0.91% PL, 2.06% PC

Yield based on crude lecithin:
0.91% PL, 2.06% PC

Fig. 13.2. Flow schematic for SC-CO₂ phospholipid recovery from canola flakes (PL = phospholipids; PC = phosphatidylcholine). *Source:* Dunford, T.G., and F. Temelli, *J. Amer. Oil Chem. Soc. 72:*1009 (1995).

extract the residual phospholipid content of the meal using extraction conditions of 68.3 MPa and 80°C and varying the ethanol content of the supercritical fluid phase from 5 to 20 mol%. They confirmed what others have found, that there was an increase in the phospholipid content of the extract, which could be optimized at a 15 mol% addition of ethanol to the SC-CO_2. Extensive analysis of the phospholipid-laden fractions by Bollman's reagent, phosphorus analysis by inductively coupled plasma spectroscopy, and high-performance liquid chromatography showed that phosphatidylcholine and phosphatidylethanolamine could be recovered at levels exceeding 90% from the deoiled meal. The more polar phospholipids in the soybean flakes could also be obtained at 70% recoveries using 20 mol% ethanol in SC-CO_2. These high recoveries could only be obtained by using both a cosolvent and the higher extraction pressures and temperatures quoted above.

Conclusions

This chapter has shown that many attributes can be imparted to the residue, such as seed meals, by extraction with supercritical fluids. The obvious absence of objectionable solvent residues in the case of processing with CO_2 is very attractive and applies equally to both extractable and nonextractable components. We have also seen how treatment with SC-CO_2 can produce an unaltered residual, equivalent to those obtained via other processing techniques. On the other hand, examples have been cited where treatment with SC-CO_2 can alter enzyme activity, thereby affecting the end use properties of the extracted substrate.

Substrates remaining after partial or exhaustive extraction can also be reextracted with SC-CO_2 and an appropriate cosolvent. This often results in the recovery of a high-value product that can be utilized in a variety of ways while retaining the nutritive value of extracted product. As researchers become even more facile in the application of pressure and supercritical fluids, new substrate morphologies may be generated that have different chemical as well as dissolution properties, thereby opening up other application opportunities for these supercritical fluid–processed materials.

References

1. Bridgeman, P., *J. Biol. Chem. 19:*511 (1914).
2. Grant, E.A., R.B. Dow, and W.R. Franks, *Science 94:*616 (1941).
3. Laufert, M.A., and R.B. Dow, *J. Biol. Chem. 140:*509 (1941).
4. Curl, L., and E. Jansen, *J. Biol. Chem. 184:*45 (1950).
5. Johnson, F.H., and D.H. Campbell, *J. Biol. Chem. 163:*689 (1946).
6. Stahl, E., E. Shutz, and H.K. Mangold, *J. Agric. Food Chem. 28:*1153 (1980)
7. Friedrich, J.P., and G.R. List, *J. Agric. Food Chem. 30:*192 (1982).
8. Friedrich, J.P., G.R. List, and A.J. Heakin, *J. Am. Oil Chem. Soc. 59:*288 (1982).
9. List, G.R., J.P. Friedrich, and D.D. Christianson, *J. Amer. Oil Chem. Soc. 61:*1849 (1984).
10. List, G.R., J.P. Friedrich, and J. Pominski, *J. Amer. Oil Chem. Soc. 61:*1847 (1984).

286 F. Favati et al.

11. Friedrich, J.P., U.S. Patent 4,446,923 (1984).
12. Friedrich, J.P. G.R. List, and G.F. Spenser, in *Proceedings of the Seventh International Conference on Jojoba,* edited by A.R. Baldwin, American Oil Chemists' Society, Champaign IL, 1989, pp. 165–172.
13. Zosel, K., German Patent 2,005,293 (1974).
14. Roselius, W., O. Vitzhum, and P. Hubert, German Patent 2,127,643 (1974).
15. Vitzhum, O., and P. Hubert, German Patent 2,127,642 (1975).
16. Barbeau, W.E., *Ital. J. Food Sci. 4:*213 (1990).
17. Hung, S.C., and J.F. Zayas, *J. Food Sci. 56:*1216, 1223 (1991).
18. Weder, J.K.P., *Cafe Cacao Thé 34:*87 (1990).
19. Weder, J.K.P., *Food Chem. 15:*175 (1984).
20. Weder J.K.P., *Z. Lebensm. Unters. Forsch. 171:*95 (1980).
21. Eldridge, A.C., J.P. Friedrich, K. Warner, and W.F. Kwolek, *J. Food Sci. 51:*584 (1986)
22. Pubols, M.H., P.C. McFarland, A.C. Eldridge, and J.P. Friedrich, *Nutrition Rep. Int. 31:*1191 (1985).
23. McFarland, D.C., M.H. Pubols, A.C. Eldridge, and J.P. Friedrich, *Poultry Sci. 61:*1511 (1982).
24. Sessa, D.J., E.C. Baker, and J.P. Friedrich, *Lebensm.-Wiss. u. Technol. 21:*163 (1988).
25. Christianson, D.D., J.P. Friedrich, G.R. List, K. Warner, E.B. Bagley, A.C. Stringfellow, and G.E. Inglett, *J. Food Sci. 49:*229 (1984).
26. Lin, C.S., and J.F. Zayas, *J. Food Sci. 52:*1308 (1987).
27. Tömösközi, S., B. Simandi, E. Borsiczky, and J. Nagy, in *Proceedings of the World Conference on Oilseed Technology and Utilization,* edited by T.H. Applewhite, American Oil Chemists' Society, Champaign IL, 1993, pp. 438–443.
28. Wu, Y.V., J.P. Friedrich, and K. Warner, *Cereal Chem. 67:*585 (1990).
29. Wu, Y.V., J.W. King, and K. Warner, *Cereal Chem. 71:*217 (1994).
30. Dakovic, S., J. Turkulov, and E. Dimic, *Fat Sci. Technol. 91:*116 (1989).
31. Favati, F., R. Fiorentini, and V. De Vitis, in *Proceedings of the Third International Symposium on Supercritical Fluids,* Vol. 2, edited by M. Perrut, Institut National Polytechnique Lorraine, Nancy, France, 1994, pp. 305–316.
32. Calvo, L., and M.J. Cocero, in *Proceedings of the Third International Symposium on Supercritical Fluids,* Vol. 2, edited by M. Perrut, Institut National Polytechnique Lorraine, Nancy, France, 1994, Vol. 2, pp. 371–375.
33. Esquivel, M.M., I.M. Fontan, and M.G. Bernardo-Gil, in *Proceedings of the Third International Symposium on Supercritical Fluids,* Vol. 2, edited by M. Perrut, Institut National Polytechnique Lorraine, Nancy, France, 1994, Vol. 2, pp. 429–434.
34. Favati, F., L. Lencioni, V. De Vitis, D. Melfi, and R. Fiorentini, in *I Fluidi Supercritici e le loro Applicazioni,* edited by I. Kikic and P. Alessi, Centro Stampa dell'Università degli Studi di Trieste, Trieste, Italy, 1995, pp. 121–128.
35. Santerre, C.R., J.W. Goodrum, and J.M. Kee, *J. Food Sci. 59:*382(1994).
36. Hung Y.C., and M.S. Chinnan, *Peanut Sci. 16:*32-37 (1989).
37. Stahl, E. K. Quirin, and R.J. Blagrove, *J. Agric. Food Chem. 32:*938 (1984).
38. Froschl, F., R. Marr, and M. Nussbaumer, *J. Supercrit. Fluids 4:*250 (1991).
39. Giddings, J.C., M.N. Myers, and J.W. King, *J. Chromatogr. Sci. 7:*276 (1969).
40. Stahl, E., and W. Schilz, *Chem.-Ing. Techn. 50:*535 (1978).
41. King, J.W., J.H. Johnson, W.L. Orton, F.K. McKeith, P.L. O'Connor, J. Novakofski, and T.R. Carr, *J. Food Sci. 58:* 950, 958 (1993).

42. King, J.W., J.H. Johnson, and J.P. Friedrich, *J. Agric. Food Chem. 37:*951 (1989).
43. Froning, G.W., R.L. Wehling, S.L. Cuppet, M.M. Pierce, L. Niemann, and D.K. Siekman, *J. Food Sci. 55:*95 (1990).
44. Rossi, M., E. Spedicato, and A. Schiraldi, *Ital. J. Food Sci. 2:*249 (1990).
45. Arntfield, S.D., N.R. Bulley, and W.J. Crerar, *J. Amer. Oil Chem. Soc. 69:*823 (1992).
46. Sun, R., B. Sivik, and K. Larsson, *Fat. Sci. Technol. 97:*214-219 (1995).
47. Taniguchi, M, R. Nomura, M. Kamihira, M. Shibata, I. Kukaya, S. Hara, and T. Kobayashi, *J. Ferm. Technol. 65:*211 (1987).
48. Taniguchi, M., R. Nomura, I. Kijima, and T. Kobayashi, *Agric. Biol. Chem. 51:*413 (1987).
49. Favati, F., J.W. King, J.P. Friedrich, and K. Eskins, *J. Food Sci. 53:*1532 (1988).
50. Favati, F., A.M. Pisanelli, R. Fiorentini, and J.W. King, in *Proceedings of the Third International Conference on Leaf Protein Research,* Special Issue of *The Italian Journal of Food Science,* Chiriotti Editori, Pinerolo Italy, 1989, pp. 286–290.
51. Hardardottir, I., and J.E. Kinsella, *J. Food Sci 53:*1656, 1661 (1989).
52. Fujimoto, K., Y. Endo, S.Y. Cho, R. Watabe, Y. Suzuki, M. Konno, K. Shoji, K. Arai, and S. Saito, *J. Food Sci. 54:*265 (1989).
53. Temelli, F., E. LeBlanc, and L. Fu, *J. Food Sci. 60:*703 (1995).
54. Stahl, E., K.W. Quirin, A. Glatz, D. Gerard, and G. Rau, *Ber. Bunsenges Phys. Chem. 88:* 900 (1984).
55. Eggers, R., and H. Wagner, *J. Supercrit. Fluids 6:*31 (1993).
56. Temelli, F., *J. Food Sci. 57:*440, 457 (1992).
57. Dunford, T.G., and F. Temelli, *J. Amer. Oil Chem. Soc. 72:* 1009 (1995).
58. Monohar, B., G. Began, and K. Udaya Sankar, *Lebensm.-Wiss. u. Technol. 28:*218 (1995).
59. Montanari, L., J.W. King, G.R. List, and K.A. Rennick, in *Proceedings of the Third International Symposium on Supercritical Fluids,* Vol. 2, edited by M. Perrut, Institut National Polytechnique Lorraine, Nancy, France, 1994, pp. 497–504.

Chapter 14

Treatment of Food Materials with Supercritical Carbon Dioxide

Masayuki Taniguchi[a], Masamichi Kamihira[b], and Takeshi Kobayashi[b]

[a]Department of Material Science and Technology, Faculty of Engineering, Niigata University, Japan and [b]Department of Biotechnology, Faculty of Engineering, Nagoya University, Japan

Supercritical fluid extraction is a relatively new separation technique with unique characteristics that distinguish it from conventional solvent extraction. Examples of the application of supercritical and liquid carbon dioxide extraction arise in the treatment of natural products such as hops, tobacco, spices, coffee, and the aroma constituents of fruits. Perhaps the best-known uses of supercritical carbon dioxide (SC-CO_2)—the decaffeination of coffee, the preparation of hops extracts for the brewing industry, and the extraction of perfume components and unsaturated fatty acids for the cosmetic industry—have been carried out on a commercial scale with considerable success. In this chapter, new applications of SC-CO_2 with potential in the food industries will be described.

Preparation of Defatted Mustard by Extraction with SC-CO_2(1)

Mustard is added to spices to strengthen their acrid taste. However, since mustard seeds contain 35 to 45% oil, the addition of mustard causes the separation of an oil fraction from other components in the spice and results in deterioration of the spices via oxidation of the mustard oils. Therefore, mustard has generally been defatted by methods such as pressing and hexane extraction. The degree of oil removed by pressing methods with a frame and plate or expeller is lower than that from hexane extraction. However, if hexane extraction is used, the hexane remaining in the defatted mustard has to be completely removed by evaporation at high temperature. This step results in inactivation of myrosinase (thioglucosidase, EC 3.2.3.1). That enzyme is essential to catalyze the formation of the acrid component (allyl isothiocyanate, AITC) from its precursor (sinigrin) by the following reaction (2,3).

$$CH_2=CHCH_2-C \overset{\displaystyle /\!/ N-OSO_3^-}{\underset{\displaystyle \diagdown S\text{-Glucose}}{}} \xrightarrow{\text{Myrosinase}} CH_2=CHCH_2-N=C=S + HSO_4 + Glucose$$

Sinigrin   Allyl isothiocyanate

To obtain mustard of high quality for use as a food additive, it is necessary to remove the oils from native mustard seeds without lowering the myrosinase activity or sinigrin content.

For this purpose, the SC-CO$_2$ extraction process has been applied to remove oils from mustard seeds. Three kinds of mustard seeds (native, half-pressed, and fully pressed) were used in the present study. The crude fat content was 36.5% for the native seeds, 34.0% for the half-pressed seeds, and 30.5% for the fully pressed seeds.

The effect of temperature on the myrosinase activity was first investigated. No decrease of myrosinase activity in the mustard seeds was observed after incubation at 20 to 50°C for 2 h. Filippi (4) has reported that the solubility of vegetable oils in liquid and SC-CO$_2$ at high pressures, around 30 MPa, was constant, independent of extraction temperature between 20 and 50°C. Therefore, an extraction temperature of 40°C was selected for the experiments subsequently described, to allow comparison with the results previously reported for SC-CO$_2$ extraction of vegetable oils (5,6).

Figure 14.1 shows the effect of pretreating mustard seeds on the amount of oil extracted with SC-CO$_2$ at 30 MPa and 40°C. Oil was not extracted at all from native mustard seeds. On the other hand, between the half-pressed and fully pressed seeds there was little difference in the amount of oil extracted for a given amount of CO$_2$ consumed. When the ratio of oil extracted was expressed on the basis of the initial fat content of the mustard sample, the extent of oil extracted from the half-pressed seeds was lower than that from the fully pressed seeds because of the higher initial fat content in the half-pressed seeds. The husk of native seeds seems to inhibit penetration of SC-CO$_2$ into the seed, so some pretreatment to disrupt this husk is essential for extracting oil efficiently from mustard seeds. Consequently, fully pressed seeds were used throughout the remainder of this study.

Figure 14.2 shows the effect of extraction pressure on the ratio of oil extracted and the myrosinase activity of the defatted residue, when the temperature and extraction time were kept constant at 40°C and 3 h, respectively. A sharp increase in the ratio of oil extracted was observed between 10 MPa and 30 MPa. The ratio of oil extracted depended primarily upon the extraction pressure and exceeded 80% at pressures above 30 MPa. No decrease in myrosinase activity per gram of dry, defatted matter was observed for extraction pressures up to 35 MPa.

Figure 14.3 shows the rate of removal of oil and water during extraction with SC-CO$_2$ at 30 MPa and 40°C as well as the effect on myrosinase activity. The amount of oil and water extracted from the mustard increased linearly with the amount of CO$_2$ used per gram of mustard sample. However, the extraction rate of the oils and water gradually decreased after 50 g of CO$_2$ per gram of mustard seed had been used, that is, after an extraction time of 3 h. In the initial linear region of Figure 14.3, the solubility of oils in SC-CO$_2$ at 30 MPa and 40°C was estimated to be 0.48 wt%, which is approximately half the value of the solubility (1.0 wt%) of rice bran oil in SC-CO$_2$ under the same conditions (6). The water content of the mustard seeds was reduced from 8.5% to 5.2% during this 3-h period. No decrease in the myrosinase activity per gram of defatted dry matter was observed during the extraction at 30 MPa and 40°C.

Table 14.1 shows a comparison between the properties of mustard seeds prepared by different procedures. The amount of crude fat extracted by pressing alone using an expeller was very low, 16% of the initial value. By extracting the pressed seeds with

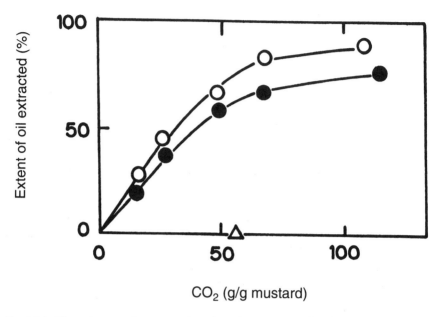

Fig. 14.1. Effect of pretreating mustard seeds. Oil was extracted with SC-CO_2 at 300 bar and 40°C. Symbols: △, native seeds; ●, half-pressed seeds; ○, fully-pressed seeds. *Source:* Taniguchi, M., et al., *Agric. Biol. Chem. 51:*413 (1987).

SC-CO_2, the crude fat content was reduced to 19% of that found in the native seeds. After extraction with SC-CO_2 at 30 MPa and 40°C for 3 h, the levels of sinigrin content and myrosinase activity on the basis of dry defatted weight were comparable to those of native seeds. Thus, extraction with SC-CO_2 makes it possible to prepare high-quality defatted mustard without lowering either the sinigrin content or myrosinase activity.

Effective Utilization of Horseradish and Wasabi by Treatment with SC-CO_2 (7)

Horseradish (*Armoracia lapathifolia* Gilib) and wasabi (*Wasabia japonica*), which belong to the same family, *Cruciferae,* contain useful compounds such as peroxidase (POD, EC 1.11.1.7) and glucosinolates. POD has been widely used as a label enzyme in clinical diagnosis and microanalysis with immunoassay. In the past few years there has been a growing demand for the enzyme, which is easily extracted with water and purified mainly from horseradish. However, glucosinolates—precursors of flavor and acrid components (isothiocyanates, ITCs)—in the residue have to be discarded after the extraction of POD (8–10). The inherent acridity and flavor of horseradish and

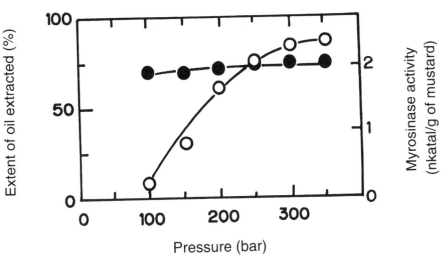

Fig. 14.2. Effect of extraction pressure on the ratio of oil extracted and myrosinase activity. Extraction with SC-CO$_2$ was carried out at a temperature of 40°C and extraction time of 3 h. Symbols: ○, ratio of oil extracted; ●, myrosinase activity. *Source:* Taniguchi, M., et al., et al. *Agric. Biol. Chem. 51:*413 (1987).

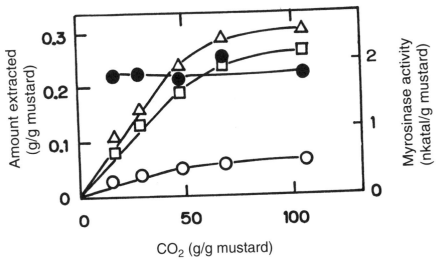

Fig. 14.3. Extraction of mustard seed oils with SC-CO$_2$ at 300 bar and 40°C at amount extracted and effect of extraction time on myrosinase activity. Symbols: △, total amounts; □, oil; ○, water; ●, myrosinase activity. *Source:* Taniguchi, M., et al., *Agric. Biol. Chem. 51:*413 (1987).

TABLE 14.1 Quality of Mustard Defatted with Supercritical CO_2 (1)

Sample	Crude fat content mg/g (%)		Sinigrin content[b] mg/g (%)		Myrosinase activity[b] nkatal/g of mustard
Native seeds	365	(100)	57	(100)	—
Half-pressed seeds	305	(83.6)	56	(98)	1.5
Fully pressed seeds treated with SC-CO_2[a]	69	(18.9)	58	(102)	2.0

[a]Extraction conditions: 300 bar, 40°C, 3 h.
[b]Based on dry, defatted matter.
Source: Taniguchi, M., et al., (1987) Agric. Biol. Chem. 51:413.

wasabi are attributable mainly to ITCs formed from corresponding glucosinolates with myrosinase (EC 3.2.3.1) in the presence of water. If the acrid and flavor compounds are separated efficiently from the plants without reduction in POD activity, the residue can be used as an enzyme source, and the extracted substances may find use as an additive to spices and foods. Moreover, the wasabi flavor is valuable, and it may be obtainable by separation from the formerly unused stems and leaves of the wasabi plant as well as from the traditionally used root.

Hence, to improve the utilization of horseradish and wasabi, the present authors investigated

1. The extraction of the ITCs formed from the glucosinolates with myrosinase
2. The effect of the ITC extraction conditions on POD activity
3. The conditions for efficient recovery of volatile ITCs with β-cyclodextrin (β-CD).

Preliminary experiments on enzymatic reaction in compressed CO_2 were carried out using the SC-CO_2 extraction apparatus with an extraction vessel of 75 cm³. The extracted substances were collected in a cold trap. Figure 14.4 shows gas chromatograms of ITCs extracted with liquid CO_2 at 20°C and 20 MPa and with SC-CO_2 at 40°C and 20 MPa from freeze-dried horseradish powder. Water, essential to the enzymatic reaction, was added at a level of 75% on a wet basis. The water content of freeze-pulverized wasabi root used in this experiment was 70%.

From horseradish only AITC (a common component in spices such as horseradish and mustard—see the preceding section) was formed, as shown in Fig. 14.4. The amount of AITC collected was limited not only by the extraction step but also by the enzymatic formation stage. Its increase with time, confirmed by the increasing peak height of AITC, as shown in Fig. 14.4a, means that liquid and SC-CO_2 at 20 MPa are utilized as nonaqueous media for enzymatic reactions, not only as solvents for extracting products formed by the enzymatic reaction.

Several ITCs, including AITC were extracted from wasabi root. AITC was a major component among the ITCs extracted, but the characteristic flavors of wasabi

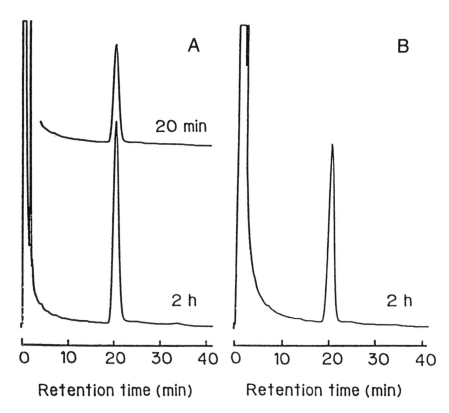

Fig. 14.4. Gas chromatograms of acrid component extracted from horseradish with (a) SC-CO$_2$ (40°C, 20 MPa) and (b) liquid CO$_2$ (20°C, 20 MPa). The peak at about 20 min corresponds to AITC. *Source:* Taniguchi, M., et al., *J. Ferment. Technol. 66:*347 (1988).

root, such as isopropyl and sec-butyl ITCs, were also extracted with liquid or SC-CO$_2$ (7). In addition to the root, the stem and leaf of wasabi are known to contain derived flavor compounds. These flavor compounds were also extracted with CO$_2$, not only from the root, but also from the stem and leaf.

Figure 14.5 shows the effect of CO$_2$ pressure on POD activity in horseradish powder with a water content of 75%. The treatment was carried out at 20°C for 2 h. No effect of CO$_2$ pressure was observed on POD activity under the conditions used. Figure 14.6 shows the effect of temperature on POD activity in the wet horseradish at CO$_2$ pressures of 0.1 and 6 MPa. POD activity decreased when the temperature was elevated above 20°C. Therefore, a temperature of 20°C was chosen for further studies. The decrease in POD activity was greater at a CO$_2$ pressure of 6 MPa than at 0.1 MPa.

To collect ITCs in a cold trap, additional cooling is necessary. Thus, recovery of the compounds by formation of inclusion complexes with β-CD was attempted. The

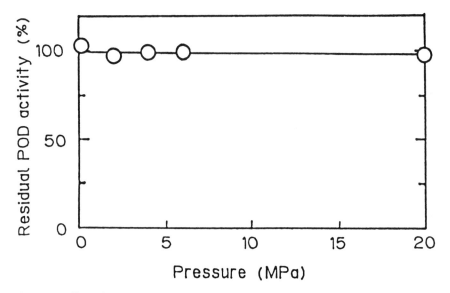

Fig. 14.5. Effect of CO_2 pressure on POD activity. Treatment conditions: temperature, 20°C; extraction time, 2 h. *Source:* Taniguchi, M., et al., *J. Ferment. Technol. 66:347* (1988).

β-CD molecule consists of seven glucose residues circularly bound; it has a hydrophobic hollow interior that allows it to play a role as a host molecule. First, the formation of the inclusion complex between β-CD and AITC was examined. Figure 14.7 shows the effect of CO_2 pressure on the formation of the inclusion complex. The collection was carried out using β-CD with a moisture content of 17.5% for 20 min. The molar ratio of AITC included in β-CD depended on the CO_2 pressure used for the collection. Two peaks were observed reproducibly at CO_2 pressures of 2.5 and 5 MPa; in the case of the inclusion complex between β-CD and geraniol (a main fragrance component in rose essential oil), only one peak was observed, at 6 MPa (11). The peak at 2.5 MPa for AITC may be attributable to other factors, such as the structure of the guest molecule.

Figure 14.8 shows the effect of the moisture content of β-CD on the formation of the inclusion complex. The ratio of the AITC included in β-CD was maximized at a moisture content of 24% at both pressures. The driving force for including the guest molecule is the stabilization in the energy that occurs when the guest molecule replaces water in the hollow of β-CD (12). In other words, β-CD including AITC is more stable in energy than β-CD containing water, which causes distortion of the β-CD cavity. Thus, water must be considered as a controlling factor in the formation of the inclusion complex.

On the basis of the results just described, SC-CO_2 extractions using an extraction vessel of 4 L were performed with three kinds of freeze-pulverized natural products:

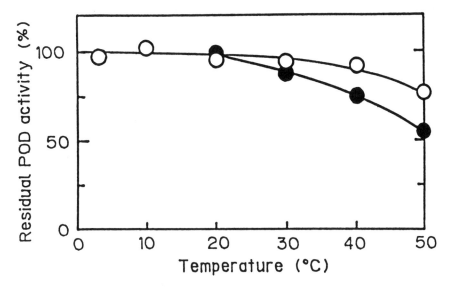

Fig. 14.6. Effect of temperature on POD activity. Extraction time was 2 h. Symbols: ○, 0.1 MPa; ●, 6 MPa. *Source:* Taniguchi, M., et al., *J. Ferment. Technol. 66:*347 (1988).

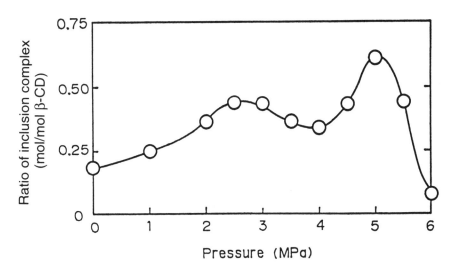

Fig. 14.7. Effect of pressure on the formation of AITC–β=CD inclusion complex. Inclusion conditions: temperature, 20°C; time, 20 min; moisture content of β-CD, 17.5%. *Source:* Taniguchi, M., et al., *J. Ferment. Technol. 66:*347 (1988).

horseradish, wasabi root, and wasabi leaf (Table 14.2). Horseradish was treated mainly for the purpose of the recovery and extraction of AITC without decreasing POD
activity in the residue after extraction. With wasabi, apart from utilization of POD, the
characteristic flavor compounds were expected to be recovered and extracted from
wasabi leaf, as well as its root. The extraction temperature selected was 37°C in order
to extract volatile flavor compounds efficiently from both parts of the wasabi.

As seen in Table 14.2, the content of sinigrin in the horseradish decreased as a result of the extraction with liquid CO_2 at 20 MPa and 20°C. The POD activity of the
horseradish residue was 85% of the initial value, which was high enough for the
residue to be used as a commercial enzyme source for POD. The sinigrin content in
the wasabi leaf as well as in the root also decreased with SC-CO_2 at 20 MPa and
37°C. The products extracted from wasabi leaf were smaller in quantity than those
from its root, but they were found to possess a very similar flavor to those from the
root in a sensory test.

Table 14.3 shows the recovery of AITC with β-CD. The pressure selected for
producing the inclusion complex was 2.5 MPa in order to maintain a stable operation.
The molar ratios of AITC included in β-CD were 0.615 and 0.506 for horseradish and
wasabi root, respectively. AITC included in β-CD from horseradish may be utilized as
an additive without further treatment. Residual POD activity after the extraction was
affected by the extraction time and water content of the sample. However, as shown in
Tables 14.2 and 14.3, the value of residual POD activity in horseradish was reproducibly high, about 90% of the initial activity. In the case of wasabi root, other ITCs

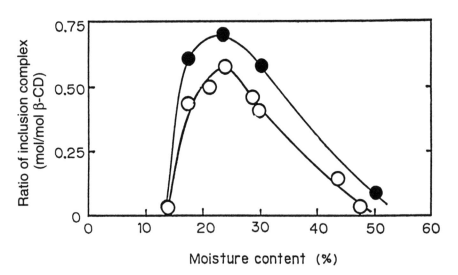

Fig. 14.8. Effect of moisture content of β-CD on the formation of AITC–β-CD inclusion
complex. Inclusion conditions: temperature, 20°C; time, 20 min. Symbols: ○, 2.5
MPa; ●, 5 MPa. *Source:* Taniguchi, M., et al., *J. Ferment. Technol. 66:*347 (1988).

TABLE 14.2 Composition of Freeze-Pulverized Horseradish and Wasabi Before and After Extraction

Sample	Total amount of sample (kg)	Water content (%)	Sinigrin content (%)	Relative POD activity (%)
Horseradish root[a]				
raw sample	1	72.0	1.76	100
residue	0.960	68.7	0.31	85
Wasabi root[b]				
raw sample	1	70.6	0.93	—
residue	0.957	67.3	0.38	—
Wasabi leaf[b]				
raw sample	1	69.5	0.37	—
residue	0.975	68.5	0.20	—

[a]Extraction conditions: 20°C, 20 MPa, 3 h.
[b]Extraction conditions: 37°C, 20 MPa, 3 h.
Source: Taniguchi, M., et al., (1988) *J. Ferment. Technol.* 66:347.

in addition to AITC were also included in β-CD; the inclusion complex mixture was examined further by gas chromatographic analysis and sensory testing.

Brewing of Sake from Rice and Rice-Koji Defatted by SC-CO₂ Treatment (13)

Rice for producing sake is generally polished to 70 to 80% of the weight of the original unpolished rice in order to reduce its lipid, protein, and mineral content. In particular, the lipid content significantly affects the quality of sake, such as fragrance and taste (14,15). However, the cost of rice as a starting material amounts to a majority of the total cost for brewing sake. To improve the utilization ratio of rice as a raw material and to reduce the lipid content of rice, lipids have been removed from 90%-polished rice (from which 10% of its original weight has been removed) by hexane extraction (16) and from 70 to 80%–polished rice by lipase treatment (17). Lipids, however, are also contained in the rice-koji (fermentation starter culture). If lipids can be removed from rice and rice-koji without the loss of enzymatic activities, by SC-CO₂ extraction, the utilization of rice will be improved and sake may be brewed with high quality. This part reports the results of extraction of lipids from polished rice with SC-CO₂ and brewing sake from the defatted rice and rice-koji.

Table 14.4 shows contents of crude fat and total lipids in rice defatted by extraction with SC-CO₂. Contents of crude fat in 80%- and 90%-polished rice decreased to about 30% of those in corresponding untreated rice. Total lipid contents of 90%- and 80%-polished rices were reduced to levels comparable to those found in 80%- and 70%-polished rices, respectively.

TABLE 14.3 Recovery of AITC and Residual POD Activity (7)

Sample	Total amount of sample packed (kg)	Amount of AITC included in β-CD		Relative POD activity (%)
		Total weight (g)	Molar ratio (mol/mol β-CD)	
Horseradish[a]	1	1.35	0.615	96
Wasabi root[b]	0.96	1.11	0.506	—

[a]Extraction conditions: 20°C, 20 MPa, 3 h. Inclusion conditions: β-CD; 30 g, 20°C, 2.5 MPa.
[b]Extraction conditions: 37°C, 20 MPa, 3 h. Inclusion conditions: β-CD; 30 g, 37°C, 2.5 MPa.
AITC: allyl isothiocyanate; POD: peroxidase.
Source: Taniguchi, M., et al., (1988) J. Ferment. Technol. 66:347.

TABLE 14.4 Treatment of Rice with Supercritical CO_2 (13)

Rice		Untreated		CO_2[a]		CO_2 + EtOH[b]	
		Crude fat (%)	Total lipids (%)	Crude fat (%)	Total lipids (%)	Crude fat (%)	Total lipids (%)
Unpolished rice		2.91	3.06	2.51	2.77	2.24	2.55
	90%	0.55	0.71	0.16	0.33	0.12	0.28
Polished rice	80%	0.063	0.29	0.023	0.23	0.024	0.18
	70%	0.039	0.21	0.022	0.20	0.019	0.15

[a]Treatment with supercritical CO_2 at 30 MPa and 40°C.
[b]Treatment with supercritical CO_2 containing 5 wt% ethanol.

Unpolished rice defatted by the SC-CO_2 treatment could not be used as a starting material for sake brewing, because the crude fat content was very high. The reduction in its total lipid content by extraction with SC-CO_2 was primarily due to removal of free oil. It was found that although SC-CO_2 dissolved unbound oil, it could not dissolve the complexed lipid matter in rice; this result is in accordance with the results reported on extraction of seed oils (5,18).

Addition of 5% ethanol to SC-CO_2 resulted in further decrease in contents of the total lipids compared with the extraction with SC-CO_2 alone. Thus, extraction with SC-CO_2 containing 5% ethanol reduced the total lipids in the three kinds of polished rice further, at a ratio of 50 mg per 100 g sample.

The present authors have reported that nine commercial enzyme preparations were not inactivated by exposure to SC-CO_2 or to SC-CO_2 containing ethanol or water as cosolvent (19). Table 14.5 shows several enzymatic activities in rice-koji defatted with SC-CO_2 alone and with SC-CO_2 containing 10% ethanol. Little decrease was observed in activities of the enzymes, except for acid carboxypeptidase, in the extraction with

SC-CO$_2$ alone. Extraction with SC-CO$_2$ containing 10% ethanol reduced enzymatic activity, especially in acid carboxypeptidase, which was selectively inactivated.

Table 14.6 shows the effect of the rice treatment on the quality of the resulting sake. Three kinds of the polished rice (70%, 80%, and 90%), defatted with SC-CO$_2$ alone or with SC-CO$_2$ containing 5% ethanol, were used. The rice-koji used in this experiment was treated with SC-CO$_2$ alone. Use of the polished rice defatted with SC-CO$_2$ resulted in

1. A decrease in the amino acids, except for the 70%-polished rice
2. Lowering of the color intensity
3. An increase in the ratio of the amount of isoamyl acetate to that of isoamyl alcohol (E/A) when compared with the ratio in sake from untreated rice—also a favorable change (20).

However, no significant improvement in the quality of sake was obtained in the experiments using rice treated with SC-CO$_2$ containing 5% ethanol.

Table 14.7 shows the effect of the rice-koji treatment on the quality of the resulting sake. When the rice defatted with SC-CO$_2$ was used, no improvements were observed for the cases of the rice-koji defatted with SC-CO$_2$ and SC-CO$_2$ containing ethanol. However, when the effect of SC-CO$_2$ treatment for rice was compared in the case using untreated dry rice-koji, the quality of the sake was also improved, regardless of the extent of polishing. From these results it was found that the SC-CO$_2$ treatment for rice was effective for improvement of quality of sake, but the SC-CO$_2$ treatment for rice-koji was ineffective.

Treatment with SC-CO$_2$ containing ethanol makes it possible to change the balance of enzymatic activities to some extent, as shown in Table 14.5. Hence, a new type of sake is expected to be brewed from the rice-koji treated with SC-CO$_2$ containing ethanol. Further work is still necessary on optimization of the treatment with SC-CO$_2$, such as ethanol concentration and water content of rice-koji, in connection with flavor and taste of sake.

TABLE 14.5 Enzymatic Activity of Rice-Koji Defatted by Supercritical CO$_2$ Extraction (13)

	Residual activity (%)[a]	
Enzyme	CO$_2$	CO$_2$ + EtOH[b]
α-Amylase	85	86
Glucoamylase	103	88
Protease (pH 3)	103	86
(pH 6)	92	74
Acid carboxypeptidase	75	26

[a]Treatment conditions: 40°C, 30 MPa, 2 h.
[b]Ethanol concentration was 10%.
Source: Taniguchi, M., et al., (1987) *J. Ferment. Technol. 65:*211.

TABLE 14.6 Effect of Rice Defatting with SC-CO_2 on the Quality of Sake[a] (13)

Type of rice		Alcohol (%)	Amino acidity (ml)	Color intensity (A_{430})	UV absorption (A_{260})	i-AmOH A (ppm)	i-AmOAc E (ppm)	E/A×100 (%)
Polishing rate (%)	Treatment							
	—[b]	16.9	2.84	0.077	21.8	409	16.2	3.95
90	CO_2[c]	16.9	2.75	0.060	20.6	417	19.5	4.68
	CO_2+EtOH[d]	18.5	2.48	0.061	20.3	412	15.3	3.70
	—[b]	16.4	2.66	0.059	19.2	378	12.0	3.18
80	CO_2[c]	17.7	2.45	0.053	18.4	297	14.1	4.76
	CO2+EtOH[d]	16.6	2.55	0.058	19.2	338	15.1	4.47
	—[b]	17.3	2.45	0.073	18.2	369	12.7	3.46
70	CO_2[c]	16.6	2.54	0.053	17.1	346	13.4	3.85
	CO_2+EtOH[d]	17.3	2.49	0.069	20.8	319	13.4	4.20

[a]The rice-koji was treated with supercritical CO_2 at 30 MPa and 40°C.
[b]Rice was not treated.
[c]Rice was treated with supercritical CO_2 at 30 MPa and 40°C.
[d]Rice was treated with supercritical CO_2 containing 5 wt% ethanol at 30 MPa and 40°C.
i-AmOH (A): isoamyl alcohol. i-AmOAc (E): isoamyl acetate.

TABLE 14.7 Effect of Rice-Koji Treatment on the Quality of Sake (13)

Type of rice		Rice-koji treatment	Amino acidity (mL)	Color intensity (A430)	E/A × 100 (%)
Polishing rate (%)	Treatment				
	—[a]	—[c]	2.85	0.079	2.28
90	CO_2[b]	—[c]	2.67	0.056	4.49
	CO_2[b]	CO_2[d]	2.75	0.060	4.68
	CO_2[b]	CO_2 + EtOH[e]	2.01	0.056	4.34
	—[a]	—[c]	2.67	0.069	2.83
80	CO_2[b]	—[c]	2.52	0.050	4.27
	CO_2[b]	CO_2[d]	2.45	0.053	4.76
	CO_2[b]	CO_2 + EtOH[e]	1.87	0.045	3.37
	—[a]	—[c]	2.46	0.070	3.45
70	CO_2[b]	—[c]	2.45	0.062	4.49
	CO_2[b]	CO_2[d]	2.54	0.053	3.85
	CO_2[b]	CO_2 + EtOH[e]	1.93	0.060	4.19

[a]Rice was not treated.
[b]Rice was treated with supercritical CO_2 at 30 MPa and 40°C.
[c]Rice-koji was not treated.
[d]Rice-koji was treated with supercritical CO_2 at 30 MPa and 40°C.
[e]Rice-koji was treated with supercritical CO_2 containing 10 wt% ethanol at 30 MPa and 40°C.
Source: Taniguchi, M., et al., (1987) J. Ferment. Technol. 65:211.

Sterilization of Microorganisms with SC-CO$_2$ (21)

Sterilization using gases (such as ethylene oxide), ultraviolet light, microwaves, radiation, superheated steam, and extrusion cooking has been extensively studied. Unfortunately, such methods cannot sterilize bioactive products, such as medicines and food, without lowering their quality. Sterilization methods using ethylene oxide or radiation are promising, but are limited to specific materials by law in Japan. Consequently, the present authors studied the sterilizing effect of SC-CO$_2$ in comparison with liquid CO$_2$ (L-CO$_2$) and gaseous CO$_2$ (G-CO$_2$). The feasibility of sterilizing thermally unstable compounds with SC-CO$_2$ was also investigated.

Table 14.8 shows the effect of the treatment conditions on *Escherichia coli* and baker's yeast. CO$_2$ in the gaseous state was used in runs 1 and 4, L-CO$_2$ in runs 2 and 3, and SC-CO$_2$ in runs 5 and 6. CO$_2$ exhibited a sterilizing effect against wet *E. coli* cells in all the experiments, and the effect was enhanced with increasing temperature and pressure. In contrast, the ratios of living cells were reduced to the order of 10^{-1} to 10^{-2} in the case of dry *E. coli* cells. For the case of baker's yeast, the sterilizing effect was observed only on wet baker's yeast cells with SC-CO$_2$. The L-CO$_2$ and G-CO$_2$ had no sterilizing effect on either wet or dry baker's yeast cells. There was a great difference in the sensitivity toward CO$_2$ between wet *E. coli* and wet baker's yeast, particularly for the experiments using gaseous and liquid CO$_2$.

Figure 14.9 shows the effect of treatment time on the sterilization with SC-CO$_2$ at 20 MPa and 35°C. Here, zero treatment time means that the sample was exposed to CO$_2$ until the pressure had been stabilized at 20 MPa (about 20 min), after which the pressure was reduced to 0.1 MPa (about 20 min). During this increase and decrease in pressure, some cells were sterilized, so the ratio of living cells became less than unity at zero treatment time. Wet cells of *E. coli* and baker's yeast were rapidly sterilized in the early stage of the treatment, but then the sterilization effect decreased. For dry

TABLE 14.8 Sterilizing Effects of CO$_2$ under Various Conditions (21)

		Ratio of living cells (−)			
		E. coli		Baker's Yeast	
	Treatment conditions[a]				
Run	Temperature (C°)	Pressure (MPa)	Wet cells[b]	Dry cells[b]	Wet cells[b]	Dry cells[b]
1	20	4	1.3×10^{-4}	0.30	0.84	1.09
2	20	10	3.1×10^{-5}	0.10	0.47	0.89
3	20	20	3.6×10^{-5}	6.6×10^{-2}	0.13	0.75
4	35	4	9.9×10^{-5}	0.11	0.73	1.08
5	35	10	6.0×10^{-5}	0.12	1.3×10^{-4}	0.65
6	35	20	7.2×10^{-6}	4.7×10^{-2}	1.3×10^{-7}	0.50

[a]Treatment time: 2 h.
[b]Water content: wet cells, 65–75%; dry cells, 5–15%.
Source: Kamihira, M., et al., (1987) *Agric. Biol. Chem. 51*:407.

cells of both of these microorganisms, the ratio of living cells was not sufficiently reduced, regardless of the time for sterilization.

Table 14.9 shows the sterilizing effect of SC-CO_2 at 20 MPa and 35°C on other
microorganisms. Four kinds of wet microorganisms (baker's yeast, *E. coli,
Staphylococcus aureus,* and *Aspergillus niger*) were sterilized by exposure to SC-CO_2
at 20 MPa and 35°C for 2 h. However, dry cells of baker's yeast and *A. niger* remained unsterilized under the same conditions. For the dry cells of *E. coli* and *S.
aureus,* the ratios of living cells were reduced to below 5% of their initial levels. The
endospores of both bacillus strains were not sterilized.

Table 14.10 shows the effects of adding ethanol and acetic acid to SC-CO_2. Ethanol
and acetic acid were selected because they are used as sterilizing reagents and are harmless [generally recognized as safe (GRAS)] even if they remain in such samples as the
starting materials for foods. With wet conidia of *A. niger* the sterilizing effect of CO_2
was enhanced by adding ethanol or acetic acid at a weight ratio of 2% or 0.5%, respectively. Dry conidia was not sterilized by SC-CO_2 alone, but the addition of ethanol or
acetic acid to SC-CO_2 made it possible to sterilize even dry conidia. The ratio of living
cells of dry conidia was reduced to the order of 10^{-6}, which corresponds to that obtained
in experiments on wet conidia with SC-CO_2 alone. For both wet and dry endospores of

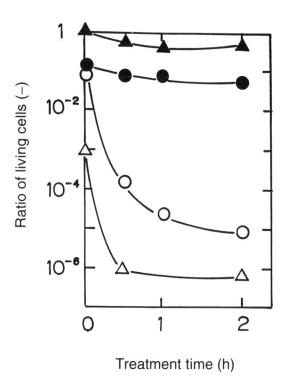

Fig. 14.9. Time courses for
sterilization with supercritical
CO_2 at 20 MPa and 35°C.
Symbols: ○, wet *E. coli*;
●, dry *E. coli*; △, wet baker's
yeast; ▲, dry baker's yeast.
Source: Kamihira, M., et al.,
*Agric. Biol. Chem. 51:*407
(1987).

Treatment time (h)

TABLE 14.9 Sterilizing Effect of Supercritical CO$_2$ at 35°C and 20 MPa (13)

	Ratio of living cells (−)	
Microorganism	Wet cells[a]	Dry cells[a]
Baker's yeast	$5.4×10^{-7}$	0.50
E. coli	$7.2×10^{-6}$	0.047
S. aureus	$1.5×10^{-5}$	0.037
A. niger (conidia)	$1.2×10^{-5}$	0.88
B. subtilis (endospore)	0.47	0.99
B. stearothermophilus (endospore)	1.07	0.80

[a]Water content: wet cells, 70–90%; dry cells, 2–10%.
Source: Kamihira, M., et al., (1987) Agric. Biol. Chem. 51:407.

TABLE 14.10 Effect of Additives on Supercritical CO$_2$ Sterilization[a] (13)

	Ratio of living cells (—)					
	Wet cells			Dry cells		
Microorganism	CO$_2$	CO$_2$+EtOH[b]	CO$_2$+ACOH[b]	CO$_2$	CO$_2$+EtOH[b]	CO$_2$+AcOH[b]
A. niger (conidia)	$1.2×10^{-6}$	$<2.3×10^{-6}$	$<7.9×10^{-6}$	0.88	$1.2×10^{-6}$	$<1.6×10^{-5}$
B. stearothermophilus (endospore)	1.07	0.62	0.43	0.80	0.49	0.43

[a]Treatment conditions: 35°C, 20 MPa. 2 h.
[b]Ethanol (EtOH) or acetic acid (AcOH) was added to CO$_2$ at a weight ratio of 2% or 0.5%, respectively.
Source: Kamihira, M., et al., (1987) Agric. Biol. Chem. 51:407.

Bacillus stearothermophilus, a sterilizing effect was observed with the addition of ethanol or acetic acid; and about 50% of the endospores were inactivated.

Table 14.11 shows the results of sterilizing enzyme preparations with SC-CO$_2$ at 20 MPa and 35°C for 2 h. After wet *E. coli* (water content: 74%) or baker's yeast (water content: 68%) was mixed with crude, dry α-amylase or lipase at a weight ratio of 9:1, the microorganisms in the enzyme preparations could be sterilized with SC-CO$_2$. Here, the initial water content of the enzyme preparation with *E. coli* or baker's yeast was 68% or 62%, respectively. *E. coli* cells in the two enzyme preparations were sterilized with SC-CO$_2$ to a living cell ratio on the order of 10^{-5}. This ratio almost corresponded to that obtained using the wet *E. coli* cells alone, as shown in Table 14.8. In the case of baker's yeast, the ratio of living cells was reduced only to the order of 10^{-3}. This may be due to a slightly lower water content (62%) compared to that used in the experiment shown in Table 14.8 (68%). The enzyme activity did not decrease for α-amylase after sterilization with SC-CO$_2$ but decreased slightly for the lipase.

The foregoing results show that SC-CO$_2$ can be used as a safe sterilizing agent as well as an alternative to a solvent for extracting lipophilic materials such as vegetable

TABLE 14.11 Sterilization of Enzyme Preparations with Supercritical CO_2[a] (21)

Enzyme	Microorganism	Residual enzymatic activity (%)	Ratio of living cells (−)
α-Amylase	E. coli	121	5.2×10^{-5}
	Baker's yeast	135	3.6×10^{-3}
Lipase	E. coli	88	8.9×10^{-5}
	Baker's yeast	78	4.7×10^{-3}

[a]Treatment conditions: 35°C, 20 MPa, 2 h.
Source: Kamihira, M., et al., (1987) Agric. Biol. Chem. 51:407.

or essential oils. As sterilization proceeds during the extraction of oils with SC-CO_2, both an oil and a defatted residue can be obtained free from microorganisms. Based on these results, the present authors applied SC-CO_2 sterilization to plasma powder (22), resulting in a greatly decreased ratio of living cells in humidified plasma powder, particularly with SC-CO_2 containing a small amount of acetic acid.

References

1. Taniguchi, M., R. Nomura, I. Kijima, and T. Kobayashi, Agric. Biol. Chem. 51:413 (1987).
2. Kojima, M., Nogyo Oyobi Engei 57:107 (1981).
3. Kawagishi, S., Nippon Shokuhin Kogyo Gakkaishi 32:836 (1986).
4. Filippi, R.P., Chem. Ind. 6:380 (1982).
5. Taniguchi, M., T. Tsuji, M. Shibata, and T. Kobayashi, Agric. Biol. Chem. 49:2367 (1985).
6. Taniguchi, M., T. Tsuji, M. Shibata, and T. Kobayashi, Nippon Shokuhin Kogyo Gakkaishi 34:102 (1987).
7. Taniguchi, M., R. Nomura, M. Kamihira, I. Kijima, and T. Kobayashi, J. Ferment. Technol. 66:347 (1988).
8. Kawagishi, S., Nippon Shokuhin Kogyo Gakkaishi 32:836 (1986).
9. Kojima, M., Y. Akahara, and I. Ichikawa, Nippon Nogeikagaku Kaishi 42:185 (1968).
10. Ohtsuru, M., Rep. Res. Inst. Food Sci. of Kyoto Univ. 38:13 (1975).
11. Kamihira, M., T. Asai, Y. Yamagata, M. Taniguchi, and T. Kobayashi, J. Ferment. Bioeng. 69:350 (1990).
12. Bender, M.L., and M. Komiyama, Cyclodextrin Chemistry, Springer-Verlag, Berlin, 1978, pp. 39–47.
13. Taniguchi, M., R. Nomura, M. Kamihira, M. Shibata, I. Fukaya, S. Hara, and T. Kobayashi, J. Ferment. Technol. 65:211 (1987).
14. Ishikawa, T., and K. Yoshizawa, Hakkokogaku 56:24 (1978).
15. Yoshizawa, K., T. Ishikawa, and K. Noshiro, Nippon Nogeikagaku Kaishi 47:713 (1973).
16. Tokumura, H., M. Shibata, S. Ohkura, T. Okada, and I. Fukaya, J. Brew. Soc. Japan 75:163 (1980).
17. Yoshizawa, K., and T. Ishikawa, J. Brew. Soc. Japan 71:975 (1975).

18. Friedrich, J.P., and G.R. List, *J. Agric. Food Chem. 30:*192 (1982).
19. Taniguchi, M., M. Kamihira, and T. Kobayashi, *Agric. Biol. Chem. 51:*593 (1987).
20. Yoshizawa, K., *J. Brew. Soc. Japan 68:*59 (1973).
21. Kamihira, M., M. Taniguchi, and T. Kobayashi, *Agric. Biol. Chem. 51:*407 (1987).
22. Taniguchi, M., H. Suzuki, M. Sato, and M. Kobayashi, *Agric. Biol. Chem. 51:*3425 (1987).

Chapter 15

Enzymatic Synthesis in Supercritical Fluids

Kozo Nakamura

Division of Agriculture and Agricultural Life Sciences, The University of Tokyo, Bunkyo-ku, Tokyo 113, Japan

Supercritical carbon dioxide ($SC–CO_2$) is an attractive solvent to food processors because it is benign with regard to human health. In addition, it exhibits excellent transport properties, such as low viscosity and high diffusivity. Solute solubilities in $SC–CO_2$ change with density and hence can easily be controlled by a variation of temperature and pressure. $SC–CO_2$ has been applied to extracting flavors and other natural substances and to removing undesirable components from the raw materials of food. It is expected that the application of $SC–CO_2$ can be extended to other unit operations, such as conducting enzymatic reactions in this unique medium (1–3).

Lipases can catalyze esterification reactions at low concentrations of water. This solvent-free process can be adapted to synthetic reactions where it is essential to separate traditional organic solvents completely from the end products (4–6). The reaction temperature of the $SC-CO_2$ process is typically high enough to melt the substrates but not so high that the enzyme is denatured, resulting in loss of activity. Therefore, use of a relatively harmless solvent, such as $SC–CO_2$ can provide a flexible and efficient enzyme-based process to the fats and oils industry.

Recent Developments

The general features of enzymatic reactions in SCF have been summarized in previous reviews (1–3), and only recent advances in this research are covered in this section. Most of the current studies deal with lipase-catalyzed reactions of esterification and transesterification (acidolysis, alcoholysis, and interesterification) in supercritical fluids, which are of interest to the food industry.

Erickson et al. (7) studied the acidolysis of trilaurin by palmitic acid to elucidate the effect of pressure on reaction rate. They used the lipase purified from *Rhizopus arrhizus*, immobilized on diatomaceous earth. Reactions were carried out batchwise in a 50-mL autoclave. Steytler et al. (8) examined the synthesis of butyl laurate from butanol and lauric acid in near-critical carbon dioxide using a crude lipase enzyme (*Candida* lipase B). The effects of temperature, pressure, and humidity were studied by comparing reaction rate profiles at a standard set of conditions ($T = 40°C$, $P = 30.0$ MPa) in a fixed-bed reactor of 120 mL internal volume. The same type of reactor (volume = 200 mL) was also used by Aaltonen and Rantakylä (9) to study the lipase-catalyzed reactions in $SC–CO_2$ for resolution of racemic ibuprofen. They tested free lipases from *Aspergillus niger* and *Candida rugosa* as well as immobilized *Mucor miehei* and *Candida rugosa* lipases.

Martins et al. (10) used a variable-volume reaction cell, consisting of a sapphire tube with a stainless steel piston, to study batch esterification of glycidol with butyric acid. The enzyme used was porcine pancreatic lipase. Marty et al. (11) also used a sapphire tubular reactor, 16 mL in volume, to study the esterification of oleic acid by ethanol with the immobilized lipase (Lipozyme, Novo,). The SC–CO_2 fluid, containing substrates and products, was circulated through the reactor, the external loop, and a sampling loop set in parallel with the external loop, permitting samples to be taken at discrete intervals. This reaction system allowed a whole kinetic curve to be obtained in one single run.

Continuous acidolysis or esterification has also been studied using immobilized lipase. Nakamura (12) used a column reactor packed with the immobilized lipase (Lipozyme IM20, Novo) to study the effect of temperature, pressure, moisture, and residence time on the acidolysis of triolein by stearic acid. The column size of the reactor was 0.5 cm i.d. and 5 to 20 cm length. Miller et al. (13) also studied the continuous acidolysis of trilaurin by myristic acid using a column reactor; the enzyme packed in the reactor was the lipase isolated from *Rhizopus arrhizus* (Sigma Chemical Co.,), immobilized on porous aminopropyl glass beads (75 Å pore size, Sigma), and the carbon dioxide, saturated with the substrate and water was fed to the reactor. Dumont et al. (14) studied the continuous synthesis of ethyl myristate using a column reactor (0.46 cm i.d. × 40 cm length) packed with Lipozyme TM (Novo); a small column, partially filled with water, was connected between the feed pump for the CO_2 and the reactor, to saturate the CO_2 with water.

Stability of Enzymes

The stability of enzymes as well as their activity depend on the concentration of water (1). Recently Dumont et al. reported that the effect of water on the activity of an immobilized lipase was reversible unless the concentration of water exceeded a specific value (14). The critical moisture was 0.15% wt/wt or 55 mM at 12.5 MPa and 323 K, which was less than the solubility (0.22% wt/wt) estimated by Chrastil's correlation equation (15). The conversion decreased as the moisture in SC–CO_2 was replaced by dry SC–CO_2. Dumont's study confirmed the irreversible loss of enzymatic activity by measuring the activity of the enzyme withdrawn from the reactor.

Marty et al. (11) also showed the effect of water on the activity of Lipozyme in SC–CO_2. They measured the residual activity of the enzyme that had been in contact with moist SC–CO_2 for one day at 13 MPa and 313 K. The loss of activity became significant when the added water concentration was above 200 mM. They suggested that water might lead to undesirable side reactions, such as hydrolysis of proteins or changes in their conformation. Miller et al. (13) measured the operational stability of immobilized lipase in contact with dry SC–CO_2 for 80 h at 9.5 MPa and 308 K but did not observe any change in the activity. This result indicated that pressure had minimal effect on the stability of the enzyme, because the SC–CO_2 pressure was far less than the inhibitory stage, which sets in over several hundred MPa (16). Interestingly, there have been no long-term experiments to test the operational stability of immobilized lipase in SC–CO_2.

Because the water adsorbed on the immobilized enzyme may directly relate to the stability of the enzyme, several researchers initiated the measurement of the adsorption isotherms of water between the support and the SC–CO_2. Figure 15.1 shows the examples of those adsorption isotherms reported on Lipozyme, which is hydrophilic with high specific surface area (14,17). According to Marty et al. (11), the commercial enzyme contained 8% water, which corresponded approximately to the water concentration critical to the enzyme stability in SC–CO_2 for the case of continuous synthesis of ethyl myristate (14). Marty et al. showed the effect of temperature, pressure, and ethanol on the adsorption isotherms of water for Lipozyme. The adsorption decreased with an increase in pressure, because the solvation effect can be more significant in SC–CO_2 than the vapor pressure effect. On the other hand, the effect of temperature was similar to that of gas adsorption; that is, the adsorption decreased with increase in temperature under the same condition of moisture in SC–CO_2. Ethanol in SC–CO_2 exhibited an "entrainer effect" for water and reduced the adsorption of water on the immobilized enzyme. This entrainer effect seems to be superficial and actually seems to arise due to the competitive adsorption of alcohol and water on the support particles.

Kinetics of Enzymatic Reactions in SCF

The study of kinetics can be useful for elucidating the characteristics of supercritical fluid as a solvent for reactions and to contribute to the design of the reactor.

Effect of Mass Transfer

Heterogeneous reactions on immobilized enzymes are influenced by inter- and intraparticle mass transfer as well as by mixing in the reactor. Figure 15.2 illustrates some important conclusions of mass transfer with the phenomena of solvent clustering and adsorption, which are related to the kinetics of enzymatic reactions in SCF. Erickson et al. (7) estimated the Thiele modulus (the ratio of the characteristic reaction rate to the characteristic *internal* diffusion rate) and the Damköhler number (the ratio of the characteristic reaction rate to the characteristic *external* diffusion rate) for a transesterification reaction in SC–CO_2, using diffusion coefficients reported in the literature (7). The Thiele modulus and the Damköhler number were found to be 10^{-2} and 10^{-5}, respectively, allowing Erickson et al. to conclude that mass transfer effects could be neglected in this case.

Miller et al. (13) examined the reaction rate by using packed-bed reactors of different lengths, which made it possible to change the flow velocity and to keep the residence time constant. They found no effect of superficial flow velocity on the observed transesterification rate. Miller et al. further alluded to the possible effect of intraparticle mass transfer limitation by estimating the observed modulus Φ, which yielded an

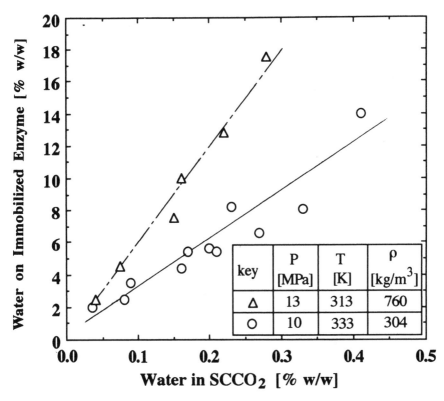

Fig. 15.1. Equilibrium partition of water between supercritical carbon dioxide and immobilized enzyme. *Source:* Dumont, T., D. Barth, and M. Perrut, in *Proceedings of the 2nd International Symposium on Supercritical Fluids,* edited by M.A. McHugh, Johns Hopkins University, Baltimore, MD, 1991, pp. 150–153.

estimated reaction rate similar to that obtained experimentally. Internal mass transfer effects were studied by Steytler et al. (8), who measured the reaction rate profiles of an ester synthesis, using immobilized lipase of two sizes: 3 mm and 1 mm. The rate profiles were found not to differ between the different sizes of porous glass beads when the same volume of beads—that is, the same amount of lipase—was used. This means that the reaction rate, which was low in this case, is not limited by the internal mass transfer.

Effect of Temperature

Steytler et al. (8) also reported that the activity of *Candida* lipase B (Biocatalysts Ltd., Wales) in $SC-CO_2$ (30 MPa) did not increase monotonously in the range from 20–60°C, but exhibited a maximum at about 40°C. Miller et al. found a similar temperature dependence for the pseudo-first-order rate constant for interesterification in

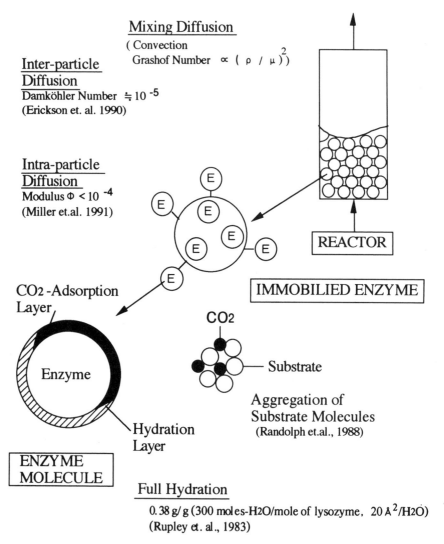

Mixing Diffusion
(Convection
Grashof Number $\propto (\rho / \mu)^2$)

Inter-particle Diffusion
Damköhler Number $\fallingdotseq 10^{-5}$
(Erickson et. al. 1990)

Intra-particle Diffusion
Modulus $\Phi < 10^{-4}$
(Miller et.al. 1991)

REACTOR

CO_2-Adsorption Layer

IMMOBILIED ENZYME

CO_2

Enzyme

Substrate

Aggregation of Substrate Molecules
(Randolph et.al., 1988)

Hydration Layer

ENZYME MOLECULE

Full Hydration
0. 38 g/g (300 moles-H_2O/mole of lysozyme, 20 Å^2/H_2O)
(Rupley et. al., 1983)

Fig. 15.2. Immobilized enzyme reaction and mass transfer in SCF.

SC-CO_2; they theorized that a loss of enzyme activity could be caused by thermal denaturation, causing the rate constant to decrease with increasing temperature from 35–40°C at 9.58 MPa (13). As mentioned before, the stability of enzymes in SC-CO_2 depends on the moisture. The thermal denaturation of enzyme could be avoided when the concentration of water in SC-CO_2 is below the saturation level, and the reaction rate increased with increase in temperature (9).

Effect of Pressure

Miller et al. (13) analyzed the effect of pressure on acidolysis between trilaurin and myristic acid in SC–CO$_2$ by excluding the effect of substrate concentration. Figure 15.3 corresponding to Figure 8 in the original paper (13), shows that the pseudo-first-order rate constants for the overall reaction and for interesterification both decrease as pressure rises, but the overall reaction rate constant decreases more than the rate constant for transesterification. According to Miller et al., the suppression of hydrolysis relative to transesterification could be the result of particular phenomena in SC–CO$_2$: changes in enzyme-solvent interactions that depend on the density-dependent properties of the solvent. Steytler et al. also examined the effect of pressure (15,30,50 MPa) on the conversion of batch esterification reactions (8) and showed that reaction rate increases with an increase in pressure. They mentioned that both enhanced solvation and improved distribution of components toward the SCF phase could occur with increasing density of SCF and improve the reaction rate.

On the other hand, the initial reaction rate of esterification between ibuprofen and propyl alcohol was reported not to depend on pressure (9). Adverse effects of pressure on the reaction rate for transesterification between trilaurin and palmitic acid have also

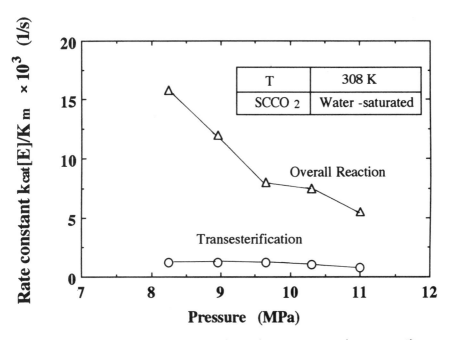

Fig. 15.3. Effect of pressure on the pseudo-first-order rate constants for interesterification and the overall reaction. *Source:* Adapted from Miller, D.A., H.W. Blanch, and J.M. Prausnitz, *Ind. Eng. Chem. Res. 30:*939 (1991).

been reported (7); Erickson et al. modeled this unfavorable effect by using the mole fractions of the reactants in place of the molar concentrations in the rate law.

Effect of Water

Water is a substrate of hydrolysis, so the use of excess water results in poor synthesis of esters or transesterification. However, enzymes need a small amount of water to maintain their active conformation. Water in a SCF functions also as modifier and changes the solubility of substrate and product indirectly, influencing an enzymatic reaction. It is important, therefore, to find the optimum water concentration in the synthetic reaction where the concentration of water is comparable to that of substrates. First, the optimum water concentration depends on the type of reaction, because water is produced in esterification and consumed in hydrolysis. Water is neither produced nor consumed in transesterification if no side reaction ever occurs. Transesterification, however, proceeds by formation of acyl-enzyme complex, and the reaction rate can be higher at the water concentration, where the side reaction of hydrolysis also takes place. The optimum water concentration also depends on factors such as substrate, enzyme, and carrier as well as on the reaction conditions of temperature, pressure, and substrate concentration.

Miller et al. (13) conducted the transesterification of trilaurin by myristic acid at 9.5 MPa and 308 K. They observed a decrease in the reaction rate with increase in water content of the $SC–CO_2$, partly due to the decreasing solubility of trilaurin (13). On the other hand, the pseudo-first-order rate constant, calculated by dividing the reaction rate by the initial trilaurin concentration, remained constant as the water content changed, as shown in Figure 15.3.

Dumont et al. (14) reported that the maximum reaction velocity of esterification of myristic acid by ethanol in a stirred reactor appeared at water concentrations ranging from 25 to 55 mM. The experimental conditions were as follows:

Ethanol = 50 mM

Myristic acid = 10 mM

Lipozyme = 20 mg

Reactor volume = 100 mL

$P = 12.5$ MPa

$T = 323$ K

The reaction time was not reported.

Marty et al. (11) used the optimum water content of the immobilized enzyme (Lipozyme) in place of water concentration in the medium in their study on the esterification of oleic acid by ethanol at 13 MPa and 313 K. They obtained sharp, bell-shaped curves as shown in Figure 15.4 in both $SC–CO_2$ and n-hexane. The optimum water content, 10% wt/wt, corresponds to about 1.4 g/L $SC–CO_2$, that is, 77 mM. They explained that with increasing water content, negative effects on the enzyme's activity could occur because of hydrophilic hindrance of the hydrophobic substrate on its way to the enzyme sites.

Aaltonen and Rantakylä (9) showed the effects of water concentration and propanol concentration in SC–CO$_2$ on the production of esterified ibuprofen; both effects were similar to the bell-shaped curves. The enantiomeric purity of the *S*(−)-form ester, however, was not dependent on water concentration, propanol concentration, or lipase concentration.

The effect of water on enzyme activity can also depend on how the water is added to the reactor. Steytler et al. (8) demonstrated a clear difference between the method in which water was added directly to the enzyme and that in which water was introduced above the catalyst bed after the reactor was loaded. Enhanced activity was observed only when the water was not in direct contact with the enzyme. This result is supposedly due to the reverse reaction (ester hydrolysis), caused by some portion of the enzyme being highly hydrated.

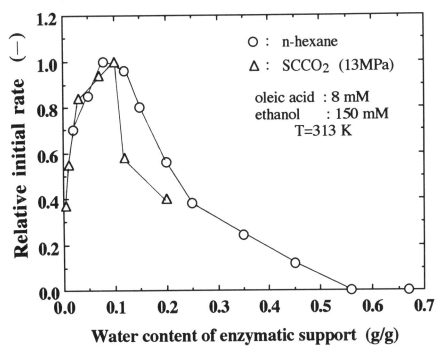

Fig. 15.4. Influence of water content of enzymatic support on enzyme activity in SC–CO$_2$ (13 MPa, 313 K) and *n*-hexane (313 K). *Source:* Marty, A., W. Chulalakananukul, R.M. Willemot, and J.S. Condoret, *Biotechnol. Bioeng. 39*:273 (1992).

Comparison of SCF with Organic Solvents

Kinetic studies have been advanced to compare SCF with organic solvents. Miller et al. (11) obtained the maximum velocity of their transesterification by application of the pseudo-first-order rate equation to the initial rate of reaction and calculated the apparent turnover numbers k_{cat} using the estimated Michaelis constant. The value of k_{cat} in SC–CO$_2$ was found to be 0.1 sec^{-1} at 9.5 MPa and 308 K at trilaurin concentration maintained at 2.9×10^{-3} M, whereas k_{cat} in cyclohexane was 0.3 s^{-1} when trilaurin concentration was 7.8×10^{-3} M. Considering that these turnover numbers were possibly determined at unsaturated conditions, Miller et al. suggested that the rates of reaction in SC–CO$_2$ and cyclohexane would be roughly similar.

Marty et al. (10) analyzed the kinetic data on esterification of oleic acid by ethanol using the rate equation predicted by a Ping-Pong Bi-Bi mechanism. Kinetic parameters obtained in SC–CO$_2$ and n-hexane are shown in Table 15.1. The apparent affinity constant of enzyme for oleic acid [K_m(ol)] was smaller in SC–CO$_2$ than in n-hexane, whereas the other affinity constant [K_m(eth)] was larger in SC–CO$_2$ than in n-hexane. Marty et al. (10) explained this result qualitatively by comparing the solubility and the diffusivity of each substrate in the two media. Experimental initial velocities were definitively shown to be similar in both media, provided the oleic acid concentration is the same in the two media for unsaturated SC–CO$_2$.

Continuous Transesterification Reaction

The present author has recently reported (20) the acidolysis reaction between triolein and stearic acid, catalyzed by the commercial immobilized enzyme (Lipozyme IM20, Novo), a 1,3-regiospecific lipase. Both the product composition and experimental conditions were reported.

Equilibrium Composition of the Product

The lipase is a hydrolytic enzyme; however, mono- and diglycerides can be produced even though the reaction is conducted with limited moisture. In this case, the total reaction consists of multiple reaction pathways, each of which can proceed through for-

TABLE 15.1 Kinetic Parameters Obtained in SC–CO$_2$ (13 MPa, 40°C) and n-Hexane at 40°C (11)

	V_m (mmol/min•g)	K_m (ol) (mM)	K_m (eth) (mM)	K_i (mM)
SC–CO$_2$	14	170	1600	65
n-hexane	23	450	600	60

Average relative error between model and experimental values: 5.3% and 4.2% with a maximum about 15.7% and 14.3% in SC–CO$_2$ and n-hexane, respectively.

mation of an acyl-enzyme complex. The present author calculated only the equilibrium concentrations of the product by developing equations based on the following assumptions:

1. The substrate specificity of lipase is the same for each of the fatty acids, such as stearic acid and oleic acid
2. Efficiency of water as an acyl-acceptor is K times as much as that for transesterification.

Making these two assumptions permits the residual triglyceride, T, and the extent of interesterification, E, to be calculated at the equilibrium state of the reaction (9):

$$T_\infty = \frac{(OOO + SOO + SOS)_{equil}}{(OOO)_{initial}} = \frac{(2C_t + C_s)(C_t + C_s)}{(2C_t + C_s + KC_w)(C_t + C_s + KC_w)} \quad (15.1)$$

$$E_\infty = \frac{(SOO + 2SOS)_{equil}}{3 \times (OOO + SOO + SOS)_{equil}} = \frac{2}{3} \times \frac{C_s}{2C_t + C_s} \quad (15.2)$$

where C_t, C_s, and C_w are the initial concentration of triolein, stearic acid, and water, respectively; [SOS] denotes a triglyceride with oleate (O) residue in the 2 position and stearate (S) residues in the 1 and 3 positions; and so on. The productivity is defined by Equation 15.3, and its equilibrium value can be also expressed in terms of the initial concentration of substrates (9):

$$P = \frac{(SOO + SOS)}{OOO_{initial}} \times \frac{C_t F}{M} \quad (15.3)$$

where F is the flow rate and M is the amount of immobilized enzyme packed in the column reactor.

Reactor and Experimental Condition

The continuous reactor developed for the transesterification is shown schematically in Figure 15.5. The reactor, 10 in the figure, is a small column, 4.6 mm in diameter, having various lengths of 50, 100, 150 and 200 mm, thereby allowing the effect of flow velocity on conversion to be related to the residence time in the enzyme bed. The amount of immobilized lipase (particle size 440 μm) was 0.3 g in the shortest column and increased in proportion to the column length. One of the substrates, stearic acid, was contacted with flowing carbon dioxide in the vessel at a temperature to give the desired solubility in SC–CO_2. The other substrate, triolein, was fed separately by the syringe pump (2 in the figure) to the mixer, 9, and was mixed homogeneously with the SC–CO_2/stearic acid phase. The concentration of water in the CO_2 was controlled by feeding it manually through injector 8, at regular intervals. Product was recovered in a trap, 14, connected to the exit of the pressure regulator, 11, and the flow rate of carbon dioxide was measured at atmospheric pressure by a gas meter, 15.

Reaction temperature was changed from 313 K to 343 K at 20 MPa. The range of flow velocities achieved corresponded to Reynolds numbers from 2 to 8 and residence times from 15 to 100 sec. The feed concentrations of triolein and stearic acid were less

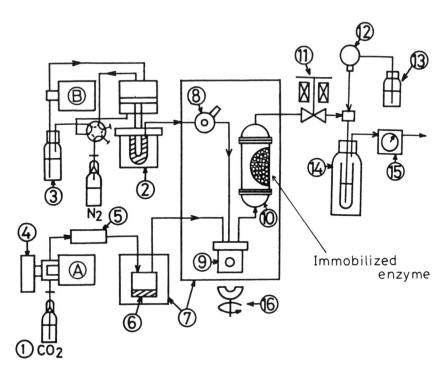

A.B.HPLC Pump	9.Mixing Cell
2.Syringe Pump	10.Column Reactor
3. I.P.A. Tank	11.Pressure Regulator
4.Cooling Unit	12.Pump
5.Molecular Sieves 13X	13.Ethanol
6.Stearic Acid	14.Trap
7.Const. Temp. Bath	15.Flow Meter
8.Injector(water)	16.Stirrer

Fig. 15.5. Scheme of experimental apparatus.

than 10 mM, and the concentration of water was changed from 2 to 20 mM during the experiment.

Experimental Results

A finite time was required for product elution to come to the steady state after the feed of substrates was started. When the residence time was kept constant, the steady state was attained much later as the flow velocity became smaller. The unsteady state seemed to be related to the adsorption of products and the distribution of water on and in the immobilized enzyme. The production rate of stearic acid–containing triglycerides (productivity) arrived at the same equilibrium value, irrespective of the flow velocity; hence, there could be no effect of external mass transfer on the reaction rate, as reported similarly by other researchers (7,14).

Effect of Water Concentration. An example of the effect of water concentration on the experimental results is shown in Figure 15.6. These results were obtained at 323 K

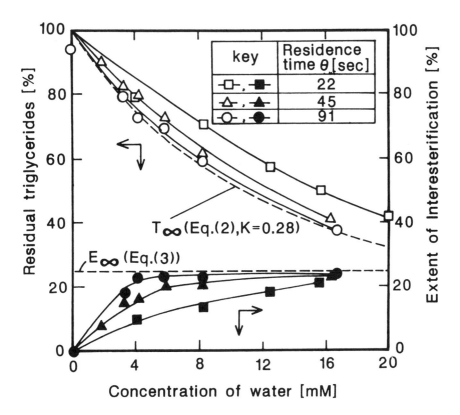

Fig. 15.6. Effect of water on residual triglyceride and extent of interesterification.

and 29.4 MPa, using feed concentrations of triolein and stearic acid of 2.6 to 2.7 mM
and 3.3 to 3.5 mM, respectively. Hydrolysis and acidolysis were both promoted with
increase in water concentration, indicating that there had to be an optimum water
concentration. The residual triglyceride and the extent of transesterification shown in
Figure 15.6, were attained at the steady state of the continuous reaction and were
equivalent to those of the equilibrium state estimated by Equations 15.1 and 15.2,
when the water concentration was large. As reported by Marty et al. (11), Figure 15.7
shows the existence of an optimum water concentration, which shifted to larger values
with an increase in the flow velocity (decrease in residence time). The higher water
concentration is preferable to promote the reaction at a shorter residence time; hydrolysis should be suppressed at longer residence times by the use of a low water
concentration.

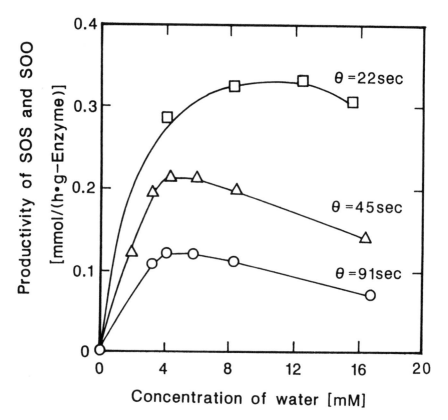

Fig. 15.7. Effect of water on productivity of triglycerides SOS and SOO.

Effects of Temperature and Pressure. The effect of temperature was examined under the same condition of pressure (29.4 MPa) and substrate concentration. The product distribution—that is, the selectivity of reaction—did not differ between 323 K and 343 K. Other reactions were conducted with different combinations of pressure and stearic acid concentration, while the concentrations of triolein and water were kept constant. Unexpectedly, a combination of low substrate concentration and low pressure resulted in slightly better productivity than the opposite combination at the short residence time of 37 s (Fig. 15.8). This result could be related to the combined effects of selectivity and reaction rate, as well as the effect of pressure on rate constants (14). The increase in pressure leads to a better partition of water to the SC–CO$_2$ phase, so that hydrolysis was suppressed by drying of the immobilized enzymes resulting in an increase in the production of triglycerides. In addition, it may become difficult for substrates and products to diffuse into the dried immobilized enzyme, and thus the reaction rate may become less at high substrate concentration and pressure.

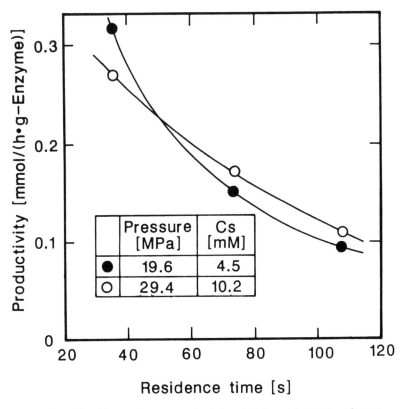

Fig. 15.8. Effect of residence time on productivity of triglycerides SOS and SOO (triolein: 1.7 mM, water: 9.7 mM).

The production of a cocoa butter substitute obtained by enzymatic transesterification in SC–CO_2 was found to be about 0.3 kg/h-kg-immobilized enzyme (Lipozyme IM20), leaving 76% residual triglyceride. Unfortunately, there is no published data on a solvent-free process for comparing. It is hoped that new immobilized enzymes will be developed in the near future that are suitable for use in SC–CO_2 leading to the use of supercritical fluids in an integrated system of reaction and separation.

References

1. Nakamura, K., *Trends in Biotechnol. 8:*288 (1990).
2. Aaltonen, O., and M. Rantakylä, *CHEMTECH 21:*240 (1991).
3. Randolph, T.W., H.W. Blanch and D.S. Clark, in *Biocatalysts for Industry,* edited by J.S. Dordick, Plenum Press, New York and London, 1991, pp. 219–237.
4. Luck, T., T. Kiesser, and W. Bauer, in *Proceedings of the World Conference on Biotechnology for the Fats and Oils Industry,* edited by T.H. Applewhite, American Oil Chemists' Society, Champaign, IL, 1987, pp. 343–345.
5. Ergan, F., M. Trani, and G. André, *Biotechnol. Lett. 10:*629 (1988).
6. Yamada, T. in *Proceedings of the Asian-Pacific Biochemical Engineering Conference,* 1990, (APBiochEC '90), the Korean Institute of Chemical Engineers and the Society of Chemical Engineers, Japan, 1990, pp. 453–456.
7. Erickson, J.C., P. Schyns, and C.L. Cooney, *AIChE J. 36:*299 (1990).
8. Steytler, D.C., P.C. Moulson, and J. Reynolds, *Enzym. Microb. Technol. 13:*221 (1991).
9. Aaltonen, O., and M. Rantakylä, in *Proceedings of the 2nd International Symposium on Supercritical Fluids,* edited by M.A. McHugh, Johns Hopkins University, Baltimore, MD, 1991, pp. 146–149.
10. Martins, J.F., S.F. Barreiros, E.G. Azevedo, and M. Munes da Ponte, in *Proceedings of the 2nd International Symposium on Supercritical Fluids,* edited by M.A. McHugh, Johns Hopkins University, Baltimore, MD, 1991, pp. 406–407.
11. Marty, A., W. Chulalaksananukul, R.M. Willemot, and J.S. Condoret, *Biotechnol. Bioeng. 39:*273 (1992).
12. Nakamura, K., in *Proceedings of the Asian-Pacific Biochemical Engineering Conference,* 1990, (APBiochEC '90), the Korean Institute of Chemical Engineers and the Society of Chemical Engineers, Japan, (1990), pp. 480-483.
13. Miller, D.A., H.W. Blanch, and J.M. Prausnitz, *Ind. Eng. Chem. Res. 30:*939 (1991).
14. Dumont, T., D. Barth, and M. Perrut, in *Proceedings of the 2nd International Symposium on Supercritical Fluids,* edited by M.A. McHugh, Johns Hopkins University, Baltimore, MD, 1991, pp. 150–153.
15. Chrastil, J., *J. Phys. Chem. 86:*3016 (1982).
16. Morild, E., *Adv. Protein Chem. 34:*91 (1981).
17. Chi, Y.M., K. Nakamura, and T. Yano, *Agric. Biol. Chem. 52:*1541 (1988).
18. Rupley, J.A., E. Graton, and G. Careri, *Trends in Biochem. Sci.8:*118 (1983).
19. Randolph, T.W., D.S. Clark, H.W. Blanch, and J.M. Prausnitz, *Science 239:*387 (1988).
20. Nakamura, K., in *Supercritical Fluid Processing of Food and Biomaterials,* edited by S.S.H. Rizvi, Blackie Academic and Professional, 1994, pp. 54–61.

Chapter 16

Basic Principles and the Role of Supercritical Fluid Chromatography in Lipid Analysis

P. Sandra[a] and F. David[b]

[a]Laboratory of Organic Chemistry, University of Gent, Gent, Belgium, and [b]Research Institute for Chromatography, Kortrijk, Belgium.

Lipid analysts have always been on the forefront of separation science, and the development of chromatographic techniques has often been catalyzed by their research activities. For example, the first application of gas chromatography by James and Martin, the inventors of the technique, was the analysis of fatty acids (1). Today, the analysis of lipids and their derivatives is one of the most important and successful applications of modern chromatography.

According to Christie (2), lipids can be defined as "fatty acids and their derivatives, and substances related biosynthetically or functionally to these compounds." Two classes of lipids can be distinguished (2–4): *simple* and *complex* lipids. Simple lipids yield, after saponification, at most two types of primary hydrolysis products per mole. Examples are triacylglycerols and sterol esters, which give, on saponification, only a fatty acid component and a sterol or glycerol, respectively. Hydrolysis of complex lipids, on the other hand, yields three or more primary hydrolysis products. An example is the phospholipid phosphatidylinositol, which yields glycerol, phosphoric acid, inositol, and two fatty acids on saponification.

Lipid mixtures are often extremely complex in their composition. Fatty acids differ in carbon number, in their degree of unsaturation, and even in the position of their double bonds. For example, a natural triacylglycerol C_{54}, with 54 fatty acid carbons, can be any of the following:

SSS   (0 NUFA)
SSO   (1 NUFA)
SSL
SSLn
SOO   (2 NUFAs)
SOL
SOLn
SLL
SLLn

We thank Unilever Research Laboratorium, Vlaardingen, The Netherlands, for financial support of our lipid research program.

SLnLn
OOO
OOL
OOLn
OLL
OLLn
OLnLn
LLL
LLLn
LLnLn
LnLnLn

where NUFA = number of unsaturated fatty acids and S stands for stearic or octade-
canoic acid, O for oleic or 9-octadecenoic acid, L for linoleic acid or 9,12-octadeca-
dienoic acid, and Ln for linolenic or 9,12,15-octadecatrienoic acid. This large number
of combinations also gives rise to positional isomers on the glycerol moiety (e.g.,
SOS, SSO) and even to optical isomers (e.g., SSO and OSS).

It would therefore be naïve to think that one chromatographic technique can resolve
all compounds in oils and fats and unravel the complex lipid profiles. A wide variety of
analytical techniques are currently applied in laboratories responsible for fat and oil
quality control as well as process analysis. The most important of these are capillary gas
chromatography (CGC) and high-performance liquid chromatography (HPLC).

Initially, simple and complex lipids were analyzed by gas chromatography after
saponification and derivatization into the fatty acid methyl esters (FAMEs). This
method is still routinely used. Capillary GC separations on biscyanopropyl stationary
phases provides a wealth of information that cannot be obtained with other chromatographic techniques (5,6). In the past, simple lipids such as mono-, di-, and triacylglycerols, sterol esters, and wax esters could not be analyzed quantitatively as molecular
species by GC. The introduction of thermostable, inert capillary columns and stationary phases, and nondiscriminative injection techniques has extended the applicability
of CGC to these high-molecular-weight compounds (7–13).

The main advantages of CGC for the analysis of simple lipids are the very high
efficiency, the speed of analysis and the large variety of sensitive detectors and selective
spectroscopic techniques. Today, high-temperature CGC is routinely applied. On apolar
columns of the dimethylsilicone type, lipids are separated according to their carbon
number (CN separations) and to the number of unsaturated fatty acids (NUFA) separations (7). On polarizable diphenyldimethylsilicone phases, lipids are also separated
according to the different combinations of saturated and unsaturated fatty acids in the
triglycerides (8). For the analysis of triglycerides, problems can be encountered with oils
containing large amounts of highly unsaturated triglycerides such as trilinolenin
(LnLnLn), which tend to polymerize in the column at the high temperatures (350°C)
needed for their elution. For quantitation of such lipids, calibration is necessary.

The analysis of complex lipids is impossible with CGC, because these polar and ionic species decompose even when cool on-column injection techniques are applied. Therefore, HPLC is the method of choice for characterization of highly unsaturated simple lipids and for the analysis of complex lipids. Triglycerides are separated according to their equivalent carbon number (ECN) by reversed-phase HPLC on octadecylated silica (3,14–17). ECN is defined as CN – 2NDB, where CN is the carbon number and NDB is the number of double bonds. ECN separation by RP HPLC gives a completely different profile of the oil than does CGC on apolar or polarizable stationary phases. Therefore, CGC and HPLC are complementary techniques.

For fats with a highly saturated character, HPLC has the disadvantage of low solubility of such compounds as SSS in the polar mobile phases that are used. HPLC on silica impregnated with silver ions is another alternative very often used for routine analysis of oils (3). Lipids are separated by so-called *E-classes*. *E*-separations are based on the total number of double bonds and on the configuration of the double bond (*cis/trans*); positional isomers, such as SOS and SSO, can be differentiated. Columns packed with Ag-impregnated silica, however, have a relatively short lifetime.

The analysis of complex lipids such as phospholipids is normally performed by straight-phase HPLC on silica or diol phases (3,18). Separation occurs according to polarity, yielding very limited information on the fatty acid composition. HPLC prefractionation followed by CGC is often used to characterize the fatty acids in the phospholipids. An important disadvantage of HPLC for the analysis of lipids is the limited number of plates in comparison to CGC, resulting often in incomplete separations of complex lipid mixtures. Moreover, many lipids do not contain chromophoric groups and can be detected only at low wavelengths by UV detection. Universal detectors such as refractive index, light scattering and moving-wire FID are applied, but their routine use is not without problems.

Considering the foregoing shortcomings of CGC or HPLC for the analysis of simple and complex lipids, supercritical fluid chromatography (SFC) has been advocated as an alternative, or at least a promising complementary analytical method (13,19–22). Because of the nature of the fluid mobile phase, the following advantages were realized: elution of high-MW, polar, and even ionic compounds at low analytical temperatures; compatibility with universal and sensitive GC detectors; and high efficiency of analysis.

In this chapter, an overview is given on the authors' practical experience using capillary supercritical fluid chromatography (CSFC) in the analysis of lipids. Over the last few years, various classes of lipids have been analyzed by CSFC, and the possibilities of CSFC will be compared to high-temperature CGC and HPLC.

Basic Principles of Supercritical Fluid Chromatography

The possibilities of supercritical fluids as mobile phases for chromatography are directly related to the nature of a supercritical fluid. In the supercritical state a substance possesses physical properties that are intermediate between those of a gas and those of a liquid. The most interesting advantages of these special physical properties for chromatography can be summarized as follows:

1. The density of a supercritical fluid is higher than the density of a gas, resulting in higher solubility for organic compounds. High-molecular-weight compounds can be eluted at low temperatures. Hence, the analysis of thermolabile compounds also becomes possible.

2. The higher diffusivity of supercritical fluids in comparison to liquids results in faster elution times at equivalent column efficiencies.

3. The lower viscosity of supercritical fluids in comparison to liquids allows the use of longer columns, thereby providing higher plate numbers.

For a detailed description of the basic principles of supercritical fluids, the reader is referred to Refs. 23–26. In most practical uses of SFC, carbon dioxide is used as the mobile phase. This is due to its favorable critical pressure (72.9 atm) and critical temperature (31.3°C). Moreover, carbon dioxide is cheap, noncorrosive and nontoxic. An additional advantage to the use of CO_2 is that it is compatible with both GC and LC detectors, so interfacing with spectroscopic methods is relatively easy.

Both capillary columns and packed columns also can be used for supercritical fluid chromatography (23). Until now, packed-column SFC has been mostly applied for fast analysis of relatively simple mixtures. Capillary column SFC offers high efficiency for the separation of more complex mixtures. In this chapter the application of CSFC will be highlighted. Some general remarks on new trends in SFC will be noted at the end of the chapter.

In general, fused silica capillary columns, 50–100 μm i.d., and 5–25 m in length, are used for CSFC. Columns with different stationary phases are commercially available and can be used with commercial SFC equipment. The sample capacity of such columns is sufficient for the analysis of lipids. Higher efficiency can be obtained on 10 to 25 μm i.d. columns, but practical experience has shown poorer results. In the studies presented here, a Carlo Erba SFC 3000 system was used with a Valco 60nL internal loop injector and FID detection. An integral restriction was made at the column end.

What possibilities does SFC, especially CSFC, offer the lipid analyst? How does SFC compare to CGC and HPLC for lipid characterization? What is the role of SFC? To obtain a better understanding, we will have to go back to some of the fundamentals of chromatography.

The intrinsic capability of a chromatographic technique for separating compounds is reflected by the *resolution equation*:

$$R_s = \frac{\sqrt{n}}{4} \; \frac{\alpha - 1}{\alpha} \; \frac{k}{k+1} \tag{16.1}$$

in which n is the column plate number, α the selectivity factor, and k the retention factor. For large k values, the resolution is controlled by n and α.

Column Efficiency

The column plate number n can be expressed as L/h, where L is the column length and h the height equivalent to a theoretical plate. The minimum plate height, h_{min}, or the highest efficiency, occurs at the optimal mobile phase velocity u_{opt}. The values of h_{min} and of u_{opt} for several columns in CGC, CSFC, HPLC, and SFC are listed in Table 16.1.

For capillary columns, h_{min} is equivalent to the internal diameter of the columns, whereas for packed columns h_{min} approaches 2 times the particle diameter. When one compares CGC and CSFC at high density, under optimal flow conditions, columns with the same dimensions (10 m × 100 µm i.d.) theoretically offer the same plate number (100,000), but the analysis in CSFC is roughly 500 times slower. Using more accurate data, CSFC is 750 times slower when the mobile phase has a density of 0.8 g/mL mobile phase density and 200 times slower at 0.2 g/mL density. Working close to the optimal u value in CSFC would result in analysis times of several hours. As the retention time t_R is equal to $L/u(1 + k)$, operating the column at 0.1 cm/s would give an analysis time of 16 h for a compound with $k = 5$ under isoconfertic conditions.

In practice, mobile phase velocities are selected that are much higher than u_{opt} with a commensurate drop in column plate number. This is illustrated in Figure 16.1, showing an isoconfertic analysis ($P = 25$ MPa and $T = 150°C$) of cocoa butter on a 10 m × 100 µm i.d. column coated with 0.2 µm SE-52. Here, the mobile phase velocity was 4 cm/s, limiting the effective plate count of the column to 5,000. Nevertheless, a complete separation of the palmitin-olein-palmitin (POP), palmitinolein-stearin (POS), and stearin-olein-stearin (SOS), the main triglycerides in cocoa butter, was achieved.

The slopes of the h–u curves at higher velocities are significant. The curves flatten when (1) the stationary phase film thickness decreases and (2) when the temperature

TABLE 16.1 Minimum Plate Height and Optimal Mobile Phase Velocity for Several Columns in CGC, CSFC, HPLC and SFC[a]

Method	CO_2 density (g/mL)	Capillary i.d. (µm)	Particle diameter d_p (µm)	h_{min} (µm)	u_{opt} (cm/s)	N/m	N/25 cm
CGC		250	—	250	40	4,000	—
		100	—	100	60	10,000	—
CSFC	0.8	100	—	100	0.08	10,000	—
	0.4	100	—	100	0.15	10,000	—
	0.2	100	—	100	0.3	10,000	—
	0.8	50	—	50	0.1	20,000	—
	0.4	50	—	50	0.2	20,000	—
	0.2	50	—	50	0.4	20,000	—
HPLC	—	—	5	10	0.15	—	25,000
SFC	—	—	5	10	0.5	—	25,000

[a]Values deduced from theoretical h–u plots.

increases. This is the reason why most CSFC applications have been performed on thin-film columns (<0.3 μm) and at relatively high temperatures (90–200°C). The effects of temperature on the resolution and on the retention factors of triglycerides have been studied by Proot et al. (21).

Very often, chromatograms show much sharper peaks and apparently higher efficiencies than in Figure 16.1. This is achieved by *density programming*—the CSFC analog of temperature programming in CGC or gradient elution in HPLC. The conditions in Figure 16.1, however, show the highest resolution for CN separations; triglycerides can be separated, to a difference of one carbon.

It is clear that as far as efficiency is concerned, CGC has to be preferred over CSFC. For example, the application of CSFC and CGC for quality control of milk chocolate is shown in Figs. 16.2*a* and *b*. High-temperature CGC is preferred for routine analysis of the lipid mixture in the chocolate industry because of its higher efficiency and much shorter analysis time (30 min in CSFC *vs.* 3.5 min in CGC). This analysis allows the determination of the quantity of milk fat in chocolate (asterisk-marked peaks) as well as the presence of cocoa butter equivalents (11). Table 16.2 compares the quantitative results obtained by CSFC and CGC with respect to the contents of POP, POS, and SOS, the cocoa butter lipids in the milk chocolate, shown in

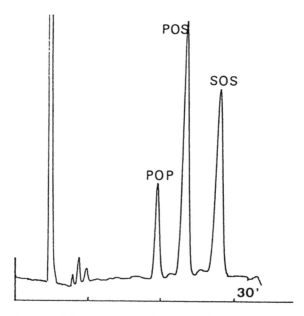

Fig. 16.1. Triglyceride analysis of cocoa butter. Column: 10 m × 100 μm i.d., 0.2 μm SE-52; temperature: 150°C; pressure: 25 MPa.

Figs. 16.1 and 16.2. The data are very similar, illustrating that no decomposition occurs in the CGC analysis of these lipids.

Further developments in CSFC should be directed to the optimization of very small internal-diameter columns (10 μm i.d.), offering high efficiencies in short analysis times. At present, technological difficulties are encountered in the practical application of such columns.

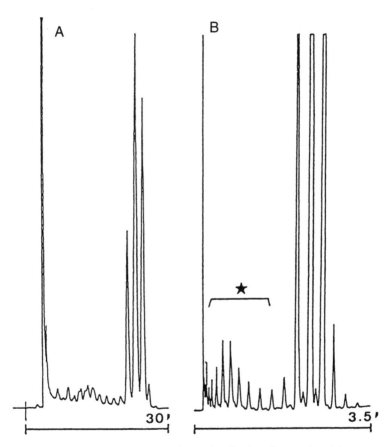

Fig. 16.2. CSFC and HT-CGC analyses of milk chocolate triglycerides. *(a)* CSFC: Column: 10 m × 100 μm i.d., 0.2 μm SE-54; temperature: 170°C; pressure: program 19 MPa (10 min) to 29 MPa at 0.5 MPa/min. *(b)* UT-CGC: Column: 6 m × 250 μm i.d., 0.1 μm OV-1; temperature: program 290–350°C at 20°C/min; carrier gas: hydrogen at 0.1 MPa. *Source:* Sandra, P., et al., in *Proceedings of the Seventh International Symposium on Capillary Chromatography,* edited by D. Ishii, et al., The University of Nagoya Press, Japan, 1986, p. 650.

TABLE 16.2 Cocoa Butter Triglycerides in Milk Chocolate

	POP	POS	SOS
CSFC	18.77%	45.63%	35.50%
CGC	18.01%	46.37%	35.62%

Comparing SFC and HPLC, one finds that the speed of analysis is roughly 4 times faster for SFC. At u_{opt}, the efficiencies are the same or slightly higher for SFC at lower fluid densities. This is an important argument for selecting SFC for applied analysis, if the same selectivity can be obtained. Recently Berger et al. (28) demonstrated apparent efficiencies of 200,000 plates or more in 30 min, using coupled columns in SFC. This result opens some interesting and new perspectives for the analysis of lipid mixtures.

Column Selectivity

Over the years CGC and HPLC have been extensively evaluated and optimized for lipid analysis. The separation mechanism in CGC is based on vapor pressure differences and selective interactions with the stationary phase. CGC, however, offers limited selectivity, because only the stationary phase can be modified; its good resolution is mainly due to column efficiency. In HPLC, retention is based on solubility and selective interactions of the solutes in both the stationary phase and the mobile phase; although HPLC lacks the high efficiency of CGC, very specific lipid separations can be realized by using the high selectivity it offers through a suitable combination of phases. CSFC and SFC are often claimed to offer more selectivity than CGC and HPLC. Notwithstanding theoretical considerations, what can be achieved with CSFC in applied analysis? Hardly any selectivity can be introduced *via* the mobile phase, CO_2, because doing so would preclude the use of the "universal" FID detector. Schwartz et al. (29) used an interesting combination of modifiers (0.3% formic acid and 0.15% water) for FID detection in packed-column SFC. The same mobile phase composition has been evaluated in our laboratory for CSFC, but no contribution to solute selectivity has been observed. Light scattering detection, as described by Hoffman and Greibrokk (30), is another possibility but has not been fully exploited extensively in lipid analysis.

A number of stationary phases have been evaluated in CSFC for triglyceride separations, including methyl silicone, methylphenyl (5%) silicone, methylphenyl (50%) silicone, methylphenyl (25%) cyanopropyl (25%) silicone, and Carbowax 20M. With the exception of very good CN separations on apolar phases, the selective interactions with the stationary phase are mostly offset by the solubility of the lipids in the $SC-CO_2$ mobile phase, and the lack of high efficiency. This is illustrated in Fig. 16.3, showing the analysis of tripalmitin (PPP), tristearin (SSS), triolein (OOO), trilinolein (LLL), and trilinolenin (LnLnLn) on a methylphenyl (25%) cyanopropyl (25%) silicone phase (22). The resolution achieved is not high enough to separate complex samples such as those encountered in applied analysis. Within the separation space of the two extreme CN 54 species (SSS and LnLnLn), 18 other CN 54 species can occur in oils and fats. To our knowledge, the best and most complete separation of triglycerides using CSFC has been reported by Richter et al. (31). In that study, soybean oil was analyzed on a 10 m × 50 µm i.d. column coated with a 0.25-µm Carbowax 20M film. Separation occured according to the degree of unsaturation in the triglycerides, but the resolution was much poorer than those obtained with CGC (8) or HPLC (15). In order to unravel the complexity of unsaturation in lipids, additional work is needed in the area of stationary phase selectivity in CSFC. NUFA, ECN, or E separations achieved to date using CSFC are inferior than those currently achieved using CGC or HPLC.

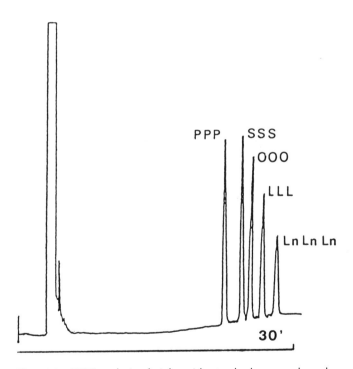

Fig. 16.3. CSFC analysis of triglyceride standards on a polar column. Column: 10 m × 100 µm i.d., 0.1 µm OV-225; temperature: 150°C; pressure: program 15–30 MPa at 0.5 MPa/min. (*Source:* Sandra, P., et al., in *Proceedings of the Seventh International Symposium on Capillary Chromatography,* edited by D. Ishii et al., The University of Nagoya Press, Japan, 1986, p. 650.)

Therefore, the role of CSFC and SFC in the routine analysis of oils and fats is presently restricted to:

1. CN separations of lipid mixtures containing solutes of high molecular weight
2. Analysis of thermolabile lipids species
3. Samples amenable to CGC, but for which the selectivity of the stationary phase at high column temperature is insufficient

Current applications of CSFC will be illustrated in the following section.

Applications

Analysis of Simple Lipids

Supercritical CO_2 is an excellent solvent for common lipids. Fatty acids, mono-, di-, and triglycerides, sterol esters, and other species, have been extensively studied with CSFC. Most successful applications concern CN determinations on nonpolar columns.

As an illustration, the carbon number separation of palm kernel oil by CSFC is given in Fig. 16.4. The triglycerides shown range from carbon numbers 28 to 56. On nonpolar columns the elution of the compounds is controlled by the density of the mobile phase. For the separation of the triglycerides shown in Fig. 16.4, an increasing pressure program was used. Alternatively, the density can be increased by decreasing the temperature at constant pressure (13). The reproducibility of retention data and quantitative results have been evaluated, and both are excellent: respectively, the quantitative analysis precision being within <0.1% σ_{rel} and <1% σ_{rel}. This is the same magnitude as for CGC and HPLC.

Figs. 16.5a and b show an application taken from the pharmaceutical industry. *Miglyols* are fractions of coconut oil, and the differentiation of miglyol 814 and 812 can easily be made. The lipid peaks range from 18 to 26 min and from 28 to 34 min, respectively. Frequently, the use of columns with i.d.'s of 100 μm provides better quantitation.

Fig. 16.6 shows the analysis of meat fat on a 50-μm i.d. column. A very good CN separation is obtained, and free cholesterol can also be detected with good peak shape. The reproducibility of quantitation, however, was only on the order of 3.5% σ_{rel}. For the lipids CN 52, and 54, some fine structure was observed that cannot, however, be exploited.

Fig. 16.7 shows the analysis of interesterified cocoa butter, achieved under the same conditions as Figure 16.1. Interesterification gives a random distribution of the fatty acids P, O, and S on the glycerol moiety, resulting in a much more complex profile than the triglycerides of the original cocoa butter, consisting only of POP, POS, and SOS. A comparison of Figs. 16.1 and 16.7 shows broader band widths for peaks CN 50, 52 and 54, along with the appearance of higher concentrations of diglycerides, and the appearance of the CN 48 moiety, consisting of tripalmitin (PPP).

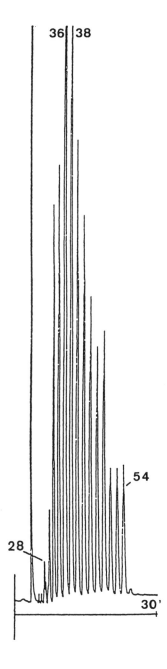

Fig. 16.4. CN separation of palm kernel oil by CSFC. Column: 10 m × 100 μm i.d., 0.2 μm SE-54; temperature: 170°C; pressure: program 19–29 MPa in 30 min.

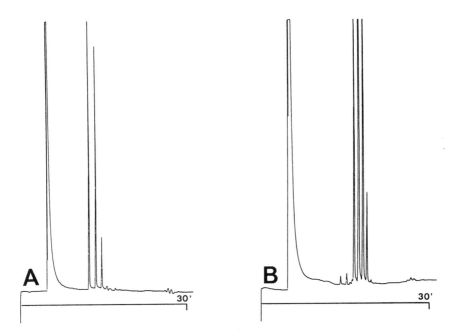

Fig. 16.5. CN separation of miglyols by CSFC. Column: 20 m × 100 μm i.d., 0.2 μm SE-54; temperature: 100°C; pressure: program 10–30 MPa at 0.5 MPa/min. (a) Miglyol 814, (b) Miglyol 812.

This interesterified cocoa butter mixture was analyzed using different stationary phases with CSFC, but the results were not as good as those attained with high-temperature CGC. Figs. 16.8a and b show the CGC results on methyl silicone and diphenyldimethyl silicone, respectively. On methyl silicone, separation in decreasing order of NUFA occurs for each CN number, because of the lower boiling point of unsaturated triglycerides relative to saturated triglycerides. On the polarizable diphenyldimethyl silicone, triglycerides having the same CN are separated in increasing order of NDB. For cocoa butter, the only unsaturated fatty acid is oleic acid, so both profiles are similar, but the fine structures in each CN are reversed.

If the triglycerides in oils and fats contain higher-carbon fatty acids, highly unsaturated fatty acids, or both, then CSFC will be a better alternative for CN determinations than CGC. Also, because of the low analysis temperature utilized, no thermal degradation or polymerization of the sample occurs. This principle is illustrated in Fig. 16.9, which shows the analysis of rapeseed oil that has a high erucic acid (C_{22}:1) content. Triglycerides as high as CN 70 were eluted perfectly. Hence, CSFC is the only alternative for the fast CN analysis of the higher-MW triglycerides. Unfortunately, further resolution for each CN number can be obtained only by HPLC (quantitative) (15) or CGC (qualitative) (8). For example, in this sample, CN number 62 includes

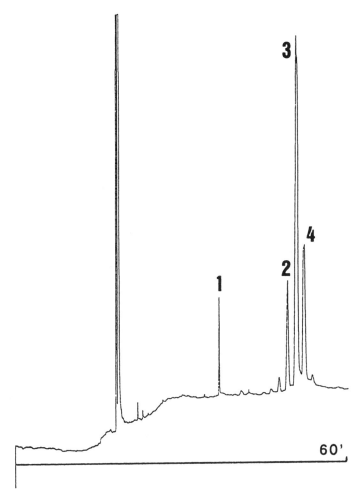

Fig. 16.6. CSFC analysis of meat fat. Column: 10 m × 50 μm i.d., 0.1 μm SE-54; temperature: 130°C; pressure: program 10 MPa (10 min), 1 MPa/min to 25 MPa, 0.25 MPa/min to 30 MPa (10 min). Peaks: 1, cholesterol; 2, CN 50; 3, CN 52; 4, CN 54.

the triglycerides dierucin-olein (ErErO), dierucinlinolein (ErErL) and dierucin-linolenin (ErErLn).

Another typical application of CSFC is for the analysis of ricinus oil (castor oil). This oil mainly contains 12-hydroxyoctadecenoic acid and is therefore relatively polar in nature. The oil has to be derivatized for CGC analysis, but with CSFC the underivatized oil can be analyzed.

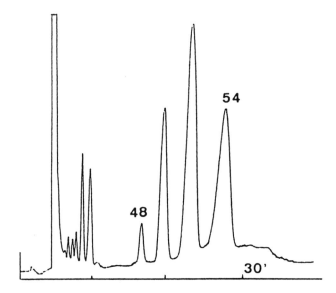

Fig. 16.7. CSFC analysis of interesterified cocoa butter.
Column: 10 m × 100 μm i.d., 0.2 μm SE-52; temperature:
150°C; pressure: 25 MPa. Chromatogram has been recorded
under the same conditions as for Fig. 16.1.

Mixtures of mono-, di-, and triglycerides can also be fractionated by CSFC. Fig. 16.10 shows the analysis of such palm oil glycerides. On a nonpolar column the mono-, di-, and triglycerides are separated according to carbon number. This analysis was performed without derivatization of the sample; using CGC, derivatization (silylation) is often needed to obtain good peak shape for the more polar mono- and diglycerides (9).

The analysis of sterols and sterol esters is another application of CSFC (32). Fig. 16.11 shows the separation of the cholesterol esters of palmitic, stearic, and erucic acid, so a more challenging separation is shown in Fig. 16.12 for pharmaceutical formulation containing testosterone esters along with di- and triglycerides. The steroid esters can easily be analyzed by CGC, but the lipid material has to be removed to generate good quantitative data and to protect the GC system. Applying CSFC, the sample can be injected without any sample preparation step, and the steroid esters and di- and triglycerides can be quantified simultaneously.

Many CSFC separations of fatty acids and fatty acid derivatives have been reported. This subject will not be discussed in this chapter, because we strongly believe that no chromatographic method can compete with CGC for the separation of fatty acids and derivatives. Some exceptions will be discussed in the section "Miscellaneous CSFC Separations."

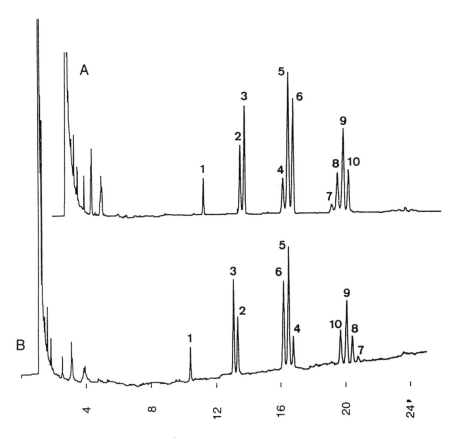

Fig. 16.8. Separation of interesterified cocoa butter by HT–CGC. (*a*) NUFA. Column: 25 m × 250 µm i.d., 0.1 µm OV-1; temperature: program 340–365°C at 1°C/min; carrier gas: hydrogen at 0.1 MPa. Peaks: 1, PPP; 2, POP; 3, PPS; 4, POO; 5, POS; 6, PSS; 7, OOO; 8, SOO; 9, SOS; 10, SSS. (*b*) NDB. Column: 25 m × 250 µm i.d., 0.1 µm diphenyl dimethyl silicone; temperature: program 340–365°C at 1°C/min; carrier gas: hydrogen at 0.1 MPa. Peaks: 1, PPP; 2, POP; 3, PPS; 4, POO; 5, POS; 6, PSS; 7.000; 8, SOO 9, SOS; 10, SSS.

Analysis of Complex Lipids

The most important class of complex lipids consists of the phospholipids. The structures of the main phospholipids are given in Fig. 16.13. The group connected to the phosphate chain defines different phospholipid categories such as phosphatidyletha-nolamine, phosphatidylserine, phosphatidylcholine, and phosphatidylinositol. Phospholipids differ in polarity—a characteristic that can be used to separate them by thin-layer chromatography or HPLC. Differentiation according to the fatty acids is normally performed by CGC after hydrolysis of the phospholipids. The analysis of

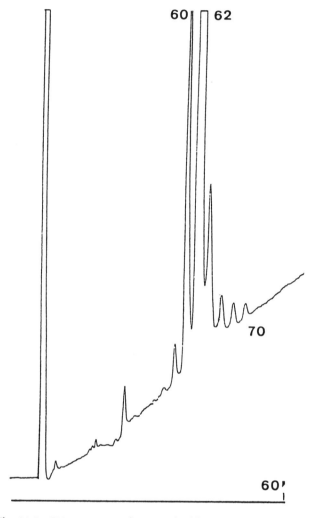

Fig. 16.9. CN separation of rapeseed oil by CSFC. Column:
20 m × 100 μm i.d., 0.2 μm SE-54; temperature: 150°C; pressure: program 15–25 MPa at 1 MPa/min, then to 35 MPa at
0.1 MPa/min.

the original phospholipid molecular species as such is impossible by CGC, because
of their thermalability (instability at high temperature): the compounds cleave at the
O–P bond to yield diglycerides. HPLC–light scattering detection (2,3) and
HPLC–MS (33) on reversed-phase columns have been applied for characterization of
the fatty acid moieties in phospholipids. A promising alternative to the high-temperature techniques is CSFC. However, phospholipids are polar and ionic, and are there-

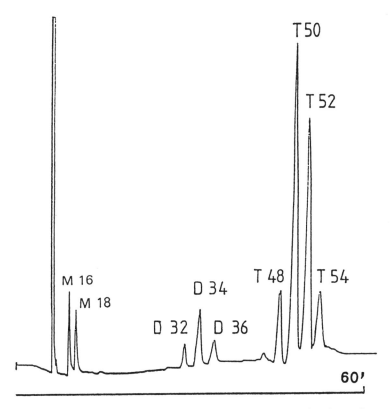

Fig. 16.10. CSFC analysis of palm oil mono- (M), di- (D), and triglycerides (T). Column: 20 m × 100 μm i.d., 0.2 μm SE-54; temperature: 150°C; pressure: program 15–25 MPa at 1 MPa/min, then to 30 MPa at 0.1 MPa/min.

fore insoluble in pure supercritical CO_2. A solution can be found in using packed columns with modifiers, but then the advantages of high resolution, inertness, and universal FID detection of CSFC are lost. To perform CSFC, the phospholipids have to be derivatized into less polar molecules. This can be achieved by methylation of the phosphoric acid group with diazomethane, followed by acylation of the amino or hydroxyl functions with trifluoroacetic acid anhydride. The quaternary ammonium group in phosphatidylcholine can also be demethylated using thiophenol (34). The derivatized phospholipids are soluble in supercritical CO_2 and can be analyzed by CSFC with pure CO_2.

This is illustrated in Fig. 16.14*a* (phosphatidylethanolamine), *b* (phosphatidylserine), *c* (phosphatidylinositol) and *d* (phosphatidylcholine). The peaks correspond to different combinations of fatty acids; hence the chromatograms provide CN information. The most difficult application in this series is the analysis of phosphatidylcholines. The analysis of some phosphatidylcholine standards, shown in Fig. 16.15, indicates that the derivatization reaction occurs, yielding good peak shapes.

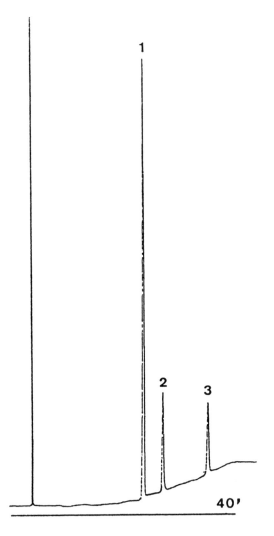

Fig. 16.11. Analysis of sterol esters by CSFC. Column: 20 m × 100 μm i.d., 0.2 μm SE-54; temperature: 150°C; pressure: program 15 MPa–5 min isobaric, then to 30 MPa in 60 min. Peaks: 1, cholesterol palmitate; 2, cholesterol stearate; 3, cholesterol erucate. (Source: Hill, H.H., in *SFC Applications 1989*, edited by K.E. Markides and M.L. Lee, Brigham Young University Press, Provo, UT, 1989, p. 117.)

The presence of the phosphate group in the eluting solutes could be confirmed by NPD detection. Because of the increasing importance of phospholipids in food and pharmaceutical products, this analysis may prove an important and unique application of CSFC.

Miscellaneous CFSC Separations

Lipids have often been used as test analytes to illustrate improvements in CSFC, rather than practical lipid analysis. Nevertheless, some unique CSFC separations have been described. The analysis of hydroxy fatty acids without derivatization has been

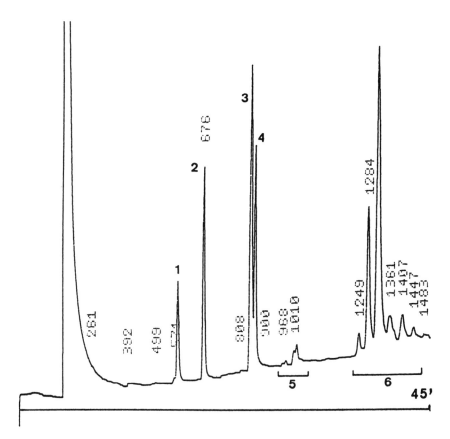

Fig. 16.12. Analysis of steroid esters in a pharmaceutical formulation. Column: 20 m × 100 µm i.d., 0.2 µm SE-54; temperature: 100°C; pressure: program 15 MPa–5 min isobaric, to 25 MPa at 0.5 MPa/min. Peaks: 1, testosterone propionate; 2, testosterone isocaproate; 3, testosterone decanoate; 4, testosterone phenylpropionate; 5, diglycerides; 6, triglycerides (main peak CN 54). (*Source:* Hill, H.H., in *SFC Applications 1989*, edited by K.L. Markides and M.L. Lee, Brigham Young University Press, Provo, UT, 1989, p. 117.)

reported by Doehl et al. (35). Artz showed an excellent separation of methyl linolenate hydroperoxides on a 20 m × 50 µm i.d. column coated with SB-cyanopropyl-50 (36). Holzer, et al. (37,38) have achieved separations of glycerol tetra ethers isolated from bacteria. This analysis is not possible by CGC because of the high molecular weights of the ethers, whereas HPLC lacks sufficient sensitivity.

Hawthorne and Miller (39) separated wax components with CSFC and identified them using quadrupole MS with only slight modification to the normal CGC/MS configuration. Although high-temperature CGC can be applied to analyze these wax sam-

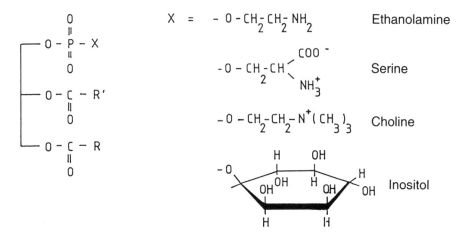

Fig. 16.13. Structures of the most important phospholipids.

ples such as bayberry wax, beeswax, and carnauba wax, CSFC offers more selectivity than CGC, even on the same stationary phase, because selectivity decreases at the high temperatures needed to elute all of the wax components. For example, Fig. 16.16 shows the analysis of beeswax on a 10 m × 50 µm i.d. column, coated with methyl silicone. The homologous series marked with an asterisk coelutes with and is masked by the following series in CGC. Different batches of waxes can also be differentiated. Figs. 16.17a and b show the relevant part of two carnauba wax batches. Qualitatively, both chromatograms appear similar, but the quantities of individual wax esters differ.

CSFC can also be used as a screening method for unknown oil samples and lubricants. Fig. 16.18 shows the analysis of an oil that was found to contain lipids and a mineral oil. The analysis of the total oil as such shows three distinct groups: mineral oil, diglycerides, and triglycerides (Fig. 16.18a). The oil was fractionated in three different fractions (nonpolar, medium polar, and polar) by solid phase extraction on silica. The nonpolar fraction contains the mineral oil and the triglycerides (Fig. 16.18b); the medium polar fraction, mono- and diglycerides (Fig. 16.18c). The CSFC analysis of the polar fraction gave no peaks. After derivatization with trifluoroaceticanhydride, fatty amines could be detected (40). Fig. 16.19 shows the resultant chromatogram.

Several unique CSFC applications of lipid-derived chemicals have never been published, for proprietary reasons. These involve the separation of esters formed when

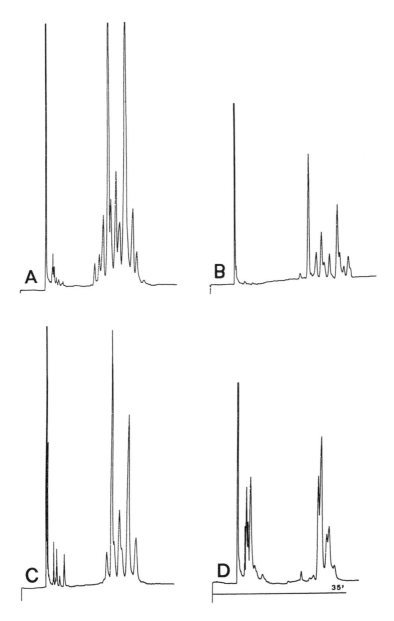

Fig. 16.14. CSFC analysis of phospholipids. Column: 20 m × 100 µm i.d., 0.2 µm SE-54; temperature: 130°C; pressure: program 18 MPa—5 min isobaric, to 28 MPa in 30 min. (*a*) phosphatidylethanolamine; (*b*) phosphatidylserine; (*c*) phosphatidylinositol; (*d*) lecithin (phosphatidylcholine).

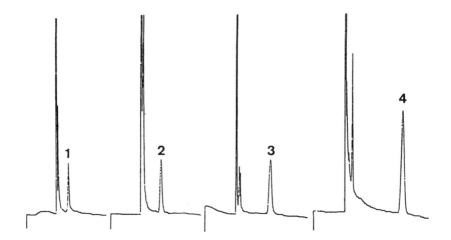

Fig. 16.15. Analysis of phosphatidylcholine standard compounds by CSFC. Column: 20 m × 100 μm i.d., 0.2 μm SE-54; temperature: 130°C; pressure: 25 MPa. Peaks: 1, dilauryl-phosphatidylcholine; 2, dimyristyl-phosphatidylcholine; 3, dipalmitylphosphatidylcholine; 4, distearyl-phosphatidylcholine.

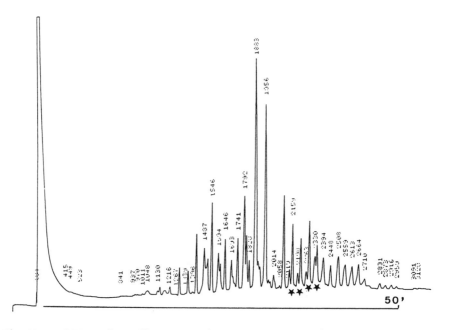

Fig. 16.16. CSFC analysis of beeswax. Column: 10 m × 50 μm i.d., 0.1 μm SE-54; temperature: 100°C; pressure: program 10 MPa–7 min isobaric, then to 30 MPa at 0.4 MPa/min.

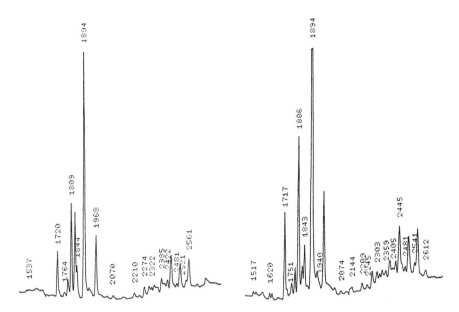

Fig. 16.17. Relevant part of the CSFC analysis or two different carnauba wax batches. Column: 10 m × 50 µm i.d., 0.1 µm SE-54; temperature: 100°C; pressure: program 10 MPa–7 min isobaric, then to 30 MPa at 0.4 MPa/min.

palmitic, stearic, and oleic acids are reacted with linear or cyclic poly oils, and of oxidized and polymerized oils and fats. Compounds having molecular weights as high as 4000 daltons can be eluted with CSFC. Some possible separations achieved by CSFC for this MW range can be found in Refs. 41 and 42, which discuss the analysis of a sucrose polyester by CSFC–FID and CSFC–MS.

Conclusion and Future Directions

For the analysis of lipids, each chromatographic technique has its own possibilities and limitations; therefore, the techniques are complementary to each other. CSFC is ideally suited for CN separations of mono-, di-, and triglycerides, sterol esters; waxes; phospholipids, etc. The separation power (efficiency and selectivity) is, however, insufficient to resolve complex lipid mixtures according to their NUFAs, NBDs, ECN and E-numbers. New developments are expected in SFC using packed and micropacked columns, and from using open tubular columns with extremely small internal diameters. Currently, the introduction of packed columns with apparent efficiencies of

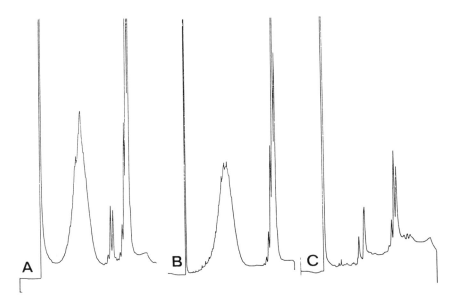

Fig. 16.18. CSFC analysis of an unknown oil. Column: 20 m × 100 μm i.d., 0.2 μm SE-54; temperature: 150°C; pressure: program 15 MPa–to 30 MPa at 0.25 MPa/min. (*a*) total oil; (*b*) nonpolar fraction (MO: mineral oil; TG: triglycerides); (*c*) Medium polar fraction (MG: monoglycerides; DG: diglycerides).

200,000 plates, in combination with diode array detection and nitrogen-phosphorus detection, opens new perspectives for the separation of phospholipids, according to their polarity and C-number.

References

1. James, A.T., and A.J.P. Martin, *Biochem. J. 50:*679 (1952).
2. Christie, W.W., *Lipid Analysis,* 2nd edn., Pergamon Press, Oxford, UK, 1982.
3. Christie, W.W., *High Performance Liquid Chromatography and Lipids,* Pergamon Press, Oxford, UK, 1987.
4. Christie, W.W., *Gas Chromatography and Lipids,* The Oily Press, Ayr, Scotland, 1989.
5. Sandra, P., F. David, and G. Dirickx, *J. High Resol. Chrom. & Chrom. Comm. 11:*256 (1988).
6. Vannieuwenhuyze, F., and P. Sandra, *Chromatographia 23:*850 (1987).
7. Geeraert, E., P. Sandra, and D. De Schepper, *J. Chrom. 279:*287 (1983).
8. Geeraert, E., and P. Sandra, *J. High Resol. Chrom. & Chrom. Comm. 8:*415 (1985).
9. Geeraert, E. in Sample *Introduction in Capillary Gas Chromatography,* edited by P. Sandra, Hüthig Verlag, Heidelberg, 1985, p. 159.
10. Hinshaw Jr., J.V., and W. Seferovic, *J. High Resol. Chrom. & Chrom. Comm. 9:*731 (1986).
11. Geeraert, E., and P. Sandra, *J. Am. Oil Chem. Soc. 64:*100 (1987).

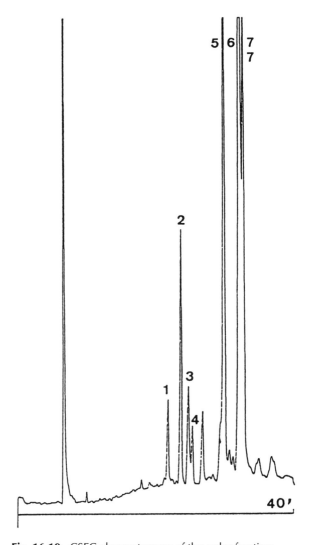

Fig. 16.19. CSFC chromatogram of the polar fraction of the oil in Fig. 16.18 (aliphatic amines, as their acyl derivatives). Column: 20 m × 100 µm i.d., 0.2 µm SE-54; temperature: 100°C; pressure: program 10 MPa–to 20 MPa at 0.2 MPa/min. Peaks: 1, $C_{16:0}NH_2$; 2, $C_{17}NH_2$ (internal standard); 3, $C_{18:1}NH_2$; 4, $C_{18:0}NH_2$; 5, $C_{16:0}NH(CH_2)_3NH_2$; 6, $C_{18:1}NH(CH_2)_3NH_2$; 7, $C_{18:0}NH(CH_2)_3NH_2$. (*Source:* David, F., and P. Sandra, *J. High Resol. Chrom. and Chrom. Comm.* 11:897 (1988).

12. Termonia, M., F. Munari and P. Sandra, *J. High Resol. Chrom. & Chrom. Comm. 10:*263 (1987).
13. Sandra, P., F. David, F. Munari, G. Mapelli, and S. Trestianu, in *Supercritical Fluid Chromatography,* edited by R.M. Smith, The Royal Society of Chemistry, London, UK, 1988, p. 137.
14. Shukla, V.K.S., W. Schiotz Nielsen, and W. Batsberg, *Fette Seifen Anstrichm. 85:*274 (1983).
15. Geeraert, E., and D. De Schepper, *J. High Resol. Chrom. & Chrom. Comm. 6:*123 (1983).
16. Fiebig, H.-J., *Fette Seifen Anstrichm. 87:*53 (1985).
17. Frede, E., *Chromatographia 21:*29 (1986).
18. Becart J., C. Chevalier, and J.P. Bresse, *J. High Resol. Chrom. 13:*126 (1990).
19. Chester, T.L., *J. Chrom. 299:*424 (1984).
20. White, W.M., and R.K. Houck, *J. High Resol. Chrom. & Chrom. Comm. 8:*293 (1985).
21. Proot, M., P. Sandra, and E. Geeraert, *J. High Resol. Chrom. & Chrom. Comm. 9:*189 (1986).
22. Sandra, P., M. Proot, and E. Geeraert, in *Proceedings of the Seventh International Symposium on Capillary Chromatography,* edited by D. Ishii, K. Jinno and P. Sandra, The University of Nagoya Press, Japan, 1986, p. 650.
23. McHugh, M., and V. Krukonis, *Supercritical Fluid Extraction, Principles and Practice,* Butterworth, Stoneham, MA, 1986.
24. Charpentier, B.A., and M.R Serenants, *Supercritical Fluid Extraction and Chromatography: Techniques and Applications,* edited by B.A. Charpentier and M.R. Sevenants, ACS Symposium Series 366, American Chemical Society, Washington DC, 1988.
25. Bartle, K.D., in *Supercritical Fluid Chromatography,* edited by R.M. Smith, The Royal Society of Chemistry, London, UK, 1988, p. 1.
26. Lee, M.L., and K. Markides, *Analytical Supercritical Fluid Chromatography and Extraction,* Chromatography Conferences, Inc., Provo, UT, 1990.
27. Schoenmakers, P.J., in *Supercritical Fluid Chromatography,* edited by R.M. Smith, The Royal Society of Chemistry, London, UK, 1988, p. 102.
28. Berger T.A., and H.M. Wilson, in *Proceedings of the Fourteenth International Symposium on Capillary Chromatography,* edited by P. Sandra and M.L Lee, FISCC, Miami, FL, 1992, p. 783.
29. Schwartz, H.E., P.J. Barthel, S.E. Mohring, T.L. Yates, and H.H. Lauer, *Fresenius Z. Anal. Chem. 330:*204 (1988).
30. Hoffman, S., and T. Greibrokk, *J. Microcol. Sep. 1:*35 (1989).
31. Richter B.E., M.R. Andersen, D.E. Knowles, E.R. Campbell, N.L. Porter, L. Nixon, and D.W. Later, in *Supercritical Fluid Chromatography and Applications,* edited by B.A. Charpentier and M.R. Sevenants, ACS Symposium Series 366, American Chemical Society, Washington, DC, 1988, p. 179.
32. Hill, H.H., in *SFC Applications 1989,* edited by K.L. Markides and M.L. Lee, Brigham Young University Press, Provo, UT, 1989, p. 117.
33. Kim, H.-Y., and N. Salem, Jr., *Anal. Chem. 59:*722 (1987).
34. Knapp, R.D., *Handbook of Analytical Derivatization Reactions,* Wiley, New York, 1979, pp. 118 and 391.
35. Doehl, J., and T. Greibrokk, *J. Chrom. 392:*175 (1987).

36. Artz, W.E., in *SFC Applications 1989,* edited by K.E. Markides and M.L. Lee, Brigham Young University Press, Provo, UT, 1989, p. 99.
37. DeLuca S.J., K.J. Voorhees, T.A. Langworthy, and G.U. Holzer, *J. High Resol. Chrom. & Chrom. Comm. 9:*182 (1986).
38. Holzer, G.U., P.J. Kelly, and W.J. Jones, *J. Microbiol. Methods 8:*161 (1988).
39. Hawthorne, S.B., and D.J. Miller, *J. Chrom. 388:*397 (1987).
40. David F., and P. Sandra, *J. High Resol. Chrom. & Chrom. Comm. 11:*897 (1988).
41. Chester, T.L., D.P. Innis, and G.D. Owens, *Anal. Chem. 57:*2243 (1985).
42. Pinkston, J.D., and D.J. Bowling, in Proceedings of the *Fourteenth International Symposium on Capillary Chromatography,* edited by P. Sandra and M.L. Lee, FISCC, Miami, FL, 1992, p. 758.

Chapter 17

Supercritical Fluid Chromatography for the Analysis of Oleochemicals

T.L. Chester

The Procter & Gamble Company, Miami Valley Laboratories, P.O. Box 8707, Cincinnati, Ohio 45253-8707.

An empirical awareness of the differences between the properties of liquids and of gases begins for nearly everyone in early childhood. It seems a natural consequence of these obvious differences that gas chromatography (GC) and liquid chromatography (LC, or high-performance liquid chromatography, HPLC) will, in turn, have very different characteristics and instrumentation requirements.

The properties of a supercritical fluid are not as obvious as those of liquids and gases and are much harder to rationalize. A supercritical fluid can bridge the properties of liquid and gas without an obvious phase transition. At pressures above its critical pressure, a liquid cannot be distinguished from its corresponding supercritical fluid as the temperature is varied through the critical temperature. Similarly, at temperatures above the critical temperature, a gas cannot be distinguished from its corresponding supercritical fluid as the pressure is varied through the critical pressure. Thus, it is possible to change a material from liquid to gas or vice versa without a phase transition, by going around the critical point. Similarly, supercritical fluid chromatography (SFC) incorporates many of the characteristics of GC and LC.

Simply stated, SFC (1,2) is a form of column chromatography in which the mobile phase is a supercritical fluid. However, it is possible, and sometimes preferable, to operate an SFC instrument with a liquid or gaseous mobile phase. This choice simply depends on the choice of the mobile phase temperature and pressure in relation to its phase behavior. GC and LC operate according to the same fundamental rules, although different empirical descriptors have evolved to explain each, within the limits of the characteristics of the respective mobile phases. The empirical distinctions between these techniques merge as the mobile phases enter the supercritical region from either a GC or an LC starting point. It is often incorrectly assumed that SFC simply duplicates capabilities already available in GC or LC. Actually, there is a unique and new combination of capabilities that the specific properties of SFC make available. In this chapter we will examine these capabilities and some of their applications in SFC.

The Scope of SFC among Separation Options

Imagine a typical, new analysis need in your area of responsibility. Let us further imagine that this need involves a fat; a fat-derived substance; something associated with fats, such as emulsifiers and surfactants; or a fat-soluble additive, impurity, or by-product in your sample. In most cases an analyst considering options thinks of applying a chemical separation technique either first or quite early in the analysis scheme. The resolving power of an appropriate separation technique is often sufficient to solve the problem all by itself. In addition, when more information is required, application of a separation technique first can simplify the problem by allowing subsequent techniques to be selectively applied to the uncertain fractions of the sample. Let us consider common separation analysis options and briefly examine how and when supercritical fluid chromatography can be used to advantage in these situations. Figure 17.1 shows a rough scoring scheme comparing SFC, GC, HPLC, and thin-layer chromatography (TLC). The intent of the figure, rather than drawing focus to the particular scores presented, is simply to indicate that the combination of attributes is unique for each technique and that each has special abilities for solving particular kinds of problems.

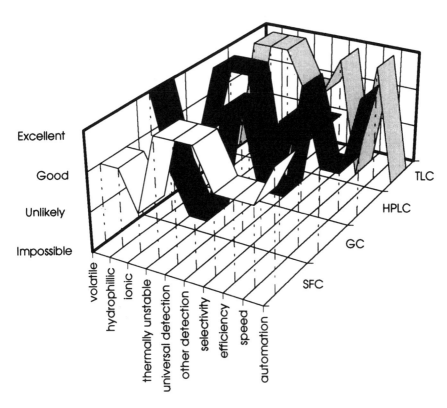

Fig. 17.1. Comparison of some of the features of SFC, GC, HPLC, and TLC. We may argue over some of the specific scores assigned to the techniques here, but most users will agree that each technique has a unique distribution of strengths.

Thin-Layer Chromatography

TLC is one of the most useful separation techniques for "fatty" problems. In practice, no other technique can match it in providing information in a short period of time. Water and other polar solvents can be incorporated into the mobile phase, making the technique amenable to polar and ionic solutes when necessary. TLC also has the great advantage that the total sample remains literally in full view—nothing gets "lost in the system" except highly volatile solutes. (However, they are often not of importance or interest in "fatty" problems.) Visualization of the spots often requires charring the plate or treating it with reagents after the separation is completed. Indirect detection methods such as fluorescence quenching, are also possible. Quantitation can be done with good accuracy and precision in many situations by the skilled practitioner, but is (perhaps unfairly or naively) still considered an art by many. Also, despite gains in automation in recent years, TLC is still comparatively labor-intensive and undesirable for large studies with hundreds or thousands of repetitive analyses. For these reasons, many people often prefer HPLC.

Liquid Chromatography

HPLC can be highly automated and is considered amenable to very large studies. Appropriate solvents can usually be found in the lab for nearly any problem involving fatty, hydrophilic, or ionic samples. However, solvent acquisition and disposal costs are major items in both HPLC and TLC lab budgets. HPLC injection is more precise than TLC spotting, and detection in HPLC is much more precise than in TLC. Relative standard deviations better than 1% are routinely produced in HPLC analyses. However, sensitive, detectable features must be present on all the solutes of interest, clearly distinguishing them from the chromatographic mobile phase. The detector and mobile phase must be carefully matched to the properties of the solutes for a successful analysis.

UV detection is most commonly used in HPLC. Solute derivatization (either in the sample preparation or post-column) can be done to enhance detectability, somewhat similar to TLC visualization techniques. (Derivatization before injection may also be used to modify the chemistry of selected solutes and improve the separation.) If several solutes of interest are present with very different detection features, a complete sample analysis may not be possible in one attempt by a single HPLC approach. If no easily detectable features exist on some or all of the solutes of interest, HPLC may not work at all unless a bulk-property detector (such as refractive index or evaporative light scattering) can provide the desired sensitivity without imposing any significant, new limitations. There is no convenient, inexpensive, and sensitive universal detector for HPLC.

Gas Chromatography

There certainly is a good, universal detector for GC: the flame ionization detector (FID). It is easily capable of detecting subnanogram quantities of solutes (in narrow peaks). Combined with an injection volume of several microliters, sub-part-per-million determinations are easily possible without any special effort. GC analyses can be highly automated for processing large numbers of repetitive analyses. Analysis precision in GC is similar to that in HPLC, especially when the injection is automated. GC can also produce high separation efficiency, needs relatively short analysis times, and does not require organic solvents for the mobile phase. However, what GC does require of the sample components is a combination of sufficient thermal stability *and* volatility to perform the analysis. Derivatization can be performed prior to (or during) injection to reduce solute polarity and increase volatility to expand the scope of GC applications.

The requirement of elevating the temperature to achieve the necessary solute vapor pressure is not always compatible with the thermal stability of the solutes of interest or with that chromatographic system itself. This is especially true if the analysis problems are difficult ones. Every gas chromatographer at one time or another has pondered the possibilities of operating 100 or 200°C higher than is actually feasible.

Unmet Needs

None of the foregoing three techniques is sufficient for all possible separation needs or even for all separations of fatty materials. TLC is in a class by itself. Because of its speed, ease of use, informing power, and the ability to observe the complete sample, TLC is superb for investigating special problems involving small numbers of samples. However, when higher accuracy and precision are required, or when a high level of automation is desired to support large numbers of routine analyses, other options are sought. GC is best for thermally stable volatiles. HPLC is versatile, accurate, and precise when detection is not a problem. However, these two techniques combined cannot cover all the separation needs routinely occurring in most laboratories. Solutes lacking thermal stability or volatility and lacking a conveniently detectable feature in the condensed phase are troublesome, to say the least. A significant fraction of "fatty" analysis problems have these characteristics. Neither GC nor HPLC offers the combination of a low-temperature, high-efficiency separation combined with universal detection. SFC, when configured with an FID, does.

This reasoning provides the fundamental justification for establishing SFC capability in an analytical laboratory. In addition, SFC offers more detection options than any other form of chromatography. Many detectors used in column chromatography have been used successfully in SFC. Bulk property detectors such as refractive index (RI) do not work particularly well, because the bulk properties of a supercritical fluid mobile phase are not well fixed and are usually programmed in the course of an analysis. However, with so many "good" detector options available, it seems no one has missed utilizing RI. Other benefits, such as the reduction in waste solvent, faster analysis times, different selectivities, and higher efficiency (compared with HPLC), will be exploited once the basic SFC capability is installed. Derivatization can be more versatile in SFC than in GC (3). Better blocking groups are available and can be used without concern over making a volatile derivative. Furthermore, these derivatives do not have to be stable at high temperatures, because as will be seen later, low temperatures may be chosen. SFC can also be used to great advantage in situations in which it complements or verifies analytical data obtained by other techniques (4). Several examples of how SFC can uniquely solve problems will be presented later in this chapter, but this ability to complement and verify other techniques should not be overlooked.

Characteristics of SFC

SFC is routinely performed on both packed and open-tubular columns. The operating characteristics of SFC are easily understood starting from either a GC or an HPLC perspective and predicting or observing what will happen as the nature of the mobile phase is changed slowly. Historically, open-tubular SFC has most often been described from a GC perspective, packed-column SFC from an LC perspective. Only the most fundamental characteristics will be described here since the literature already contains numerous descriptions and reviews; for example, see Ref. (1).

Solute Partitioning, Elution, and Mobile Phase Strength

Let us start from a GC perspective and explore what will happen if the gaseous mobile phase is replaced with one exhibiting stronger chemical properties—in particular, dispersion forces. In GC, solutes are partitioned exclusively by their vapor pressures. The mobile phase is nothing more than an inert carrier. However, if the mobile phase were suddenly to exhibit attractive forces for the solutes, the effect will lower solute retention.

If a column were operated with CO_2 as the mobile phase at typical GC temperatures and pressures, solute retention would be quite similar to what would be seen with any other GC carrier gas. However, the solute–mobile phase attractive forces are roughly proportional to the mobile-phase density. If the density of the supercritical CO_2 mobile phase were raised toward liquid values, a decrease in retention of all the solutes will be observed. If temperature programming were used, the effect would be to shift the elution temperatures to lower values, compressing the chromatogram and creating more room at the higher temperatures for higher-boiling solutes (5), as shown in Fig. 17.2. Thus, SFC can be very GC-like, with the added benefit of arbitrarily lowering the elution temperatures by means of the mobile-phase density.

If the mobile-phase density is made high enough, the "normal" solute vapor pressure will become negligible in the partitioning process for all but the most volatile solutes. This allows the chromatographer the freedom to set the column temperature to any convenient value. It may be kept low when thermally labile samples are analyzed. With a CO_2 mobile phase, temperatures as low as about 35°C can be used. When thermally stable samples are analyzed, the temperature can be set high to maximize diffusion and improve the column efficiency per unit time. In some cases, as will be seen later, it is desirable to set the temperature to a particular intermediate value to fine-tune the selectivity. In all of these cases, mobile-phase strength can be indepen-

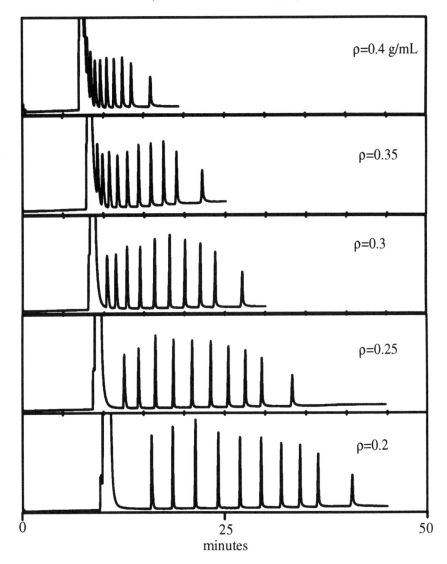

$\rho=0.4$ g/mL

$\rho=0.35$

$\rho=0.3$

$\rho=0.25$

$\rho=0.2$

0   25   50

minutes

Fig. 17.2. GC can be viewed as a special case of SFC in which the mobile-phase solvation strength is negligible. With an appropriate mobile phase, in this case CO_2, increasing the mobile-phase density in a series of GC-like, temperature-programmed experiments lowers the elution temperature of every solute. The column is a 10 m × 50 μm i.d. DB-1 (methyl), used with a tapered-capillary restrictor and a flame-ionization detector. The initial temperature, 90°C, was held for 1 min, then programmed at 4°C/min. For each chromatogram the densities were fixed at the values shown by making appropriate pressure adjustments during the course of the temperature program. The solutes are the even *n*-hydrocarbons from carbon number 16–32 and 36.

dently and continuously adjusted by means of the external pressure and its effect on the mobile-phase density.

The usual practice of programmed elution in SFC is isothermal pressure (or density) programming. Since the mobile-phase strength varies as a function of time in isothermal density-programmed SFC, this form of the technique vaguely resembles gradient elution in HPLC. One important difference is that even though the strength is programmed and (as will be seen later) a spatial strength gradient may result from the pressure drop along the column length, the chemical composition of the mobile phase remains constant in this form of SFC. Actual gradient elution (that is, mobile phase composition programming) is sometimes practiced in SFC and is now appearing as a built-in capability on some commercial instruments.

Diffusion Coefficients and Speed of Analysis

Solute diffusion coefficients are higher in supercritical fluids (10^{-4} to 10^{-3} cm^2/s) than in liquids (approximately 10^{-5} cm^2/s), but lower than in gases (10^{-1} cm^2/s). The optimum mobile-phase velocity in any chromatography column is directly proportional to the solute diffusion coefficient in the mobile phase. Thus, optimum velocities are faster if a supercritical fluid mobile phase is used in a particular column than if liquid is selected. If a gas is used, the optimum velocity is even faster. High-diffusion carrier gases such as helium and hydrogen will produce higher optimum velocities than will nitrogen or CO_2 in a chromatography column.

This comparison assumes that everything else (column dimensions, temperature, analyte, etc.) is unchanged. However, such a comparison, while theoretically illustrative, is impractical. Typical GC and HPLC analytes are very different in their properties. Conditions to elute a given solute by each technique on just one column would be much different from the optimal parameters for each technique. So, in general, we can say SFC is faster than HPLC whenever they are performed similarly, and slower than GC whenever they are performed similarly; but when specific situations involving different equipment in each technique are compared, the order may change. Although speed is important, detection options have, so far, provided more of the motivation to select SFC for an analysis.

Temperature and Pressure Range

SFC (and LC) performance can usually be improved by raising the column temperature to the highest value possible. In low-pressure gases binary diffusion coefficients are proportional to the absolute temperature to a power between about 1.5 and 2 (6). Liquid diffusion coefficients are usually assumed to be directly proportional to temperature. In supercritical fluids, both temperature and pressure will have pronounced effects on solute diffusion coefficients. Empirical estimation methods for diffusion coefficients are fairly crude. However, it is likely that diffusion coefficients will have, conservatively, at least a first-order dependency on temperature at constant density. Thus, by operating at the highest temperature possible, the optimum velocity is increased. Since chromatographers usually operate at velocities higher than optimum to begin with, this increase in optimum velocity translates either into shorter analyses or higher column efficiency for a given maximum analysis time.

However, with higher temperature comes another limit. The equipment in use will have a maximum pressure rating, and increasing the temperature will lower the maximum mobile-phase density that the equipment is capable of producing at that new temperature. Increasing the temperature at any fixed pressure will increase the retention of solutes when their ordinary vapor pressures are insignificant in the partitioning process. Therefore, increasing the temperature at a fixed maximum pressure will decrease the range of analyzable solutes. Temperature and pressure ranges in SFC instruments should be matched, and improvements in either should be made in concert with the other while considering the mobile-phase compressibility (7).

Efficiency and Pressure Drop

Optimum column efficiency is not a mobile-phase property. It depends on the physical dimensions of the column and packing (if the column is packed). Thus, it is theoretically possible to use any column optimally with any mobile phase. So, there is nothing magical about any effect that a supercritical fluid has on *column* efficiency (at least not for short columns).

Relative peak widths are a different story. Pressure drops, ΔP, in packed columns are predicted to a first approximation by the equation

$$\Delta P = \frac{v \eta L}{K^\circ} \qquad (17.1)$$

where v is the mobile-phase velocity, η is the mobile-phase viscosity, L is the column length, and K° is the permeability of the column packing (8). Viscosities of supercritical fluids are typically an order of magnitude lower than those of liquids. Thus, pressure drops can be much lower in SFC than in HPLC. This means that for a particular pressure source available for delivering the mobile phase, SFC columns can be longer than HPLC columns, with the same diameter and packing, thus providing more theoretical plates and narrower peaks relative to their retention times.

Berger found that typical HPLC columns, used with supercritical fluid mobile phases, required 1/3 to 1/5 the pressure drop while operating at 3 to 10 times higher velocities compared to operation with liquid mobile phases (9). He was able to use packed columns up to 2.2 m in length (much longer than permitted by the pressure drop with liquid mobile phase) and achieved 260,000 theoretical plates. Programming was not used. However, these should still be considered "apparent" theoretical plates, because the pressure drop, 160 bar in this case, may actually help narrow the peaks.

When a pressure drop exists longitudinally along an SFC column, it sets up a spatial mobile-phase density gradient and a spatial mobile phase velocity gradient. Because of the correlation between mobile-phase density and its strength, the density gradient is, in effect, a mobile-phase "strength" gradient. It resembles the spatial gradient established along an HPLC column during gradient elution. So, even though no programming is performed in this SFC example, the leading edge of a peak is always situated in mobile phase slightly lower in pressure, density, and strength than the trailing edge. Thus, at any particular instant, the trailing edge of a peak would tend to have a smaller local value of k' (that is, the ratio of the amount of solute in the stationary phase to the amount in the mobile phase over an infinitesimally narrow volume slice across the column at the point of interest) and a somewhat faster velocity than the leading edge relative to the moving mobile phase at each point. This represents a restoring force tending to recombine a broadened peak. The effect of the strength gradient on local k' values tends to narrow the peaks in space and in time, because the peak edges move toward each other relative to the mobile phase. The broader the peak, the stronger becomes this restoring effect, because the peak intersects a longer segment of the spatial gradient.

Somewhat balancing this focusing effect is the velocity distribution of the mobile phase across the column length. Since the mobile phase is decompressing and losing density as it travels down the column, the velocity has to increase somehow to keep the mobile-phase mass flow rate constant at all points along the column. Thus, the leading edge of a peak is located in slightly faster-moving mobile phase than the trailing edge. This represents a mechanism tending to broaden the solute peaks in space, but not in time. Thus, solutes whose partitioning is strongly dependent on density will be narrowed by the spatial strength gradient, while those, if any, with weak or no dependency of partitioning on density would be broadened in space by the velocity gradient but would not be broadened in time.

Most solutes are expected to exhibit some compression of their peak widths in time as a result of interacting with a spatial pressure gradient. Thus, the apparent SFC column efficiency already demonstrated, although extremely impressive and practically useful, does not actually represent *column* characteristics, even when no programming is performed. Great care must be taken in reporting column efficiency in packed-column SFC, especially when pressure gradients are large.

Pressure drops in open-tubular columns are much lower than in packed columns. Similar benefits from spatial gradients have not been reported for open-tubular columns.

Packed or Open-Tubular Columns?

Most of the SFC work reported in the last decade has used open-tubular columns. The biggest practical advantage is not so much the geometry but the nature of the stationary phases available. Open-tubular columns are available with highly inert, low-retention stationary-phase films. The quality of these columns makes a great deal of SFC work possible using a CO_2 mobile phase without any modifiers. FID detection is fairly routine in open-tubular SFC for solutes with molecular weights up to several thousands. This often represents genuine, new analysis capability for the chromatographer, not just a new solution to an old problem.

Selectivity. Stationary phases used in open-tubular SFC tend to mimic similar stationary phases in GC in terms of selectivity, although some pressure or temperature dependency results, especially with polar stationary phases (5,10,11). This is actually an advantage, because in SFC the temperature can be set almost independently of the other parameters. The selectivity can often be coarsely selected with the choice of stationary phase, then fine-tuned with the temperature. This feature is unique to SFC and can be exploited to optimize separations as shown in Fig. 17.3. As this figure suggests, separations at two or more temperatures, differing by at least 20°C, should be performed in analyzing an unknown mixture.

Open-tubular SFC has already claimed a place alongside HPLC and GC, particularly by offering the unique combination of temperature and detection options already mentioned. Open-tubular SFC is very well suited for oleochemical applications and is often the only practical separation alternative to normal-phase HPLC when that technique has no easy detection options for a particular problem. The columns are compatible with aqueous samples. However, even though SFC can directly analyze monoglycerides and free fatty acids, it is usually not capable of analyzing very strongly hydrophilic solutes, such as carbohydrates and amino acids, directly with pure CO_2 mobile phases. Derivatization of these solutes to improve their solubility in CO_2 is a very effective way of extending SFC–FID benefits to many highly polar solutes and even to some ionic solutes.

Packed-column SFC is currently reemerging as a practical analysis option. This reemergence is fueled by the efficiency and speed advantages mentioned earlier and by the reduction in waste organic solvent compared to HPLC. The full impact of this advantage is unclear, because both solvent disposal restrictions and costs are currently increasing. Packed-column SFC, while offering much in waste reduction, is not completely waste-free. It usually requires some addition of organic modifier to the CO_2 mobile phase, because the column packings available today are much more retentive than open-tubular SFC columns. The high surface area and activity of porous silica-based packings, even with bonded phases and end capping, contribute a solute-adsorption retention mechanism involving the silica in addition to the solute—bonded phase interactions. Therefore, a modifier is often required to increase the mobile phase strength above what is available with CO_2 alone and to provide mobile phase components more capable of displacing polar solutes from adsorption sites on the stationary phase. Unfortunately, the use of most organic modifiers is not compatible with FID, limiting the analyst to HPLC-like detectors.

Even with this restriction, packed-column SFC will compete very favorably with normal-phase HPLC. Recent reports of high-temperature HPLC (12–14) suggest that researchers working with this technique are moving toward SFC in search of new benefits, particularly lower viscosity and higher diffusion. Simply substituting CO_2 for the weak component in the mobile phase and moving the temperature above the mixture critical temperature would add the benefits of pressure control of the mobile phase strength, faster analyses, smaller pressure drop per unit column length, higher efficiency with the possibility of using longer columns, and peak focusing via spatial

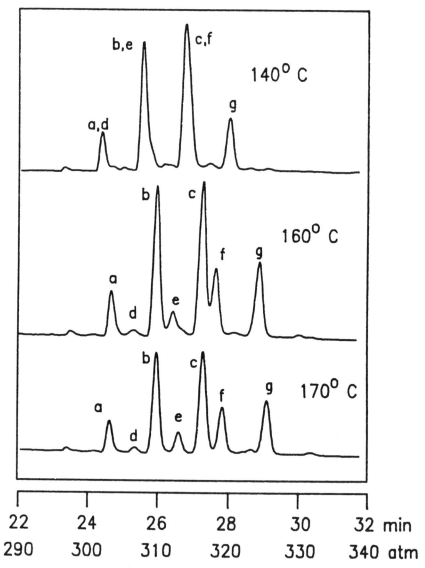

Fig. 17.3. Selectivity in SFC is often temperature-dependent, especially when polar stationary phases are used. This figure illustrates how small temperature changes affect the selectivity of several unknown solutes. The column is a 10 m × 50 μm i.d. SB-biphenyl-30 used with unmodified CO_2 as the mobile phase; a tapered-capillary restrictor; and FID.

gradients without mobile phase composition programming. It is conceivable that normal-phase HPLC will merge with SFC in practice and that CO_2 will become as common in normal-phase HPLC-SFC as water is in reverse-phase HPLC.

Basic Instrumentation

Complete treatments of SFC instrumentation are readily available in literature (1,15). The goal of this section is simply to provide an introduction or review of the most important points. The mobile phase must be pressurized and delivered to the column. The column outlet pressure must be kept elevated, either by a flow restrictor or a regulator. Samples must be injected into the mobile phase in a narrow plug. Finally, a means of detecting the separated solutes is required.

Pumping and Pressure Control

The mobile phase is usually pumped at or below ambient temperature as a liquid, because pumping is easier and more reliable when the fluids are less compressible. Syringe pumps have been used almost exclusively in SFC over the last decade, especially for open-tubular SFC, because of their excellent freedom from pressure pulses. Reciprocating pumps will also work if some provision is made by the manufacturer (or user) for the extremely volatile mobile phases normally encountered. Ordinary HPLC pumps have been used in some cases with the pump heads refrigerated to increase pumping efficiency.

A key difference between SFC and HPLC is that the SFC column must be kept at elevated pressure at its outlet. Since the mobile-phase density depends on pressure, it is desirable to control the pressure not only at the column outlet but everywhere along the column length. Of course, this is not possible. Usually, the pressure is controlled at one end of the SFC column while the pressure at the other end is determined by the pressure drop required to produce the desired mobile-phase flow. For example, when a flow restrictor interfaces a column to a low-pressure detector, the column inlet pressure is regulated. If the column outlet (or nondestructive detector outlet) pressure is fixed with a regulator, the pump is usually operated in a controlled-flow mode.

Restrictors

The three most often used SFC restrictors are compared in Fig. 17.4. The commercially available restrictors most often encountered are the porous frit restrictor (16) and the short-tapered restrictor (also known as the ground, polished, or integral restrictor) (17). The frit restrictor is a plug of porous ceramic material deposited and fixed in the end of a fused-silica tube. The frits may be from about 0.5 to 2 cm in length and may be shortened by the user to increase the flow. The short-tapered restrictor is prepared by abruptly sealing the end of a fused-silica tube with a microtorch, then grinding away the end until a small opening develops. The outlet orifice internal diameter (i.d.) is usually less than 1 μm, and the taper length is in the range of 0.5 to 1 mm. Tapered capillary restrictors have also been used successfully (18) but are not commercially available. The most successful of these restrictors are drawn by a robot under computer control (19). The outlet i.d. is usually 2–4 µm with a taper length of 1–2 cm. Although the protective polyimide coating is burned off the tapered section during the drawing operation, these restrictors are quite sturdy if care is taken not to scratch the exposed surface. For example, the tapered end can be turned in a right-angle bend with a 0.5 cm radius without harm.

All three of these restrictors are well suited to analysis conditions available on commercial instruments. The frit restrictor is sturdy, resists plugging because of its multiple flow paths, and is a good all-around choice. Its high-molecular-weight performance is poorest, but this is usually not a problem except with high-pressure research instruments. (Commercial instruments, at the time of this writing, do not have the pressure range to reveal many significant problems with frit restrictors.) The short-tapered restrictor is most easily plugged because of its small, single orifice. However, it is also sturdy and provides excellent performance, even for quite high-molecular-weight solutes, as long as the samples are filtered. It also has excellent resistance to velocity variations with changing pressure (20). The thin-walled tapered capillary restrictor is a good compromise in many respects. It has excellent high-molecular-weight performance, presumably because its thin walls efficiently transfer heat into the mobile phase in the decompression zone. Even though there is only one flow path, the larger outlet orifice makes it much less likely to plug than the short-tapered restrictor.

Injection

Samples for SFC analyses are usually dissolved in an appropriate liquid and injected onto the SFC column in a fashion vaguely resembling HPLC injection. A four-port,

Frit Restrictor   Short-tapered Restrictor   Tapered-capillary Restrictor

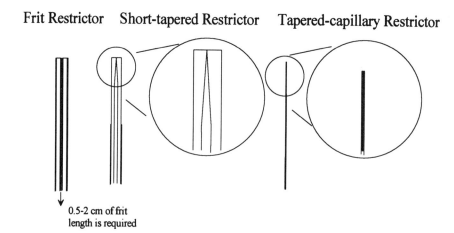

0.5-2 cm of frit
length is required

Fig. 17.4. This figure gives a schematic representation of the last 2 mm of three restrictors used in SFC to interface column outlets to low-pressure detectors such as the FID and the mass spectrometer. The details shown represent the last 0.5 mm of length.

high-pressure valve with a fixed, internal volume (selectable between 0.06 and 0.5 µL) is used as the injection device. The valve volume is determined by the depth of a scored channel on the valve rotor. The rotor is replaced to change the volume.

Injection of only 0.5 µL of liquid into a 50-µm i.d. column would completely fill about a 0.25-m length of the column with liquid and would flood a much longer length in the process of separating the injection solvent from the solutes. Injection volumes of this size generally cause no serious problems with packed-column systems, but they can create unacceptable band broadening in space in open-tubular columns. There are several simple ways to deal with this problem.

The injected volume can be dynamically split at the inlet (21) as in Fig. 17.5a. The flowing sample is simply divided between two flow paths. This is mechanically simple but has several disadvantages (as does any sample splitting approach):

- The limits of detection are increased in terms of concentration in the injected sample.
- Flow-splitting injection is usually not reliable for quantitation.
- The split ratio is never really known under the injection conditions of the real sample.
- Viscosity and diffusion differences between the real samples and external standards may yield different split ratios and lead to wrong quantitative answers.
- Split ratios may not be the same between different components of the same solution, and each may vary from one solution to the next;

thus, even the use of internal standards may not guarantee accurate quantitation.

It is also possible to split the moving sample volume without a flow splitter by rapidly actuating the valve from the "load" to the "inject" position and then quickly returning to "load" as in Fig. 17.5b (22). This is known as the time-split technique.

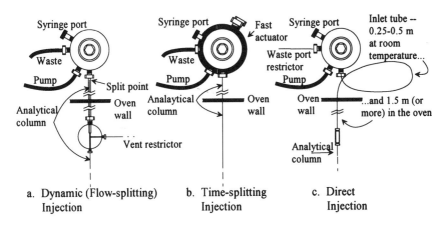

a. Dynamic (Flow-splitting) Injection   b. Time-splitting Injection   c. Direct Injection

Fig. 17.5. Three common methods for sample introduction to open-tubular SFC columns using internal-loop valves. A fast valve actuator is needed for the time-splitting technique (*b*) but not for the other two techniques.

Helium-powered pneumatic actuators can cycle the injection valve with sampling times as low as tens of milliseconds. Quantitative analysis is much better in this mode, especially with internal standards, but viscosity differences between samples and external standards may still cause uncertainty in some cases. The effective sample size injected is still about as small as with flow splitting, so concentration-based detection limits are about the same as with flow splitting.

The simplest and most effective injection method is direct injection. One implementation is illustrated in Fig. 17.5c. The entire contents of the valve are transferred to the column, but the flooding is handled either through a mixing volume (23, 24) or by the use of an uncoated inlet tube to separate the solvent from the solutes, followed by a focusing step to accumulate the solutes in a small band at the head of the analytical column (25, 26). Typically, a 2 to 4 m inlet tube of the same i.d. as the column is attached to the injection valve to receive the sample volume, with the first 0.25–0.5 m remaining out of the column oven. This length of room-temperature inlet tube is chosen to provide enough volume to transfer the liquid from the valve into a well-swept volume before the sample front is heated. The remaining length of inlet tube serves to separate the solvent from the solutes. This can be done in the shortest possible distance by selecting an injection solvent, injection pressure, and column temperature that will ensure a liquid-gas phase separation between the injection solvent and the mobile phase. A liquid film is dynamically formed on the inlet tube and removed by evaporation. The injection solvent does not penetrate the system in a liquid phase beyond the flooded zone. Focusing of the solute peaks prior to or during transfer to the analytical column is accomplished either by the solvent effect or simply by accumulating the solute at the head of the column. This is accomplished by phase-ratio focusing after the injection solvent is removed from the inlet tube. This technique takes advantage of the lower solute k' values on the uncoated precolumn relative to the analytical column. Solutes are transported to the head of the analytical column using mobile phase strong enough to transport solutes only in the absence of stationary phase. Solutes reaching the stationary phase accumulate there (25,26).

For analyses of samples at concentrations below about 1 ppm, larger effective injection volumes are required when an FID or other detector with similar sensitivity is used. For open-tubular columns, solvent-venting injection is the solution. There are several approaches. The simplest is a solid-phase injector described by Lee et al. (27). In it the liquid solvent is deposited on the end of a platinum needle. The liquid is evaporated with a dry gas; then the needle is inserted through a pressure lock into a volume well swept by the pressurized mobile phase. The injection volume can be as high as the operator would like, depending only on the number of cycles of deposition and evaporation performed before the needle is inserted into the mobile phase. The volume per cycle depends on the surface tension of the liquid sample but is typically about 1 µL. Quantitative performance of this injector has not been well established at the time of this writing but is expected to be highly technique-dependent.

Solvent venting can also be done through a valve, as shown by Berg and Greibrokk (28) and Berg et al. (29). Typically, the solvent and solutes are separated in a short precolumn and the solvent is vented through a valve connecting the precolumn and the analytical column. The valve is then closed, the system is pressurized, and the solutes are transferred to the analytical column. In another venting approach, Koski et al. accumulated solutes from 1–3 μL injections on a packed capillary while venting the solvent, then back-flushed the solutes onto an open-tubular SFC column (30).

A solvent-venting approach with great promise for open-tubular SFC was described by Cortes et al. (31,32). In their method, liquid sample, either from an injection valve or cut from an HPLC column, is diverted into a precolumn, where the solvent and solutes are separated. The precolumn is connected to the analytical column through a restrictor interface with a vent. The liquid can be evaporated and vented while the solutes remain in the precolumn; then the solutes are transferred through the restrictor and deposited near the inlet of the analytical column in a small band. The chromatogram is begun with pressurization of that region. Preliminary reports by the authors have indicated excellent performance, with injections of tens of microliters and relative standard deviations below 1%. However, at the time of this writing, an injector with these capabilities is not commercially available.

With any of these injection modes, precision depends on completely filling the valve loop. Because the dimensions are small and the injection solvents are often volatile, and because the valve may be inadvertently heated by waste heat from the oven, or heated purposely to aid transfer of poorly soluble solutes (33), vapors may prevent the complete filling of the sample loop with liquid. Filling efficiencies as low as 25% have been observed (34). This problem is easily overcome with a moving-plunger injection technique combined with a short restrictor on the waste port of the injection valve (as shown in Fig. 17.5c). A typical 25 μL syringe has an inside diameter of about 0.7 mm (0.028 in). Its crosssectional area is only 0.38 mm^2 (6.2 × 10^{-4}in^2). Thus, a modest but continuous force on the syringe plunger (for example, about 20 N, or 5 lb) becomes a pressure exceeding 500 atmospheres working against the waste-port restriction. This pressure is more than sufficient to prevent evaporation of the most volatile solvent and ensure complete loop filling. Precision around 1% RSD should be expected for mid-concentration-range standards with this loop-filling technique combined with direct injection.

Applications

Most applications chosen for SFC analysis so far have been ones requiring a combination of instrumentation properties, where neither GC nor LC is sufficient. Many oleochemical and related analyses fit into this arena. They are often too high in molecular weight for successful GC but require separation or detection capabilities not easily accomplished using HPLC. We will illustrate several of the capabilities of SFC with selected examples. This is not intended to represent the full range of SFC contributions for oleochemical analysis problems to date, but simply to show some of the possibilities. SFC applications from 1989–1991 are compiled in two fundamental reviews (35,36).

Triacylglycerols and Related Materials

There will continue to be some controversy over the choice between GC and SFC for the analysis of triacylglycerols. GC, and particularly high-temperature GC, is certainly more capable than SFC of producing information in minimum time. This derives directly from the much faster diffusion coefficients of solutes when evaporated into hydrogen or helium than when solubilized in carbon dioxide as discussed earlier. However, there is a good deal of uncertainty regarding how high temperatures can be allowed to go in GC before highly unsaturated solutes are thermally affected. GC could further improve, for example, if columns were developed that were capable of eluting and resolving these materials at lower temperatures. However, this is not likely based on current knowledge of GC, and if it is accomplished, the basic problem with thermal stability will not be completely solved. Instead of incremental improvements, SFC offers a discontinuous leap. The cost is longer analysis time.

Kallio et al. first separated triacylglycerols from Baltic herring by Ag+ TLC, then further separated the most polar fractions by open-tubular SFC (37). They reported acyl carbon numbers (ACN) as high as 62 in the most polar fraction, with docosahexanoic acid (22:6ω-3) and eicosapentanoic acid (20:5ω-3) being the most abundant fatty acids. Thus, triacylglycerols containing from 10 to 17 double bonds are present if two or three of these fatty acids are incorporated into molecules ranging up to ACN 62.

In situations where lower SFC temperatures are not required, the technique may still be chosen, either to avoid uncertainty with respect to the thermal stability of unknown solutes or just for convenience. For example, there is little or no degradation of performance with the continuous analysis of triacylglycerols by SFC, because degradation products are not formed in the injector, and essentially everything injected is washed out of the system with the analysis.

France et al. examined abused and storage-damaged vegetable oils using both packed and open-tubular SFC, as shown in Fig. 17.6 (38). Furthermore, these workers found that with packed-column SFC, neat oil could be injected without sample preparation. This allowed the trace analysis of early-eluting species in oils composed primarily of triacylglycerols.

Similarly, Hannan and Hill analyzed lipids in onion seed and investigated changes in the lipids with accelerated aging (39). One of their experiments is illustrated in Fig. 17.7. In studies like this, SFC offers a big advantage, with its ability to analyze solutes ranging from free fatty acids to triacylglycerols (or heavier lipids if necessary) in a single analysis without derivatization, and with little worry that a solute peak of interest will degrade on column or that an later-eluting peak will degrade and produce interfering materials.

Simultaneous Determination of Fatty Acids, Sterols, and Sterol Esters

This combination of materials has a wide range of boiling points and would require very high temperatures for elution of the least volatile components by GC. Again, the

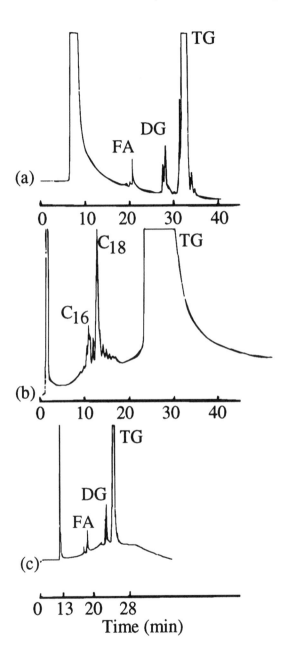

Fig. 17.6. SFC–FID chromatograms of oil from storage-damaged soybeans, performed with CO_2 mobile phase on three different columns: (*a*) Deltabond Octyl with a water-saturator precolumn, (*b*) Omni-Pac μPRN-300, and (*c*) open-tubular SB octyl column. DG = diglycerides, TG = triglycerides. *Source:* Adapted from France, J.E., et al., *J. Chrom. 54*:271 (1991), with permission.

Time (min)

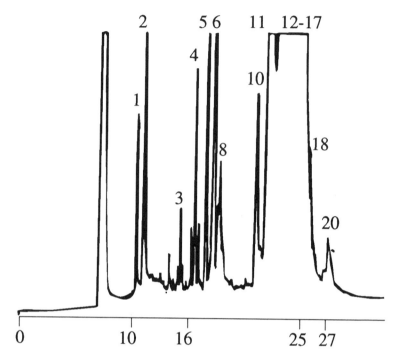

Fig. 17.7. SFC–FID chromatogram of onion-seed lipids. Palmitic (peak 1) and linoleic (peak 2) acids increased in concentration with seed aging, while tocopherol (peak 3) and some of the larger-molecular-weight components decreased. A 20 m × 50 μm i.d. SB-methyl-100 (Lee Scientific, Div. Dionex, Salt Lake City, UT) column was used with pressure-programmed CO_2 at 150°C. *Source:* Adapted from Hannan, R.M., and H.H. Hill, *J. Chrom.* *547*:393 (1991), with permission.

uncertainty of thermal stability can be completely eliminated by the use of SFC and the selection of a relatively low column temperature.

Figure 17.8 shows the open-tubular SFC separation of a supercritical fluid extract of hamster feces (40). This separation was done using a methylsilicone stationary phase. Preliminary work had indicated that an elution temperature in the range of 200–240°C was necessary with this particular column to separate sterol esters and triglycerides. However, at this temperature, the low-boiling free fatty acids were not adequately retained at the lowest possible pressure allowed by the instrumentation, so simultaneous pressure-temperature programming was performed to give the necessary combination of retention for the early-eluting materials and the needed selectivity for the later-eluting peaks.

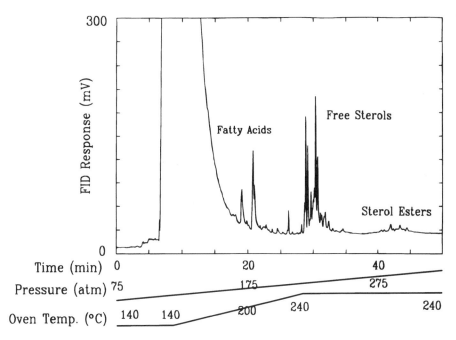

Fig. 17.8. SFC–FID of lipids extracted from hamster feces. CO_2 mobile phase was used with a 10 m × 50 μm i.d. SB-methyl-100 column. *Source:* Pinkston, J.D. et al., *J. High Resol. Chrom. 14*:401 (1991), with permission.

Alcohol–Reaction Mixture Residue

High-molecular-weight components are often produced as byproducts in commercial syntheses and can be found concentrated in precipitates of many chemical manufacturing processes. Figure 17.9 (upper trace) shows the separation of an alcohol–reaction mixture residue accomplished by open-tubular SFC (41). This chromatogram, detected with an FID, revealed late-eluting peaks that had not been detected in earlier separations of the same sample by GC. The identifications of these late-eluting peaks were made using online SFC–MS. Peak matching is more difficult in SFC–MS than in GC–MS because it is difficult to reproduce the mobile-phase velocity (and its dependence on pressure) exactly whenever two different restrictors are involved. However, the reconstructed ion chromatogram, Fig. 17.9 (lower trace), is similar enough to the FID chromatogram to establish the needed correlations. This work identified products containing more than 50 carbon atoms, made by multiple condensations of starting material and smaller byproducts. In cases like this GC lacks sufficient range to solve the problem, and HPLC is inappropriate because an adequate detector does not exist.

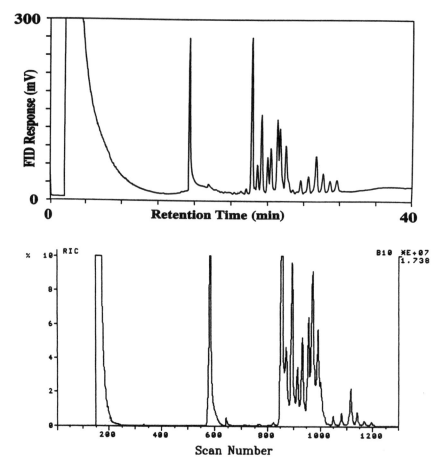

Fig. 17.9. SFC–FID (upper trace) and SFC-MS (lower trace) chromatograms of an alcohol–reaction mixture residue. Absolute retention times rarely compare well between two SFC instruments when restrictors are used. However, the relative retention times and peak elution order usually compare well and allow peak-by-peak matching.

Ethoxylates

SFC–FID is superb at analyzing ethoxylated materials. Figure 17.10 shows chromatograms of polyethylene glycol mixtures in two different molecular weight ranges (42). Although it is not absolutely necessary, derivatization of the terminal hydroxyl groups [for example, by reaction with bis-trimethylsilyltrifluoroacetamide (BSTFA) to make trimethylsilyl ethers] improves the peak shapes. If the samples are relatively water-free, the derivatization can be accomplished in one step simply by adding sufficient

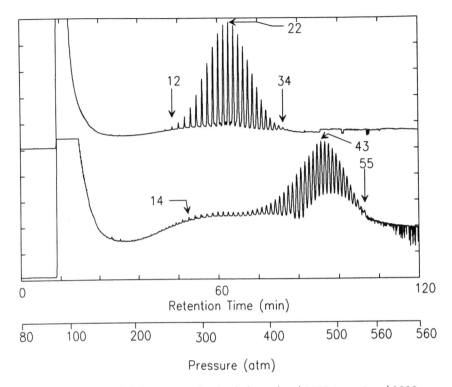

Fig. 17.10. Trimethylsilyl derivatives of polyethylene glycol 1000 (upper) and 2000 (lower).The numbers shown indicate the degree of polymerization. CO_2 was used with a 10 m × 50 μm i.d. SB-biphenyl-30 open-tubular column. Detection was by FID. *Source:* Chester, T.L. et al., *Anal. Chem. 62:*1299 (1990), with permission.

BSTFA, sealing the container, and warming. (These chromatograms also indicate that additional mass range can be made available in open-tubular SFC by using higher pressures. At the time of this writing, commercial instruments are not available in the United States for producing pressures above 415 atmospheres.)

Many nonionic surfactants are also easily prepared and analyzed using the same procedure. Figures 17.11 and 17.12 show SFC–FID chromatograms of two mixtures of alkyl ethoxylates. Neodol 23-3 consists of ethoxylated n-C_{12} and n-C_{13} alcohols, with an average degree of ethoxylation of 3. Neodol 45-11 consists of ethoxylated n-C_{14} and n-C_{15} alcohols averaging 11 ethoxylates. Figure 17.11 clearly reveals the two separate envelopes of ethoxylates, one for each of the two raw materials present. However, the mixture used in Fig. 17.12 was purposely formulated to be difficult to interpret. In this case the C_{14} and C_{15} alcohols averaged only 7 ethoxylates. (The suffix T indicates the surfactant had been topped.) The SFC procedure was not able to separate all the peaks present in either case, so the next best thing was done to complete the analysis. Selectivity was adjusted to cause the coelution of one peak from each of the two series. In both chromatograms the $C_{12}E_n$ and the $C_{15}E_{n-1}$ peaks coelute almost perfectly, beginning with the $C_{12}E_2$ and the $C_{15}E_1$ peaks. This left enough room between the series of coeluting peaks to baseline-resolve the C_{13} and C_{14} ethoxylates over nearly all of the chromatogram. Thus, the ethoxylate distribution in both starting materials can be estimated, even in Fig. 17.12. Using similar SFC capability, Johnson et al. determined the ethylene oxide distribution in ethoxylated alcohols and developed a model from the analytical data to calculate relative ethoxylation rate constants for base-catalyzed ethoxylations of normal octanol (43).

Glycoside Surfactants and Glycolipids

Lightly alkylated polyglycosides are surfactants. These materials are of interest because they can be made from cheap, renewable resources and because they are expected to be biodegradable. Like carbohydrates (44), these surfactants are easily

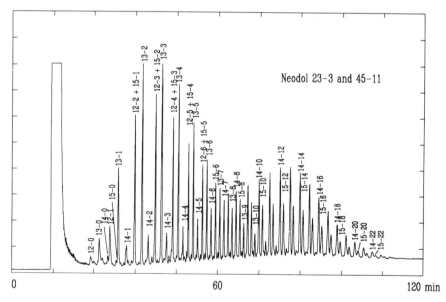

Fig. 17.11. Open-tubular-column SFC–FID separation of a nonionic surfactant mixture. Neodol 23-3 consists of n-C_{12} and n-C_{13} alcohols ethoxylated to an average value of 3. Neodol 45-11 consists of n-C_{14} and n-C_{15} alcohols ethoxylated to an average value of 11. The labels on the peaks indicate the specific alcohol and the degree of ethoxylation for that peak. For example, 13-2 is C_{13} alcohol with 2 ethoxylate units.

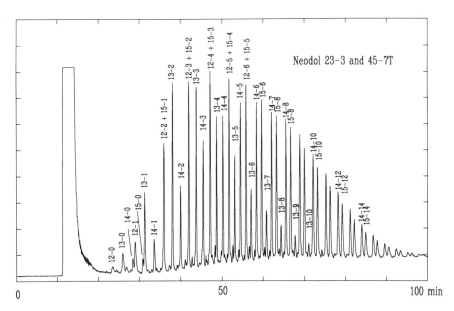

Fig. 17.12. A more complete surfactant mixture (see Fig. 17.11 and text).

derivatized, and the silyl derivatives produce very informative SFC chromatograms, as shown in Fig. 17.13. The peak identities in this chromatogram were assigned only from the retention behavior shown, combined with an independent GC analysis of the fatty acid distribution from a hydrolyzed sample.

Reinhold has pioneered the use of SFC in the challenging area of glycolipids. SFC with FID and mass spectrometric detection has convincingly proven its unique value in applications like that in Fig. 17.14 (45). Sheep urine was trimethylsilylated and analyzed by SFC-MS. The figure only shows the reconstructed ion chromatogram; however, twelve different N-linked glycans were identified, as indicated in the figure.

The versatility of the derivatization-SFC–FID approach allows analyses of materials ranging from unsubstituted to fully alkylated oligosaccharides. Partially hydrolyzed starches are being increasingly used as fat substitutes in processed foods (46). If this trend continues, chemists interested in the fatty properties of foods will be needing new analysis capabilities.

Conclusion

There is a growing realization among separation scientists that column chromatography is somewhat generic—that is, all the various techniques represent parts of just one

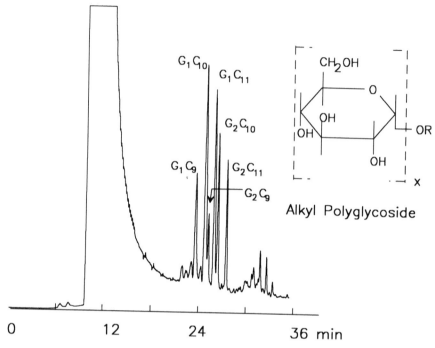

Fig. 17.13. Open-tubular SFC–FID separation of a silylated alkylpolyglycoside. The peak labels indicate the number of glucose units (G) and the number of carbon atoms (C) in the attached alkyl chain.

larger process. Martire (47,48) and Ishii (49) have both proposed the unification of gas, supercritical fluid, and liquid chromatography. Martire has focused on developing a unified theory applicable to LC, SFC, and GC, whereas Ishii combined the individual techniques in one analysis on a single column. Other workers have also performed unified chromatography, combining GC and SFC (50,51).

Our current instrumentation creates artificial boundaries separating GC, SFC, and LC. Beneath this façade lies one unified chromatography. Low-pressure SFC becomes GC. High-temperature LC becomes SFC. However, today we still must use instruments limited primarily to one of these realms.

Many reasons to try SFC were presented here. However, the value of any technique is measured by the individual user in the applications laboratory with the analysis of samples and the solution of chemical problems. SFC has unique combinations of capabilities and high problem-solving value for those willing to take on a "new" technique.

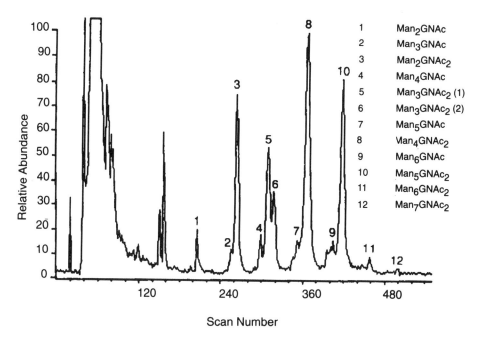

Fig. 17.14. Open-tubular SFC chromatogram of trimethylsilylated N-linked glycans isolated from sheep urine. Detection was by online chemical-ionization mass spectrometry. This figure is a reconstructed ion chromatogram. Full scans, also taken in real time, were used to make the identifications shown. Man = mannose, GNAc = N-acetyl glucose. For example, peak 3 represents a compound containing two mannose units and two N-acetyl glucose units. *Source:* Sheeley, D.M., and V.N. Reinhold, *Anal. Biochem.* *193:*240 (1991), with permission.

References

1. *Analytical Supercritical Fluid Chromatography and Extraction,* edited by M.L. Lee and K.E. Markides, Chromatography Conferences, Inc., Provo, UT, 1990.
2. Chester, T.L., *J. Chrom. Sci. 24:*226 (1986).
3. Cole, L.A., J.G. Dorsey, and T.L. Chester, *Analyst 116:*1287 (1991).
4. Chester, T.L., L.J. Burkes, T.E. Delaney, D.P. Innis, G.D. Owens, and J.D. Pinkston, in *Supercritical Fluid Extraction and Chromatography,* edited by B.A. Charpentier and M.R. Sevenants, ACS Symposium Series, American Chemical Society, Washington, DC, 1988, pp. 144–160.
5. Chester, T.L., and D.P. Innis, presented at The Fourth International Symposium on Supercritical Fluid Chromatography and Extraction, Cincinnati, Ohio, May 20–22, 1992, pp. 79–80.
6. Reid, R.C., J.M. Prausnitz, and B.E. Poling, *The Properties of Liquids and Gases,* 4th edn., McGraw-Hill, New York, 1987, p. 596.
7. Chester, T.L., and D.P. Innis, *J. High Resol. Chrom. and Chrom. Comm. 8:*561 (1985).

8. Halasz, I., in *Modern Practice of Liquid Chromatography*, edited by J.J. Kirkland, Wiley-Interscience, New York, 1972, pp. 325–240.

9. Berger, T.A., presented at The Fourth International Symposium on Supercritical Fluid Chromatography and Extraction, Cincinnati, Ohio, 1992, pp. 7–8.

10. Jones, B.A., T.J. Shaw, and J. Clark, *J. Microcol. Sep. 4:*215 (1992).

11. Chester, T.L., D.J. Bowling, D.P. Innis, and J.D. Pinkston, *Anal. Chem. 62:*1299 (1990).

12. Antia, F.D., and C. Horvath, *J. Chrom. 435:*1 (1988).

13. Erni, F., *J. Chrom. 507:*141 (1990).

14. Liu, G., N.M. Djordjevic, and F. Erni, *J. Chrom. 592:*239 (1992).

15. Chester, T.L., in *Analytical Instrumentation Handbook*, edited by G.W. Ewing, Marcel Dekker, New York, 1990, pp. 843–81.

16. Richter, B.E., presented at the Pittsburgh Conference and Exposition, Atlantic City, NJ, March 10–14, 1986, p. 514.

17. Guthrie, E.J., and H.E. Schwartz, *J. Chrom. Sci. 24:*236 (1986).

18. Chester, T.L., *J. Chrom. 299:*424 (1984).

19. Chester, T.L., D.P. Innis, and G.D. Owens, *Anal. Chem. 57:*2243 (1985).

20. Hentschel, R.T., and J.D. Pinkston, presented at the Fourth International Symposium on Supercritical Fluid Chromatography and Extraction, Cincinnati, Ohio, May 20–22, 1992, pp. 83–84.

21. Peaden, P.A., J.C. Fjeldsted, M.L. Lee, S.R. Springston, and M. Novotny, *Anal. Chem. 54:*1090 (1982).

22. Richter, B.E., D.E. Knowles, M.R. Andersen, N.L. Porter, E.R. Campbell, and D.W. Later, *J. High Resol. Chrom. and Chrom. Comm. 11:*29 (1987).

23. Hirata, Y., M. Tanaka, and K. Inomata, *J. Chrom. Sci. 27:*395 (1989).

24. Hirata, Y., and K. Inomata, *J. Microcol. Sep. 1:*242 (1989).

25. Chester, T.L. and D.P. Innis, presented at the Third International Symposium on Supercritical Fluid Chromatography and Extraction, Park City, UT, January 14–17, 1991, pp. 27–28.

26. Chester, T.L. and D.P. Innis, presented at the European Symposium on Analytical Supercritical Fluid Chromatography and Extraction, Wiesbaden, Germany, December 4–5, 1991, p. L-03.

27. Lee, M.L., S.H. Page, S.R. Goates, I.J. Koski, E.D. Lee., A. Liu, and I. Ostrovsky, European Symposium on Analytical Supercritical Fluid Chromatography and Extraction, Wiesbaden, Germany, December 4–5, (1991), p. L-04.

28. Berg, B.E., and T. Greibrokk, *J. High Resol. Chrom. and Chrom. Comm. 12:*322 (1989).

29. Berg, B.E., A.M. Flaaten, J. Paus, and T. Greibrokk, *J. Microcol. Sep. 4:*227 (1992).

30. Koski, I.J., K.E. Markides, B.E. Richter, and M.L. Lee, *Anal. Chem. 64:*1669 (1992).

31. Cortes, H.J., R.M. Campbell, R.P. Hines, and C.D. Pfiffer, *J. Microcol. Sep. 4:*239 (1992).

32. Cortes, H.J., R.M. Campbell, R.P. Hines, and C.D. Pfiffer, PCT International Application #92/05852 (1991).

33. Andersen, M.R., *LC•GC 6:*566 (1988).

34. Chester, T.L., and D.P. Innis, *J. Microcol. Sep. 1:*230 (1989).

35. Chester, T.L., and J.D. Pinkston, *Anal. Chem. 62:*394R (1990).

36. Chester, T.L., J.D. Pinkston, and D.E. Raynie, *Anal. Chem. 64:*153R (1992).

37. Kallio, H., T. Vauhkonen, and R.R. Linko, *J. Agric. Food Chem. 39:*1573 (1991).

38. France, J.E., J.M. Snyder, and J.W. King, *J. Chrom. 540:*271 (1991).

39. Hannan, R.M., and H.H. Hill, *J. Chrom. 547:*393 (1991).
40. Pinkston, J.D, T.E. Delaney, D.J. Bowling, and T.L. Chester, *J. High Resol. Chrom. 14:*401 (1991).
41. Pinkston, J.D., and D.J. Bowling, in *Hyphenated Techniques in Supercritical Fluid Chromatography and Extraction,* edited by K. Jinno, J. Chrom. Lib. 53:25 (1992).
42. Chester, T.L., D.J. Bowling, D.P. Innis, and J.D. Pinkston, *Anal. Chem. 62:*1299 (1990).
43. Johnson, A.E., P.R. Geissler, and L.D. Talley, *J. Am. Oil Chem. Soc. 67:*123 (1990).
44. Chester, T.L., and D.P. Innis, *J. High Resol. Chrom. and Chrom. Comm. 9:*209 (1986).
45. Sheeley, D.M., and V.N. Reinhold, *Anal. Biochem. 193:*240 (1991).
46. Thayer, A.M., *Chem and Eng. News 70 (24):*26 (July 15, 1992).
47. Martire, D.E., *J. Liq. Chrom. 10:*1569 (1987).
48. Martire, D.E., *J. Liq. Chrom. 11:*1779 (1988).
49. Ishii, D., and T. Takeuchi, *J. Chrom. Sci. 27:*71 (1989).
50. Liu, Y., and F.J. Yang, *Anal. Chem. 63:*926 (1991).
51. Robinson, R.E., D. Tong, R. Moulder, K.D. Bartle, and A.A. Clifford, *J. Microcol. Sep. 3:*403 (1991).

Chapter 18

Supercritical Fluid Chromatography of Trace Components in Oils and Fats

William E. Artz

Department of Food Science, University of Illinois, 382 Agr. Eng. Sci. Bldg., 1304 W. Pennsylvania Ave., Urbana, IL 61801

Fatty Acids, Mono- and Diglycerides

Supercritical carbon dioxide is an excellent solvent for nearly all the trace components found in oils because of the nonpolar nature of supercritical carbon dioxide. The discussion of the supercritical fluid chromatography trace components in oils and fats will include the constituent components of triglycerides, i.e., free fatty acids, monoglycerides, diglycerides, and a limited number of their oxidation products. Steroids and the fat-soluble vitamins, including the carotenoids, will also be covered. Excluded from this chapter will be additives and toxicants in food oils and fats, e.g., herbicides, pesticides, etc., and components of microbial origin, such as the mycotoxins. A comprehensive review on the SFC separation of natural products was published in 1991 by M. Lubke (1), which included an extensive list of references as well as detailed tables on the chromatographic conditions employed for each reported separation.

The vaporization of fatty acids and their derivatives are often used in conjunction with chromatographic studies of new stationary phases, instrumentation, etc. As a result, there are numerous examples of supercritical fluid chromatographic (SFC) separations of compounds, such as fatty acids. The primary advantage of SFC over GC is that fatty acid derivatization is not required. If derivatization is used, it is usually preferable to use capillary GC because of the significantly greater resolution that can be obtained relative to SFC. To separate saturated fatty acids (e.g., C4:0 to C24:0) with SFC a nonpolar stationary phase will suffice (100% polymethylsiloxane, e.g., SB-methyl-100, Lee Scientific, Div. Dionex, Salt Lake City, UT). However, to attain separation of either unsaturated fatty acids or a mixture of saturated and unsaturated fatty acids, a more polar stationary phase is required (25 to 50% cyanopropyl polymethylsiloxane) (2). For example, to separate C16:0; C18:0; C18:1, *cis*-9; C18:2, *cis*-9,12; C18:2, *trans*-9,12; C13:3, *cis*-9,12,15; and C:18:3, *cis*-6,9,12, a stationary phase of 50% cyanopropyl polymethylsiloxane (12 m × 50 μm) was used (2). Carbon dioxide at 62°C was held at a density of 0.16 g/mL for 10 min, then ramped at 0.05 g/mL-min to 0.35 g/mL; finally at 0.05 g/mL the density was ramped at 0.007 g/mL-min to a final density of 0.55 g/mL.

The concentration of free fatty acids, mono- and diglycerides, as well as the concentration of low-molecular-weight oxidized volatile and polymeric components, increases with time in deep-fat frying oils. The presence of these moieties in cooking oils can be used to monitor their quality. Packed as well as capillary columns can be used to separate free fatty acids from mono-, di-, and triglycerides in such mixtures. Hellgeth et al. (3) separated a mixture of saturated fatty acids with a 15 cm × 4.6 mm i.d. column containing 5 μm PRP-1 (Hamilton, Reno, NV) at 40–90°C from 2600 to 4500 psi. Packed column performance can be significantly improved with the addition of a small percentage of water to the carbon dioxide (4). For each of the seven different packed-column stationary phases examined, resolution was substantially improved with the addition of water. Packed-column SFC has also been used to monitor the quality of thermally-abused frying oils by France et al. (5). They used a precolumn of alumina to add high concentrations of water (2–3 mol%) to the CO_2 mobile phase to affect the separation of free fatty acids from triglycerides.

Hydrolytic rancidity is particularly noticeable in dairy products because of the presence of small, volatile fatty acids, which have very intense odors. For example, the characteristic flavor of blue cheese, which is due to such short-chain organic acids such as butyric, caproic, and caprylic, would be extremely objectionable in a butter matrix. Lipase will rapidly cause a substantial increase in the free fatty acid content of butter, even at freezing temperatures. Fortunately, lipase is relatively heat sensitive; pasteurization of milk will completely inactivate lipase. To separate the free fatty acids contained in milk fat or butter (C4:0 or butyric acid, C6:0, C8:0, C10:0, C12:0, C14:0, C16:0, C16:1, C18:0, C18:1 and C18:2 or linoleic acid) a 17 m × 50 μm i.d. column at 80°C containing a stationary phase of 25% cyanopropyl polymethyl siloxane (SB-cyano-25, 17 m × 50 μm i.d.) was used (Fig. 18.1). The separation was accomplished using a density program ramp, starting at 0.2 g/mL, for 10 min, followed by an asymptotic program having a final set point density of 0.6 g/mL, and a $^1/_2$ rise time of 12 min. A total program time of 60 min was used in this separation. Saturated fatty acids (C_{18} to C_{30}) have been separated on a 20 m × 100 μm column with a stationary phase of DB-1 (J&W Scientific, Inc., Rancho Cordova, CA) (6). In this case, the SFC was held at a 10-min initial isobaric period, and then the pressure was increased from 1000 psi to 3600 psi over a 15-min interval, then held isobarically at 3600 psi.

Fatty acids were also chromatographed after being extracted from whole wheat flour (7) using a modified in-line supercritical fluid extraction (SFE) cell. Samples of technical grade fatty acids were also examined as part of this study. Oleic, linoleic, and linolenic acids were separated with an SFC capillary column with a stationary phase of SB-cyanopropyl-50, 40 m × 100 μm (Fig. 18.2).

Sugiyama et al. (8) developed a rapid analysis based on the separation of hydroperoxides formed from peanut oil triglycerides with packed-column SFC (250 mm × 4.6 μm i.d., 5 μm silica gel column). The triglycerides eluted at 2.1 min, and the hydroperoxides eluted at 3.6 min. Sauer (9) separated 4-methylinolenate hydroperoxide isomers with a stationary phase of 50% cyanopropyl polymethylsiloxane (SB-cyanopropyl-50) on a 20 m × 50 μm column (40°C, UV detection at 234 nm). The separation time was excessively long, nearly 6.5 h for the hydroperoxide solutes. For further examination, these lipid oxidation products were reduced with sodium borohydride to produce a mixture of positional and *cis/trans* hydroxy fatty acid isomers. Eight

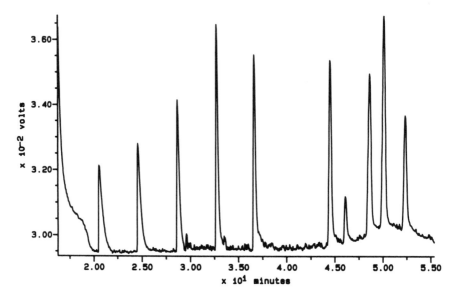

Fig. 18.1. SFC of saturated (C4:0 through C18:0) and unsaturated (C16:1, C18:1, and C18:2) fatty acids. Column: 17 m × 50 μm, SBcyanopropyl-25 at 80°C. Initial ρ was 0.20 g/mL; after a 10 min hold an asymptotic program was used with final set density of 0.60 g/mL, a $^1/_2$ rise time of 12 min, and a total program time of 60 min. FID was at 375°C. Mobile phase was CO_2.

isomers were separated (20 m × 50 μm, SB-cyano-50) and tentatively identified (Fig. 18.3). With the currently available SFC stationary phases the first choice for the separation of hydroperoxides is still HPLC. Because of the thermal instability of hydroperoxides, GC is not an option.

Hydroxy fatty acids (C_{18}, C_{20}, and C_{22}) were separated with a 20 m × 100 μm i.d. column with a stationary phase of DB-1 (10). After an initial 5 min hold at 2700 psi (185 bar), the pressure was increased to 3500 psi (240 bar) over 3 min and then held at 3500 psi (240 bar).

Chester (11) separated a mixture of mono- and diglycerides from triglycerides with capillary SFC using a 12 m × 100 μm i.d. column at 90°C with a stationary phase of BP-10. Diglycerides can be separated with either nonpolar stationary phases, such as DB-1 and DB-5 (12,13) or the slightly polar stationary phases, such as 50% phenyl-polymethylsiloxane (14).

Carotenoids

Although SFC has been applied to a wide variety of compounds and mixtures in the past few years (15–20), there have been few reports on carotenoids and only three on the separation of carotenoid *cis/trans* isomers (21–25). The first application of SFC to carotenoids was in 1968 by Giddings et al. (22), who separated α- and β-carotene. In

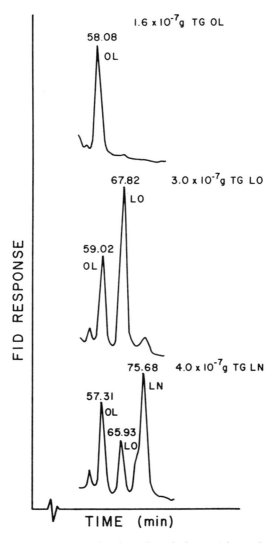

Fig. 18.2. SFE/SFC of technical grade fatty acid samples. Quantity of sample injected is given for each sample. Peak (OL) is oleic, peak (LO) is linoleic, and peak (LN) is linolenic acid. Samples were extracted at approximately 500 atm/(80 ± 3°C) for 15.0 min. Chromatography: 40 m × 100 μm SB-cyano-propyl-50 column at 100°C; pressure program: initial pressure 140 atm; 10 atm/min to 150 atm; 0.5 atm/min to 152 atm; hold at 152 atm (52 min); 0.2 atm/min to 155 atm; 0.7 atm/min to 170 atm; hold at 170 atm (19 min). FID was at 350°C. Mobile phase is CO_2.

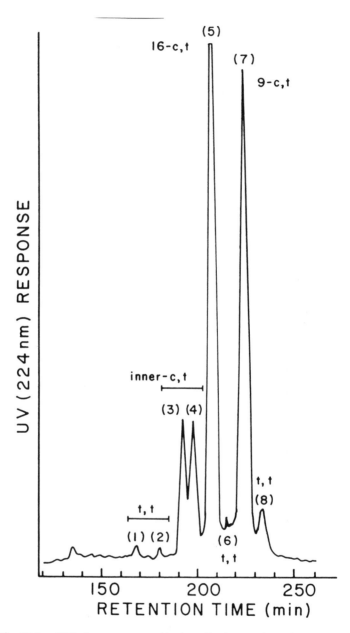

Fig. 18.3. SFC chromatogram of hydroxylinolenates from the mono-hydroperoxides produced from autoxidized methyl linolenate. Chromatography: 20 m × 50 µm i.d., SB-cyanopropyl-50, pure CO_2, 135°C, asymptotic density program; 0.192–0.193 g/mL, 20 min $1/_2$ rise time. FID at 350°C. Mobile phase is CO_2.

1987 Frew et al. (21) separated and identified several carotenoids using SFC and mass spectrometry.

Schmitz et al. (26) separated a mixture of carotenoids with capillary SFC. They were also the first to separate carotenoid *cis/trans* isomers with capillary SFC. To separate lycopene, α-carotene, and β-carotene extracted from tomatoes and carrots, a SB-phenyl-50 column (10 m × 50 μm i.d., d_f = 0.25 μm) was used with a stationary phase of 50% phenyl- and 50% polymethylsiloxane (26). For the separation of the major β-carotene *cis/trans* isomers, an SB-cyanopropyl-25 column (7 m × 50 μm i.d., d_f = 0.25 μm) was used with a stationary phase of 25% cyanopropyl- and 75% polymethylsiloxane. For the separation of α-carotene *cis/trans* isomers, an SB-cyanopropyl-25 column (17 m × 50 μm i.d.) was used containing a stationary phase of 25% cyanopropyl- and 75% polymethylsiloxane. The mobile phase was SFC grade carbon dioxide with 1% ethanol, vol/vol (Scott Specialty Gases, Plumsteadville, PA).

In addition to the solvent-extracted carotenoids from carrots and tomatoes, α- and β-carotene contained in an aqueous isolate of carotenoproteins were extracted with on-line SFE (Lee Scientific, Inc., Div. Dionex, Salt Lake City, UT, Part #012970). To extract the carotenoproteins, a small volume (10 to 25 μL) of the aqueous extract of carotenoproteins was placed on a small piece of filter paper inside the microextraction cell (27). The paper functioned as a moisture absorbent, which facilitated extraction of the carotenoids from the aqueous solution of carotenoproteins. A Lee Scientific Model β501 capillary SFC pump and oven were used with a Linear UV/VIS 204 detector (Reno, NV). β- and α-carotene extracted from the carrots were separated at 50°C under isobaric SFC conditions at 0.70 g/mL.

Carotenoids present in the tomato extract were separated at 45°C with an asymptotic density program with a $^1/_2$ rise time constant of 30 min, a termination time of 30 min, an initial density of 0.66 g/mL, and a final convergence density of 0.72 g/mL. β-carotene isomers could be separated under isoconfertic conditions using a mobile phase density of 0.66 g/mL. For the separation of α-carotene isomers, the extraction cell was used as an injection system. The extraction conditions were 50°C and 0.92 g/mL for 5 min. Column eluents were monitored at 461 nm, and the detector sensitivity was set at 0.02 AUFS for the vegetable carotenoids and β-carotene *cis/trans* SFC analyses. However, for the α-carotene analysis the eluent was monitored at 453 nm at 0.06 AUFS setting.

Figure 18.4 indicates the separation of α- and β-carotene from the carrot extract achieved with the SB-phenyl column, and Fig. 18.5 represents a separation of isomeric mixtures of acarotene with SFC on cyanopropyl polymethylsiloxane columns. Frew et al. (4) had suggested that the cyanopropyl stationary phase was the most promising for carotenoid separations with SFC. In the present author's lab, the cyanopropyl column appeared to be the best of the SFC stationary phases available for the separation of carotenoids. However, because of the severe tailing of lycopene on the cyanopropyl columns, other stationary phases were examined. Consequently, a SB-phenyl-50 column was employed for the separation of samples containing lycopene, which provided separation with minimal peak tailing.

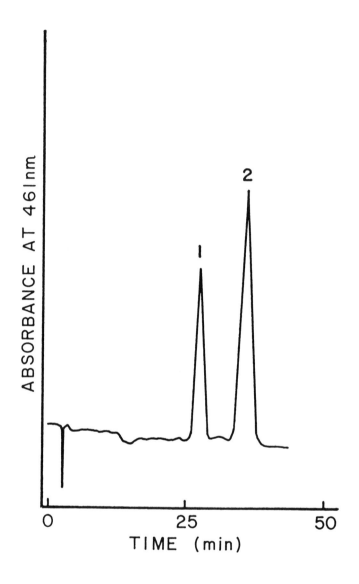

Fig. 18.4.   SFC separation of α- and β-carotene from carrots. Peak
1 is α-carotene, and peak 2 is β-carotene. Chromatography: isoconfertic conditions at a density of 0.70 g/mL, a SB-phenyl-50 column
(10 m × 50 μm i.d., d_f = 0.25 μm) the mobile phase was SFC grade
CO_2 containing 1% ethanol, v/v (Scott Specialty Gases, Plumsteadville, PA); UV/VIS detection at 461 nm (Linear, Reno, NV) 0.02 AUFS;
cell path length 250 μm.

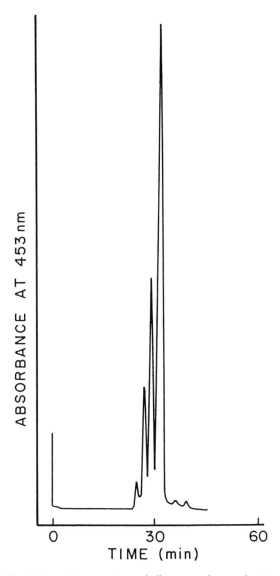

Fig. 18.5. SFC separation of all-*trans* and several *cis* isomers of α-carotene. The largest peak is all *trans* α-carotene. Chromatography: isoconfertic conditions at 0.645 g/mL, a SB-cyanopropyl-25, 17 m × 50 μm i.d., d_f = 0.25 μm; the mobile phase was SFC grade CO_2 containing 1% ethanol, vol/vol (Scott Specialty Gases, Plumsteadville, PA); UV/VIS detection at 453 nm (Linear, Reno, NV) 0.06 AUFS; cell path length 250 μm.

Both pure supercritical CO_2 and supercritical CO_2 with 1% ethanol were evaluated in the present author's lab for their effect on the separations. No appreciable difference in separation efficiency was observed in the absence of the ethanol modifier with the capillary SFC columns examined. Although no effect was expected, the CO_2 with modifier was used for on-line SFE. The effect of modifiers on packed-column SFC separations, as well as SFE, can be substantial. A polar sample solvent can significantly alter a packed-column separation, e.g., methanol. However, the effect of low to moderate concentrations of modifier on capillary SFC separations is usually negligible— as one would expect from the absence of exposed silica support sites in a deactivated capillary column.

Steroids

Markides et al. (2) separated a mixture of six different prostaglandins, which were derived from arachidonic acid. A mixture of 6 prostaglandins were separated on a 12 m × 50 μm column with a stationary phase of 50% cyanopropyl methyl-polysiloxane. Carbon dioxide at 62°C was programmed at 0.05 g/mL-min from 0.16 g/mL to 0.40 g/mL and from 0.40 to 0.87 g/mL at 0.007 g/mL-min. In addition, five sterols (progesterone, testosterone, 17-hydroxyprogesterone, 11-deoxycortisol, and corticosterone) were effectively separated with a 10 m × 100 μm column of SB-cyanopropyl-25 (CO_2, 60°C). A ramp of 3 atm/min to 150 atm was followed by a 10 atm/min ramp to 440 atm, then an isobaric period was used to separate the five sterols. A mixture of 11 prostaglandins and prostaglandin esters (28) were separated with a 6.5 m × 50 μm i.d. column and a stationary phase of oligoethylene oxide polysiloxane (100°C, density program = 0.260 to 0.525 g/mL at 0.013 g/(mL-min), then 0.525 to 0.600 g/mL at 0.020 g/(mL-min), then 5 min at 0.600 g/mL, followed by a 0.010 g/(mi-min) from 0.600 to 0.760 g/mL).

David and Novotny (29) examined a phosphorus-selective detector using a mixture of 6 steroids derivatized with dimethylthiophinic chloride. A 10 m × 50 μm column with a stationary phase of polymethylhydrosiloxane and a mobile phase of nitrous oxide was used to separate the steroids. In addition to nitrous oxide, Freon-22 (chlorodifluoromethane) has been used for the capillary SFC separation of steroids by Li et al. (30). A mixture of 6 cholesterol derivatives were separated with a capillary column of SE-52 (10 m × 100 μm). The same researchers also determined the cholesterol content of an egg yolk sample with SFE and capillary SFC (31).

Fat-Soluble Vitamins

King (32) separated both a-tocopherol and cholesterol from a fish oil concentrate with an SB-methyl-100 column (10 m × 50 μm i.d.) at 100°C. After 10 min at 0.28 g/mL, a linear density ramp to 0.66 g/mL was applied. Supercritical fluid extraction and chromatography have been used for the extraction and separation of vitamin K (33), including the analysis of a mixture of vitamin K isomers (34) with a silica gel column and a mobile phase of CO_2 and 5% acetonitrile. Nakano and Adachi (35) also found that a modifier facilitated the packed-column separation of vitamin K isomers. SFC separations of fat-soluble vitamins have been reported for vitamins A, E, and D (36,37) on reversed-phase packed columns as well as on capillary columns (DB-WAX and DB-5) (38,39).

References

1. Lubke, M., *Analysis 19:*323 (1991).
2. Markides, K.E., S.M. Fields, and M.L. Lee, *J. Chromatographic Science 24:*254 (1986).
3. Hellgeth, J.W., J.W. Jordan, L.T. Taylor, and M. Ashraf Khorassoni, *J. Chrom. Sci. 24:*183 (1986).
4. Geiser, F.O., S.G. Yocklovish, S.M. Lurcott, J.W. Guthrie, and E.J. Levy, *J. Chrom. 459:*173 (1988).
5. France, J.E., J.M. Synder, and J.W. King, *J. Chrom. 540:*271 (1991).
6. Greibbrok, T., B.E. Berg, A.L. Bilie, J. Doehl, A. Farbrot, and E. Lundanes, *J. Chrom. 394:*429 (1987).
7. Artz, W.E. and R.M. Sauer, Jr., *J. Am. Oil Chem. Soc.* (in press) (1992).
8. Sugiyama, K., T. Shikawa, and T. Moriya, *J. Chrom. 515:*555 (1990).
9. Sauer, R.M., Jr. M.S. Thesis, University of Illinois, Urbana, 1990.
10. Doehl, J., A. Farbrot, T. Greibrokk, and B. Iversen, *J. Chrom. 392:*175 (1987).
11. Chester, T.L. *J. Chrom. 299:*424 (1984).
12. White, C.M., R.K. Houck, *J. High Resol. Chrom. 8:*293 (1985).
13. Brownlee Labs, *J. Chrom. Sci. 25:*89 (1987).
14. Holzer, G., S. Deluca, K.J. Voorhess, *J. High Resol. Chrom. 8:*528 (1985).
15. Later, D.W., B.E. Richter, D.E. Knowles, and M.R. Andersen, *J. Chrom. Sci. 24:*249 (1986).
16. Later, D.W., B.E. Richter, W.D. Felix, M.R. Andersen, and D.E. Knowles, *Amer. Lab. 18 (August):*108 (1986).
17. Chang, H-C.K., K.E. Markides, J.S. Bradshaw, and M.L. Lee, *J. Chrom. Sci. 26:*280 (1988).
18. Chang, H-C.K., and M.L. Lee, *J. Chrom. Sci. 26:*238 (1988).
19. Taylor, L.T. *J. Chrom. Sci. 26:*45 (1988).
20. Lee Scientific, Inc., *J. Chrom. Sci. 25:*179 (1987).
21. Frew, N.M., C.G. Johnson, and R.H. Bromund, in *Supercritical Fluid Extraction and Chromatography Techniques and Applications,* edited by B.A. Charpentier, and M.R. Sevenants, American Chemical Society Symposium Series 366, Washington DC, 1988, pp. 208–228.
22. Giddings, J.C., L. McLaren, and M.N. Meyers, *Science 159:*197 (1968).
23. Gere, D.R., *Hewlett Packard Application Note 800-5,* Hewlett Packard, Avondale, PA, 1983.
24. Aubert, M.-C., C.R. Lee, A.M. Krstulovic, E. Lesellier, M.-R. Pechard, and A. Tchapla, *J. Chrom. 557:*47 (1991).
25. Lesellier, E., A. Tchapla, M.-R. Pechard, C.R. Lee, A.M. Krstulovic, *J. Chrom. 557:*59 (1991).
26. Schmitz, H.H., W.E. Artz, C.L. Poor, J.M. Dietz, and J.W. Erdman, Jr., *J. Chrom. 479:*261 (1989).

27. Artz, W.E., J.E. Erdman, Jr., J.M. Dietz, H.H. Schmitz, R.M. Sauer, Jr., and L. Unlu, presented at the American Oil Chemists' Society Symposium on Applications of Supercritical Fluid Chromatography, Baltimore, MD, April, 1990.
28. Koski, I.J., B.A. Jansson, K.E. Markides, and M.L. Lee, *J. Pharm. Biomed. Anal. 9:*281 (1991).
29. David, P.A., and M. Novotny, *J. Chrom. 461:*111 (1989).
30. Li, S.F.Y., C.P. Ong, M.L. Lee, and H.K. Lee, *J. Chrom. 515:*515 (1990).
31. Ong, C.P., H.K. Lee, and C.F.Y. Li, *J. Chrom. 515:*509 (1990).
32. King, J.W., *J. Chrom. Sci. 28:*9 (1990).
33. Millet, J.L., *Analysis 15(4):*38 (1987).
34. Hondo, T., N. Saito, and M. Senda, *Bunseki Kagaku 35:*316 (1986).
35. Nakano, K., and Y. Adachi, Japanese Patent #62,292,744 (1986).
36. Yarbro, S.K., and D.R. Gere, *Chromatography 2:*49 (1987).
37. Mourier, P., P. Sassiat, M. Caude, and R. Rosset, *Analysis 12:*229 (1984).
38. Board, R., D. McManigill, H. Weaver, and D. Gere, *CHEMSA* 12,21,22,24, June, 1983.
39. White, C.M., D.R. Gere, D. Boyer, F. Pacholec, and L.K. Wong, *J. High Resol. Chrom. & Chrom. Comm. 11:*94 (1988).

Chapter 19

Analytical Supercritical Fluid Extraction for Oil and Lipid Analysis

A.A. Clifford and D.F.G. Walker

School of Chemistry, University of Leeds, Leeds LS2 9JT, UK

Supercritical fluid extraction (SFE) is becoming an important technique for the preparation of samples for analysis, especially for chromatography (1–6). Usually it is more rapid, is less labor-intensive, and gives cleaner extracts than liquid extraction. Additional advantages over chemical solvent extractions include faster analysis, better selectivity, higher efficiency, and the absence of toxic solvent waste, thus reducing safety hazards. Until recently the use of SFE has generally been confined to chemical processing applications on an industrial scale, yet SFE can be performed for analytical purposes on what may be considered microscale using extraction vessels ranging from tens of milliliters down to those measuring a fraction of a milliliter.

The development of analytical SFE has predominantly occurred since 1986. The related literature has tended to be orientated toward applications, and the development of analytical SFE methods has been largely empirical; hence, the processes that control SFE are still not totally understood (7). This chapter focuses on the mechanical and chemical principles that affect the development of an analytical SFE method. In the field of lipid analysis, the range of problems can be quite large, but they appear to fall into two main categories: the measurement of oil and fat content of natural materials, and the analysis of natural toxins and pollutants in oils and fats. However, the research described in this chapter makes use of generic studies to describe effects in SFE, some of which are outside, but applicable to, the field of oil and fat analysis.

Methodology

There are essentially two ways in which the analyst can exploit the attractive properties of supercritical fluids for rapid and quantitative extraction of target analytes from a bulk matrix. The approach chosen depends on the nature of the extraction to be performed. When SFE is to be used purely for sample preparation, by collecting the extracted analytes for subsequent analysis by a variety of techniques including chromatographic, spectroscopic, and gravimetric methods, the approach is known as "off-line" SFE. In contrast, in the "on-line" (or "coupled") approach, the SFE step replaces the normal sample introduction step in an analytical separation process. For the case of chromatography the SFE step replaces the injection process, and the extracted analytes are transferred to and collected in a chromatographic injection loop, in a thermal or sorbent trap ahead of the chromatographic column, or in the stationary phase at the head of the chromatographic column itself.

Off-line extraction is inherently more simple than on-line SFE (since the analyst needs to consider only the analyte extraction and collection steps) and allows the extract to be analyzed by any appropriate technique. Off-line SFE accommodates large sample sizes better than on-line SFE and is adaptable to sample sizes ranging from a few milligrams up to hundreds of grams (8,9). Off-line SFE can be performed in two different ways: a dynamic or a static extraction. Both modes have certain advantages. Dynamic extraction allows a continuous supply of pure extraction fluid to the sample without extensive plumbing between cell and outlet, therefore minimizing loss of the analyte being extracted. Static extractions consume less carbon dioxide, requiring less pumping capacity, and known modifier concentrations can be added directly onto the sample matrix within the extraction cell.

The principal advantage of on-line SFE is that it allows quantitative transfer of all the extracted material to the analytical system without any extra sample-handling stages. The technique is experimentally more difficult to practice than off-line SFE; not only the extraction and trapping stages but also the analytical step have to be understood before analysis can be successfully completed. In a typical off-line analysis the extract is collected in 1 mL of solvent, and 1 μL is injected into the chromatograph, resulting in a loss of analyte by a factor of 10^3. Thus, on-line SFE can yield similar or better sensitivities than off-line SFE and with much smaller samples. This is particularly advantageous where trace levels are required and the sample size is small. However, it becomes a disadvantage if the sample size for extraction needs to be relatively large in order to maintain the integrity (i.e., homogeneity) of the sample. Such circumstances require further consideration of how much of the extracted material needs to be transferred to the chromatographic or other analytical step.

The Basic SFE Apparatus

The most basic system required for carrying out SFE is shown in Fig. 19.1. It consists of a solvent supply (A); a pump (B); a sub-ambient circulating bath to cool the pump head (C); an extraction cell (D), mounted in a ceramic heater tube (E); a controller for the temperature of the extraction cell (F); a restrictor connected to the outlet of the cell (G); a restrictor heater (H), to prevent extract deposition and resulting blockage in the restrictor; and a collection vial (I). In this simple unit, the pump is used to control the pressure while the restrictor controls the flow rate, or conversely, the pump can be set to control flow rate while the restrictor maintains the pressure of the extraction. Cooling of the pump head is not always essential, however, as will be discussed later. It is often done for fluids whose critical temperatures are not much above ambient, to ensure that the fluid in the pump head is always in the liquid state. Similarly, the restrictor is not always heated.

In a representative experiment the extraction cell might have a volume of 1 mL and be loaded with approximately 0.5 g of the matrix to be extracted. Carbon dioxide would be pumped at a rate of 0.5 mL per minute, metered as a liquid by the pump. The temperature would be 50°C; the pressure would be 400 bar, maintained by a

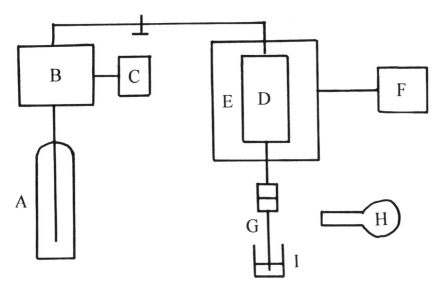

Fig. 19.1. Schematic apparatus for carrying out SFE: solvent supply (A); pump (B); pump head cooler (C); extraction cell (D), ceramic heater tube (E); heater controller (F); restrictor (G); restrictor heater (H); and collection vial (I).

restrictor consisting of a silica tube, 25 μm i.d. and approximately 12 cm in length. The effluent could be trapped in a vial containing about 3 mL of a typical organic solvent such as dichloromethane. A reference standard can then be added to the vial for quantitative analysis. The extraction would be carried out for approximately 30 min.

In addition to the dynamic mode just described, SFE can be carried out alternatively in a static mode. In this mode valves are fitted on either side of the cell. The outlet valve is closed, the cell is filled with fluid, and the inlet valve is closed. The fluid is then left in contact with the matrix for about 15 min. The valves are then opened, and the fluid in the cell is swept into the analyte trap by performing a short dynamic extraction for 15 min. Static extractions are less common than dynamic extractions but have been considered preferable for some applications.

Equipment Items

Pumps and Fluid Supply. Pumps for SFE are either those used for high-performance liquid chromatography (HPLC) or products specifically designed for SFE applications. They can be of the syringe or reciprocating-piston type. Since typical SC fluids are more compressible at ambient temperature than liquids, pump control systems are often more sophisticated than for HPLC pumps. The necessity for refilling is a disadvantage of syringe pumps, but advantages include rapid equilibration of pressure, once the pump is opened to the cell, and steady fluid delivery.

Cooling of the pump head can be carried out by pumping a water–antifreeze mixture through a jacket around the pump head from a cooling bath. Many practitioners cool the pump head even when this is not warranted. For example, it is not necessary if the fluid substance is supplied in a cylinder under pressure, as will be discussed later.

A liquid solvent is often added to a fluid as a "modifier" or cosolvent to improve its extraction properties. Such modified fluids may be delivered by one of the following methods:

1. Using a dual-pump system with a device for mixing the two fluids.
2. Using a single pump, to feed the contents from a cylinder containing premixed solvents.
3. Filling a syringe pump first with the modifier and then the major SC fluid component. Slower mixing of the components relies on turbulence during this filling process and there is a chance that the mixture will not be uniform when this method is used.
4. Using a multi-pumping setup that is part of a supercritical-fluid system in a well-equipped laboratory to fill individual syringe pumps on other systems *via* a network of tubing.

A cylinder with a dip-tube containing the fluid substance is often used to supply liquid to the pump. Because analytical SFE can concentrate impurities from the fluid substance, it has become common to use high-grade gases similar to those used for supercritical fluid chromatography. For studies in which trace amounts are to be extracted, it is essential that a blank run with an empty cell be carried out first to check impurities in the gas and subsequently the system.

Mixtures of carbon dioxide with various added proportions (5, 10%, etc.) of common modifiers are commercially available. The proportion of modifier in the fluid supplied from these cylinders increases as the contents of the cylinder are consumed, because the CO_2 gas left behind in the cylinder will contain a larger proportion of modifier. This may cause problems, because extraction conditions will change slowly. Gases and mixtures are also available that have an overpressure of helium of 100 to 130 bar, which is sufficient to maintain the CO_2 as liquid and removes the necessity of cooling the pump. Carbon dioxide supplied in this way contains about 3% of helium, which is assumed to have a negligible effect on the SFE, although this view is the subject of controversy at present (10). Cylinders of mixtures as well as those with helium head-pressure are relatively expensive, however. Of course, solvents used as modifiers are normally HPLC-grade.

The pumping system is the major cost of constructing a simple analytical SFE apparatus. Considerations in choosing a pump are as follows:

1. Whether flow rate, pressure, or both can be controlled by the pump.
2. Whether both pressure and flow rate can be measured at the pump.
3. Whether a pressure as high as 400 bar can be delivered by the pump with a maximum flow rate of at least 2 mL per min.

4. How quickly a syringe pump can be filled by the operator and returned to the desired pressure.

5. What facilities are available for cooling the pump head, to avoid the use of costly overpressure gas.

Extraction Cells and Their Use. Cells for SFE can be constructed easily from high-pressure components; however commercial cells are now available, not much more expensively, in sizes from ca. 1 to 20 mL. Cells contain a metal or ceramic frit at the outlet end to contain the matrix to be extracted. The shape and orientation of the cell can affect the efficiency of SFE from solids, as will be discussed in a later section. When conducting detailed kinetic studies prior to establishing a standard operating procedure for a specific type of analysis, vertical orientation with the fluid being supplied from the top, gives better results because there is no delay in the emergence of the extract. Conversely, when extracting a liquid, it is preferable to have the fluid enter the bottom of a vertically oriented cell that is considerably larger in volume than the liquid being extracted, in order to prevent foam being entrained and carried out of the cell. If the matrix is solid and the cell is horizontal, a glass wool plug can be used to hold the matrix in position; care must be taken in assembling the cell to keep particles away from the surfaces of the connectors.

Restrictors (Flow/Pressure Control). The simplest method of providing pressure or flow control is to use a piece of fused silica capillary tubing as a linear restrictor, which can be connected to the stainless steel outlet tube from the cell with a connector whose ferrule is made of a composite material. Tubing of 20 to 30 μm internal diameter and length of 10 to 50 cm is usual. The dimensions required are determined by experiment; the dimensions given in the representative extraction described previously can be used as a starting point.

Linear restrictors of this type are subject to blockage if the sample is wet or contains large quantities of other extractable material such as lipids or waxes. Water can freeze in the outlet, causing plugging, as it is cooled by the expansion of the fluid and the evaporation of the trapping solvent. Freezing can be avoided by heating the solvent vial, for example by placing it in a beaker of tepid water (5°C). Blockage by other materials can usually be prevented by heating the restrictor as previously described, taking care not to evaporate the trapping solvent and lose volatile analytes. Blockage quite often occurs at the beginning of an extraction, but it can be reduced by carrying out the initial extraction more slowly at a lower temperature or pressure, changing to the desired conditions later, when the larger quantities of blocking material have been removed. A coaxial metal restrictor and an adjustable restrictor, both of which are electrically heated, are available commercially and reduce plugging difficulties.

A much more expensive option is to use a commercial device with an electronically controlled orifice that maintains a back-pressure or flow rate set by the operator. These electronic valves are also not so subject to blockage (except perhaps under extreme circumstances where larger amounts of solid are precipitated). These devices often form part of commercial SFE equipment.

Analyte Collection.   One of the most widely used methods for analyte trapping is to pass the effluent through an organic solvent, as already described. A solvent compatible with subsequent analysis, such as dichloromethane, is used. Solvent trapping provides a suitable solution in one stage ready for many types of analysis. Other methods used are depressurizing in a cooled trap and depressurizing in the presence of an adsorbent. The last two methods can be less prone to the loss of volatile components as well as nonvolatile components lost as aerosols. Trapping with the use of an adsorbent has the potential for selective trapping. Whichever method is used, the possibility of loss of analytes at the trapping stage must be investigated in the overall assessment of an SFE procedure; see subsequent sections.

Commercial Equipment

A number of companies, including the large, well-known instrument manufacturers and some specialist companies, produce SFE apparatus. They regard SFE as a growth area, and their products are prominently displayed at exhibitions and conferences. Although more expensive than constructing a simple apparatus, these products offer convenience and easy control of conditions as well as such features as simultaneous extraction of a number of samples and automatic loading of samples.

Coupling of SFE to Other Techniques

There are a number of reports of "hyphenated" techniques involving SFE, and some commercial systems have been produced for this purpose. Examples are coupling to gas chromatography, capillary and packed-column supercritical fluid chromatography, and HPLC. This is an extensive subject and is covered in a later section.

General Principles of SFE

For SFE to be of use to the analytical chemist it must be quantitative. To achieve this, a procedure needs to be developed for each application on the basis of a good understanding of the extraction process, in order to avoid a number of possible difficulties. The speed of the technique and the relative ease of obtaining kinetic extraction data allow good assessment and development.

Extraction by a supercritical (or any) fluid is never complete in finite time. It is relatively rapid initially, followed by a kinetically limited region. For a successful SFE procedure, the curve of percent extracted vs. time has the form of curve A in Fig. 19.2. For a typical situation, 50% of the analyte is extracted in 10 min, but it may be 100 min before 99% is extracted. In less typical circumstances the kinetic behavior may have the form of curve B, where a short, slow period is followed by a rapid extraction and then a very slow extraction for a substantial fraction of the analyte, so that the SFE seems to have finished when only part of the analyte has been recovered.

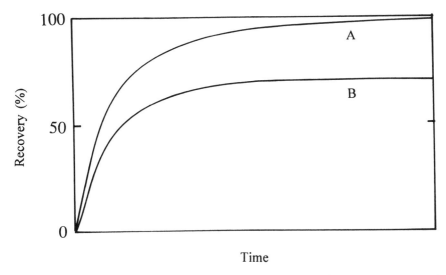

Fig. 19.2. Schematic curves of percentage extracted versus time for SFE: A: a curve with matrix effects largely absent; B: a curve showing substantial matrix effects.

It is considered that three interrelated factors influence recovery, as shown in the SFE triangle:

Solubility

For extraction to be successful, the solute must first be sufficiently soluble in the supercritical fluid. Only modest solubility may be required, particularly in trace analysis. Solubility depends on the chemical nature of the fluid, its density during SFE, and the temperature. Density, rather than pressure, is important. Raising the temperature at constant pressure will reduce density, and the net effect may be a reduction in solubility.

Information on the solubility for the analytes of interest in various fluids may be obtained from the literature (11) or inferred from reports of successful applications of SFE or supercritical fluid chromatography (SFC). If an SFC apparatus is available, experiments for the analyte/fluid systems can be carried out. The most direct method to determine whether solubility is sufficient is to deposit the analytes of interest onto filter paper and attempt to extract them under various conditions. References to solubility measurements of interest in the field of lipids are given in Table 19.1 (12–30).

Control of solubility can allow stepwise extraction. For example, the bulk of the triglycerides may be extracted from peanut meal by CO_2 alone before the analytes of interest (aflatoxins) are extracted by a CO_2-methanol mixture. If necessary, surfactants

TABLE 19.1 Published Sources for Solubility Data in CO_2 for Some Compounds of Interest for Lipids

Compound	Experimental method[a]	Temp. range (K)	Pressure range (atm)	Reference
Fatty acids:				
Behenic acid	S	313–333	80–250	12
Lauric acid	G	313	77–248	13
Myristic acid	G	313	82–249	13
	G	313–323	249	14
Oleic acid	G	313–333	200–300	14
	G	308–333	84–189	15
	S	313–333	100–250	12
Palmitic acid	G	318–338	142–575	16
	G	313	80–248	13
	G	298–313	80–187	17
	G	308–323	200–300	14
	G	310–320	112–359	18
	G	313–333	200–300	14
	S	313–333	100–250	12
	C	313	270–1900	19
Glycerides:				
Mono-olein	G	308–333	102–187	15
Tributyrin	S	313–333	100–250	12
Trilaurin	G	313	90–253	13
Trilinolein	S	313–333	80–250	12
Trimyristin	G	313	95–304	13
Triolein	G	298–333	69–197	20
	G	313–333	200–300	14
	S	313–353	80–250	12
Tripalmitin	S	313–353	80–250	12
	G	313	122–297	13
	G	298–313	86–182	17
Tristearin	G	313–333	200–300	14
	S	313–333	80–250	12
Esters:				
Behenyl behenate	S	313–333	100–250	12
Palmityl behenate	S	313–333	100–250	12
Oils:				
Canola oil	G	298–363	100–360	21
Corn germ oil	G	353	265	22
Cottonseed oil	G	313–353	476–1020	22
Jojoba oil	C	293–353	99–2568	23
Rapeseed oil	G	313–353	100–850	24
Soybean oil	C	298–353	99–2568	25
	C	293–313	148–346	26
	G	313–343	207–689	22
	G	323–333	136–681	27
Sunflower seed oil	C	313	178–691	26
Vegetable oil	G	298–328	160–480	28

(continued)

TABLE 19.1 Published Sources for Solubility Data in CO_2 for Some Compounds of Interest for Lipids (cont.)

Compound	Experimental method[a]	Temp. range (K)	Pressure range (atm)	Reference
Steroids:				
Cholesterol	S	313–353	80–200	12
Ergosterol	C	313	79–198	29
Ethinyl estradiol	C	313	79–198	29
Solasodine	C	313	79–198	29
Vitamins:				
α-Tocopherol	G	298–313	100–183	17
	S	313–353	100–250	12
Pesticides and herbicides:				
3,4-Dichloroaniline	G	313	197	30
2,4-Dichlorophenoxyacetic acid	G	313	197	30
2-[4-(2,4-Dichlorophenoxy)-phenoxy) propionic acid	G	313	197	30
2-[4-(2,4-Dichlorophenoxy)-phenoxy) propionic acid methyl ether	G	313	197	30
Linuron	G	313	197	30
Methoxychlor	G	313	197	30

[a]S = spectrometry; G = gravimetry; C = chromatography.

or complexing agents can be dissolved in a modifier to enhance solubility. Derivatization of analytes before extraction to improve their solubility will be covered in a later section.

Diffusion

The analyte must be transported sufficiently rapidly by diffusion from the interior of the matrix in which it is contained. The transport process may be simple diffusion of the solute, as in the case of a polymer, or it may involve diffusion of the fluid into the sample matrix followed by diffusion of the analyte through the fluid *via* pores in the matrix. Adsorption and desorption may take place during transport. Often the precise process will not be known, but a transport process similar to diffusion occurs. The timescale for diffusion and transport will depend on the shape and dimensions of the matrix or matrix particles. Of these the shortest dimension is of great importance, because the times depend on the inverse of the square of its value. Optimum values for the shortest dimension should be of the order of 1 mm or preferably less to ensure rapid extraction.

The Matrix

The third factor influencing SFE is the matrix (in addition to its effect on diffusion). Examples of the effect of the matrix are adsorption of analyte molecules on surface sites, trapping of molecules in viscous liquids such as polymers, and the necessity for analyte molecules being extracted from a biomatrix to penetrate cell walls. Any of these processes can be a rate-determining step in the later stages of an extraction process. These effects are considered to be responsible for the very slow final stage in some applications, as illustrated by curve B in Fig. 19.2. Often this final stage is so slow that it appears that not all of a compound is "extractable," the rest being locked into the structure of the matrix or strongly bound to its surface. However, the analyte is in fact still being extracted slowly.

Of the three factors influencing SFE, the matrix is the least well understood at present. Research into the problems that arise is continuing, both by searching for successful procedures and by modelling SFE and analyzing experimental data in terms of the models. These approaches are not helpful for cases with matrix problems, because the practicing analyst needs to extract and quantify the total amount of a particular compound present in a sample. A particular application must therefore be investigated experimentally, by one or more of the following:

1. Varying the SFE technique used (the fluid, temperature and pressure).
2. Comparing SFE with other methods.
3. Testing it with spiked samples (although spiked samples, even if aged, may not be truly representative of a real sample, especially for soils and plant and animal tissues).

With respect to method 1 it is generally helpful to raise the temperature, especially if the density is kept constant. For many difficult SFE cases the best solvents are the polar chlorinated and fluorinated methanes. The use of these compounds unfortunately removes the environmental advantage of SFE using SC-CO_2. Particular applications are discussed in a later section on method development.

Kinetic Studies of SFE

Unless the particular problem of interest to the analyst has been well studied and reported in the literature, it is well worthwhile to carry out a kinetic study on a particular application before deciding on an operating procedure. It is tempting to assume that extraction is essentially complete if it has been carried out for two consecutive equal periods of time and the second period yields only a small fraction of the compound extracted in the first period. Because of the form of the recovery curve, however, it can be seen that this assumption is not correct.

A kinetic study can be carried out using the apparatus shown in Fig. 19.1 by changing the vial containing the trapping solvent periodically during extraction. For many applications, short periods of a few minutes are appropriate at the beginning, increasing to, say, 20 min later. Appropriate times vary with the application. Recovery can then be compared with the total amount present in the matrix, found *via* another method such as liquid extraction. A curve of recovery versus time, as shown in Fig. 19.2, can then be drawn. A suitable procedure may then be developed, provided studies of this type are made for various conditions of fluid, temperature, and pressure.

Extrapolation Procedures

Even if no matrix problems are present during the extraction, the associated extraction kinetics may make extraction times excessively long. However, the extraction kinetics frequently follow an exponential form, which allows a simple extrapolation procedure to be used, thereby shortening the time required for SFE.

If extraction is carried out until the kinetically controlled region of the extraction curve is reached, the original total solute mass in the sample, m_0, can be determined from the extracted mass m_1 at time t_1, followed by extraction over two subsequent equal time periods to obtain masses m_2 and m_3 by the equation

$$m_0 = m_1 + m_2^2/(m_2 - m_3) \qquad (19.1)$$

This equation may be used to obtain m_0, provided the difference between m_2 and m_3 is large enough compared with the errors in the two quantities. This is not too serious a problem, because usually the last term in the equation is small compared with m_1.

An extreme example follows for the case of extraction from polymers. Here there is an advantage in working with the original sample pellets, for convenience as well as the possibility of loss of analyte during the grinding process. However, a fairly exhaustive extraction of polymer pellets of a few millimeters in diameter takes around 80 hours! The extrapolation procedure was therefore applied to this type of system. Table 19.2 gives data for the extraction of 2,6-di-*tert*-butyl-4-methylphenol (BHT) from standard polypropylene cylinders of ca. 3 mm in both length and diameter, with additive concentrations known to within 1% w/w. Although extraction for 8 hours yielded only 57% of the additive, an estimate of the final amount made using Equation 19.1 was only 5.2% below the given value (31).

Development and Optimization of an SFE Procedure

The initial selection of SFE conditions has often been based on the degree of solubility of the target analyte in the supercritical fluid. This is clearly a primary consideration, because successful extraction conditions are often inferred from solubility data for compounds in supercritical fluids, together with trial extraction data for spiked samples from inert matrices. However, the actual situation is not as straightforward; extraction of "real" samples quite often indicates that there must be additional circumstances governing extraction other than the extent of analyte solubility. Such a circumstance might be the degree of analyte interaction with the host matrix. Hawthorne et al. (7) termed these parameters "sample/analyte parameters" and other factors controlling SFE recoveries "experimental parameters;" some of the latter have been discussed earlier in this chapter, such as choice of fluid (and modifier if used) and

TABLE 19.2 Extrapolation to Obtain Final Quantities in the Extraction of BHT from Polypropylene (31)

Extraction times (min)	Weight extracted (µg)	Cumulative times	Weight extracted (µg)
0–20	7.1		
20–60	25.0		
60–120	45.7		
120–180	36.8		
180–240	26.8	0–240	141.4
240–300	16.4		
300–360	17.1	240–360	33.5
360–480	27.8	360–480	27.8
Total	202.7	Given total	356.8
Total from extrapolation equation			338.3
Difference between given total and extrapolation			-5.2%

[a]BHT (0.2% w/w); 178.4 mg of standard polypropylene pellets; pure CO_2 at 50°C and 400 atm.

extraction temperature and pressure (density control). Operational and mechanical considerations also dictate recoveries, such as flow rate and total volume of supercritical fluid consumed, extraction time, dead volume of the extraction cell (linked with sample size), and the effectiveness of the postextraction analyte collection system.

Two features of SFE are of paramount interest to the analyst:

1. The *extraction recovery* is the amount of solute extracted relative to the initial amount, which is known or computed from values obtained for the same sample material using a traditional extraction technique. This is most usually represented as a mass percentage.

2. The *rate of extraction* is the extraction recovery per unit time at a given velocity of the fluid through the cell. This decreases exponentially with time.

Conditions for the extraction of a specific analyte from a particular matrix clearly need to be optimized to maximize the efficiency of the procedure. The parameters optimized are those that are of most significance to the application of SFE: the pressure, the temperature, the possible addition of an organic modifier to the fluid, and the flow rate.

Influence of Pressure

King (32) found that knowledge of four basic parameters of SFE is extremely helpful in understanding solute behavior in compressed gaseous media and thus the execution of successful analytical SFE. The first of these is the *miscibility,* or the *threshold pressure* as termed by Giddings et al. (33,34), which corresponds to the pressure at which the solute partitions into the supercritical fluid and to the critical loci of mixing between the dissolved solute and the solvent gas. Next is the pressure at which the solute attains its maximum solubility. This can be approximated by Giddings's equation, which relates the solubility parameter of the gas to its critical and reduced state properties.

The third parameter is the pressure region between the miscibility and solubility maximum pressures, known as the *fractionation pressure region*. In this interval it becomes possible to regulate the solubility of one solute relative to another. Fractionation between solutes is maximized in this pressure region by differences in the physical properties of the dissolved solutes and by low concentrations in the fluid. The difference between the threshold pressure and the solubility maximum pressure can be shown for any substance that dissolves in a supercritical fluid. A good example is shown by the solubility–pressure curve of the herbicide atrazine; see Fig. 19.3*a* (35). This solute is slightly soluble in CO_2 at 100 bar (threshold pressure); as the pressure increases, the solubility increases up to the maximum value, with a particularly steep rise at 150 bar. The fluid pressure is the main parameter that influences the extraction recovery, as can be seen from the pressure–recovery curve for atrazine in Fig. 19.3*b*. Elevation of the pressure for a given temperature results in an increase in fluid density and therefore in higher solubility of atrazine. Thus, at higher pressure with improved analyte solubility, CO_2 consumption is reduced compared to a lower-pressure (decreased solubility) situation. McNally and Wheeler (36) demonstrated this when extracting diuron herbicide from soil at low pressure (110 bar) and high pressure (338 bar); the former conditions required twice the volume of CO_2 to extract 70% of the analyte. Consideration must be made when using high pressures to extract complex matrices, because the enhanced solubilities may complicate the extract containing the target analyte by enabling coextracted analytes to appear in the extract and complicate subsequent analysis. Coextracted solutes may also enhance or reduce the solubility level of the target analyte. These effects are known as synergistic effects and arise through solute/solute/solvent interactions.

Influence of Temperature

The solubility of a substance is determined by two effects: its volatility and the solvating effect of the fluid; the latter depends on the fluid density. For a constant pressure the density of a supercritical fluid such as CO_2 decreases with increased temperature, with the effect becoming more pronounced in the critical region, where the compressibility is high. Thus, the solvating effect of the fluid decreases with increasing temperature, whereas the volatility of the analyte increases. As a result, the solubility of a substance in a supercritical fluid at constant pressure first decreases with temperature as the density effect is dominant, then reaches a minimum, and then increases as the volatility effect takes over. The temperature of this minimum varies with pressure and the solute. For solutes of limited volatility the minimum is at a high temperature and may never be seen experimentally.

Thus, for a relatively nonvolatile species, increasing the temperature for a fixed fluid pressure will result in a reduction in its density and thus the solubility of the analyte in the fluid, leading to reduced extraction efficiencies. In many other examples

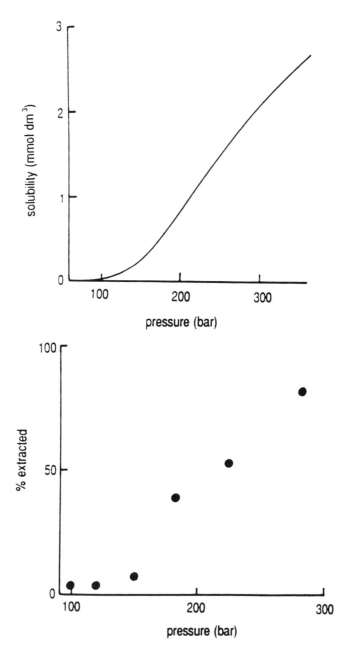

Fig. 19.3. Variation with CO_2 pressure of (a) predicted solubility (upper curve), and (b) experimental percentage recovery in unit time (15 min) for extraction of atrazine from soil at 80°C. *Source:* S. Ashraf, et al., *Analyst 117*:1697 (1992).

where the solute is more volatile, efficiencies improve despite the reduction in fluid density. In the example given by McNally and Wheeler (36), diuron recovery from soil improved to 80% from 50% at 340 atm for an increase in temperature to 100°C from 75°C. Langenfeld et al. (37) show that a large increase in temperature rather than pressure was more significant in improvement of recoveries for polychlorinated biphenyls (PCBs) and polycyclic aromatic hydrocarbons (PAHs) from environmental samples. The temperatures investigated were 50°C and 200°C, and the pressures were 350 and 650 atm. At a relatively low density (pressure, 350 atm), recoveries at 200°C compared to 50°C were extremely high and much greater than 100% (when compared with Soxhlet-derived values) for many of the analytes. These were even greater than those obtained by a density increase by raising the pressure from 350 atm to 650 atm at 50°C.

Increasing the temperature increases the vapor pressures of the analytes. Many solubility studies show the significance of temperature and the way it affects the vapor pressure. Solute solubility is increased for increases in temperature at a constant density, for example triphenylene, perylene, and anthracene in supercritical CO_2 (38,39). The use of very high temperatures is, however, precluded by the limits of some commercial apparatus and their component parts.

Addition of a Modifier

Carbon dioxide is the most popular choice of supercritical fluid and is ideal for solutes of low to moderate polarity because of its own low polarity. The molecular quadrupole of CO_2 does give rise to some interaction with polar solutes. Alternative solvents are available, but as discussed earlier, some of these have less attractive characteristics. CO_2 can be used for more polar compounds by the addition of small proportions of a polar organic solvent. The concept of adding cosolvents, also called modifiers or entrainers, to supercritical fluids has received much attention (40–45). Dobbs et al. (45) demonstrated experimentally and theoretically that polar cosolvents interact specifically with certain polar solutes in supercritical CO_2. In contrast, they showed (44) that a nonpolar modifier such as octane had approximately the same effect on polar and nonpolar solutes if their molecular weights or polarizabilities were similar. For a hydrocarbon cosolvent, analyte solubility is enhanced with an increase in chain length or solute polarizability, and the increase in attractive forces outweighs the increase in repulsive forces associated with increasing solute chain length. Foster et al. (46) showed enhanced solubilities of cholesterol in CO_2 and ethane when polar (acetone) and nonpolar (hexane) modifiers were added to each supercritical fluid, respectively.

The amount of modifier added is significant in the way it affects the solubility and subsequent extraction recoveries during SFE. Dobbs et al. (44) showed increased solubility of phenanthrene in CO_2 when the mole percent of *n*-octane cosolvent was increased for a given fluid density. They also demonstrated the same trend for the solubility of benzoic acid with sulfur dioxide as cosolvent (45). This trend is supported by numerous extraction experiments in which an increase in modifier concentration resulted in increased extraction recoveries, for example, the extraction of sulfonyl urea from soil (47), diuron from Sassafras soil (36), and PAHs from C_{18} cartridges (48). The selectivity of CO_2-ethanol for cocoa butter relative to xanthines is enhanced with increasing ethanol concentration (49), as were recoveries of aflatoxin spiked on peanut meal (50) and curcumin from turmeric using methanol-modified CO_2 (51).

Cosolvent studies require consideration of the effect of the modifier on the mixture properties. These conditions must be known or approximated to ensure that the mixture remains homogeneous (i.e., one phase) throughout the range of experiments. The phase diagram for a CO_2-methanol mixture at 50°C can be seen in Fig. 19.4 (52). Above about 95 bar the mixture is one phase at any composition; however, with smaller amounts of methanol, the pressure can be lower while still maintaining a single phase. The hatched region shows "supercritical" conditions for such a mixture.

Care should also be taken in the storage of the modifier. Two distinct values have been obtained for a solvatochromic dye in methanol (53). The less polar value was obtained from a freshly opened bottle of methanol; the more polar value, from an older, used supply. The values represent two extremes observed over several months using methanol with no special preparation. The addition of small amounts of water to methanol induced a spectral effect similar to that of the "old" methanol. This indicates that the methanol that had been opened to the atmosphere many times may have contained water, which may have altered its properties as a modifier.

Chemical Derivatization

Chemical derivatization can be viewed in a similar way to modifier addition, because in both processes an additional compound is being introduced to the fluid. Unlike the addition of polar modifiers to supercritical fluids such as CO_2, which improves the selectivity and extraction of polar species, the addition of a suitable derivatizing agent is frequently used to convert polar compounds into less polar compounds, which are then more soluble in conventional SFE solvents such as SC-CO_2. Moreover, many polar species need to be derivatized prior to chromatographic analysis of the extracts. For example, acid herbicides such as 2,4-dichlorophenoxyacetic acid require liquid solvent extraction, followed by diazomethane derivatization of the herbicides to their methyl derivatives prior to capillary gas chromatography (CGC) analysis. With *in situ* derivatization during the SFE step the herbicides can be converted to their acetyl derivatives (which also makes them easier to extract) and extracted in a single step (54).

In the same work, Hawthorne et al. (54) demonstrated quantitative (>90%) recoveries for spiked and native acetic acid and Dicamba from soil and sediment, microbial phospholipid fatty acids (as their methyl esters) from whole cells, and wastewater phenolics (as their methyl ethers) from water and C_{18} sorbent disks. For all these examples, the analytes were derivatized to less polar species using reagents such as trimethylphenylammonium hydroxide and boron trifluoride in methanol. Gholson et al. (55) showed that extraction efficiency improved for liquids when derivatizing reagents were added to the extraction solvent. In that work, derivatization of the

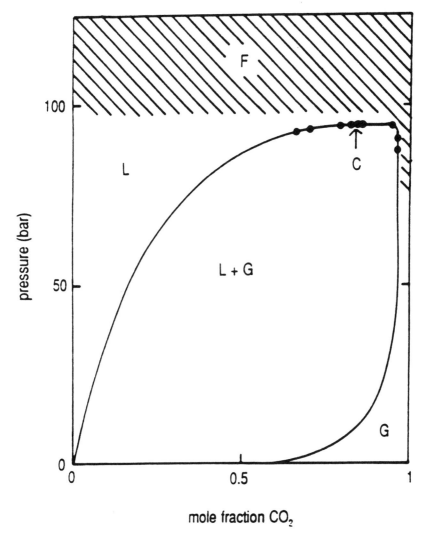

mole fraction CO$_2$

Fig. 19.4. Semischematic phase diagram for the methanol-CO$_2$ system at 50°C. Abbreviations: L: liquid; G: gas; F: one fluid phase; C: critical point. Experimental data from Ref. 52.

sample during ultrasonic agitation to form trimethylsilyl derivatives increased the recovery of organic acids, alcohols, and phenols from airborne particulates. Hills and Hill (56) have used the derivatization reagent *N*-trimethylsilylimadazole in simultaneous supercritical fluid derivatization and extraction (SFDE) of components from roasted coffee beans. Levy et al. have performed *in situ* trimethylsilylation of sucrose esters (57). White et al. have reported fatty acid methyl ester formation under supercritical fluid conditions (58). A more comprehensive study of SFDE by Hills et al. used CO_2 with tri-sil concentrate added directly to the exhaustively extracted sample matrices of roasted coffee beans, Japanese tea leaves, and marine sediment (59). Tri-sil concentrate, a mixture of hexamethyldisilane and trimethylchlorosilane, yielded many additional derivatized and nonderivatized compounds from these sample matrices.

Since 1991 there has been an increased interest in SFDE or *in situ* chemical derivatization in the literature. The extraction of calcium montanate (comprising calcium salts of C_{28}–C_{32} fatty acids, traces of C_{28}–C_{32} hydrocarbons, C_{28}–C_{32} free fatty acids and esters of these fatty acids) from nylon 6.6 can be done in one step by the addition of citric acid as a complexing agent for Ca^{2+}, converting the montanate salts into the free fatty acids, which can be extracted with 10% methanol/CO_2 (60). Lee et al. demonstrated *in situ* SFE and derivatization by producing acetylated derivatives for pentachlorophenol and related compounds from soils (61) and phenolics from sediments of pulp mill origin (62). The acetylated derivative was produced under static conditions by using acetic anhydride in CO_2 in the presence of triethylamine.

Field et al. quantitatively recovered secondary alkanesulfonate (SAS) and linear alkylbenzenesulfonate (LAS) surfactants from sewage sludges as their tetrabutylammonium ion pairs using SFE and derivatization of the analyte (63). The ion pair reagent served two purposes in this study. First, it enhanced the extraction of sulfonated surfactants in supercritical carbon dioxide by decreasing their polarity. Second, surfactant ion pairs underwent derivatization in the injection port to form sulfonate alkyl esters, which were successfully analyzed by GC/MS. King et al. (64) extracted triglycerides from oilseed samples using supercritical CO_2, which were transesterified to fatty acid methyl esters *in situ* over a solid alumina catalyst. The extracts were collected off-line in a vial or on-line in a chromatographic retention gap.

The diversity of the *in situ* derivatization SFE method is further demonstrated by Cai et al. in their determinations of butyltin and phenyltin compounds in sediment (65). These organotin compounds are alkylated by an excess of hexylmagnesium bromide (Grignard reagent), injected into the extraction cell, followed by sonication, prior to SFE. Quantitative recoveries were returned for spiked di- and triphenyltins but not for monophenyltins. Quantitative recoveries by this method were also obtained for certified reference sediments in the absence of available standard reference material for phenyltins in sediments.

Influence of Flow Rate

Flow rate can be easily controlled or changed by adjusting the degree of back pressure in the extraction cell. For the commonly used linear-type restrictors constructed of fused silica capillary tubing, changing the length or inner diameter of the tubing alters the flow rate. It is desirable to have a constant flow rate during extraction, and much effort has been directed in restrictor development to produce plug-free devices. Restrictor blockages result from the freezing of the outlet on depressurization or from analyte precipitation along the length of the restrictor, and this contributes to incomplete recoveries or poor reproducibility of recovery data. Many reported off-line extractions in which the analytes are trapped in a collection solvent are performed using these linear, fused silica restrictors (66–68). To avoid loss of analytes through impromptu blockages or breakage from silica (made brittle from the passing of certain modifiers), some workers recommend their replacement frequently (66,67,69). Controlling the restrictor temperature has been shown to reduce plugging for some types of samples (70). For the extractions of PCBs from sewage sludge and PAHs from marine sediment and mussel tissue (71) it was necessary to heat the entire linear restrictor device (end to end) in the optimized restrictor-trapping scheme.

To obtain the best efficiencies it is necessary to optimize flow rate for a given cell volume and sample. An extraction cell should be selected that minimizes dead volume in the system, allowing larger samples to be extracted with lower fluid flow rates, therefore reducing the amount of fluid needed and simplifying the collection of more volatile analytes (72). Decreasing the flow rate results in a lower linear velocity and usually in increased recoveries, as a result of extended contact time between the fluid and the sample (73). This was shown to be the case for ^{14}C-labeled linear alkylbenzenesulfonates extracted with 40 mole% methanol–modified CO_2 at 380 bar and 125°C; for the same volume of fluid, recovery was only 75.6% at a liquid CO_2 flow rate of 1.2 mL/min^{-1}, compared to 90.8% when the flow rate was decreased to 0.45 mL/min^{-1} (74). Furton and Lin (75) observed a similar result for the recovery of PCBs from different sorbents at different flow rates (0.075–0.448 mL/min). So did Kane et al. (76), who also observed this effect during studies in the effects of density and flow rate on extraction efficiency, extracting megestrol acetate as a pure compound deposited on the walls of a glass liner using a 1.5-mL cell at 55°C over a 10-min extraction period.

Lower flow rates mean longer extraction times, and it is clear that there is a correlation between the extraction flow rate and extraction time for a given cell volume. Once a flow rate has been optimized, it is necessary to conduct the extraction sufficiently long for analytes to be swept out of the cell. Thomson and Chesney (77), extracting 2,4-dichlorophenol from crop tissues, found that 10 cell volumes of supercritical CO_2 should pass through the extraction cell. Kane et al. (76) found for steroid compounds that four cell volumes were necessary. Hence, for a known flow rate and cell volume, a suitable extraction time could be determined.

The effect on recovery of high flow rates was shown by the decreased recovery of diuron from Sassafras soil (36). In this study the optimum flow rate was 5 mL/min of 10% methanol–modified CO_2. Further, this study showed that for an increase in flow rate there was also an increase in pressure drop across the extraction cell, from 6.8 atm at 2.5 mL/min to 68 atm at 7.5 mL/min. Control of this pressure drop did improve efficiencies at the higher flow rate. The overall flow rate of gaseous CO_2 leaving the restrictor must be kept at a modest level to minimize aerosol formation, excessive evaporation, or purging of trapped analytes from the collection solvent. Thus, an optimum flow rate has to be found, but a typical value is approximately 1 mL/min of compressed fluid (with extraction cells of inner diameter of approximately 1 cm). This corresponds to a gas flow rate of approximately 500 mL/min on decompression.

Influence of Cell Geometry

Cell geometry and flow rate are closely linked. Changing the volume or dimensions of the extraction cell for a given restrictor size will result in a change in flow rate, fluid linear velocity, flow pattern, and extraction efficiency. If the flow rate is fixed, the fluid linear velocity can be changed by using several cells having the same volume but different inner diameters. Higher extraction efficiencies are expected for low fluid linear velocities because of the larger contact time the fluid has with the matrix, as discussed earlier. Low fluid velocities can be brought about by employing short, broad cells.

Furton and coworkers (78–80) observed changes in recoveries in their investigation of the geometry of microextractors used for the extraction of PAHs and PCBs from specified sorbents. In one experiment 520 mg of octadecyl-bonded silica was spiked with a PAH mixture containing a range of molecular weights and ring numbers and extracted with CO_2 at 300 bar and 100°C from extraction cells of dimensions 1.0 × 1.0 cm and 7.3 × 0.37 cm; that is, the ratios of internal diameter to length of bed were 1:1 and 20:1, respectively. In each case the total volume of the sorbent bed was approximately the same, 0.79 cm^3 (78). Observed SFE efficiencies were increased by more than a factor of 2 by decreasing the microextractor cell diameter-to-length ratio from 1:20 to 1:1 for the largest PAHs studied. Similar results were reported for extraction of a mixture of PAHs and methoxychlor from the same sorbent, cell diameter-to-length ratio, and extraction conditions (79). Furton and Lin also extracted PCBs from octadecylsilane using cell geometries of 9.9 × 9.9 mm and 4.4 × 50 mm, having 1:1 and 1:11 cell diameter-to-length ratios respectively. Extractions were carried out on 510 mg of the sorbent using CO_2 at 150 bar and 60°C, a flow rate of 0.075 mL/min and a total CO_2 volume of 1.5 mL (80). Changing the cell geometry from 1:11 to 1:1 resulted in relative increases in recoveries ranging from 2% to 77%, depending on the type of PCB.

Additional studies have explored cell shape and orientation with a view to nonhomogeneous environmental samples such as sludges, soils, sediments, exhaust, and air particulates (72,81). These studies looked at extraction rates and recoveries for native PAHs from 3 g of railroad bed soil and for the flavor and fragrance compounds in 1 g of lemon peel. The conclusions drawn from this particular study were that no observable and significant differences in extraction rates were apparent from these samples when using long, narrow extraction vessels as compared with short, broad ones. This investigation also showed that, provided the vessels were loaded in such a way as to minimize dead volume, the cell orientation (horizontal versus vertical) showed no significant effects on the observed rates.

Furton and Lin (80) also showed no significant effect on the recoveries of PCBs from the adsorbent Florisil using the same cell geometries described previously. Their results showed the importance of the matrix effect in SFE: For the adsorbent system of Florisil, the analyte is retained by surface adsorption and the interactions are thought to be similar to the analyte–matrix interactions associated with soil material.

It has been shown in certain situations that reduction of the extraction cell diameter increases turbulence within the cell (as indicated by the Reynolds number), and this improves extraction efficiencies *via* increased mass transfer within the cell. This was observed for the *n*-alkanes, *n*-octadecane and *n*-eicosane, spiked onto equal volumes of alumina sorbent for cells of 6 mm and 4 mm diameter. Extractions were for 1 min with CO_2 at 80 atm and 70°C, at a constant flow rate. These extracts were cryogenically trapped at −70°C and analyzed by gas chromatography. The peak area counts recorded on elution of the alkanes from the extraction cell, for equal alumina volumes and CO_2 mass flow rates, clearly showed the advantages of the smaller cell diameter for enhancing extraction (82).

Applications and Sample Preparation

Successful SFE has been carried out on a wide variety of systems. The majority of applications have involved the extraction of solid matrices; however, air and liquids, including liquids containing suspended microorganisms, can also be extracted. *In situ* derivatization of liquid samples has been carried out prior to extraction, for example the esterification of fatty acids. The capability to extract nonsolid matrices has been particularly relevant for environmental samples, where many applications of SFE to date have been directed.

The nature of the sample matrix can have a profound effect on the results that are obtained with SFE. A knowledge only of analyte solubilities in supercritical fluids does not always allow a prediction to be made as to the effectiveness of SFE for a particular matrix (83). This is an unfortunate complication when extracting real sample matrices, such as soils or biological tissue, because predictions made on the basis of theory or results obtained for neat analytes or spikes from artificial matrices do not necessarily parallel those found for the same analyte from a natural matrix. This has led to a necessity for experimentation and method development with real samples. Many factors influence the nature of the matrix, such as the particle size, shape, surface area, porosity, moisture and level of extractable solutes, all of which can contribute to the overall analytical results. Similarly, the interactions between solutes and active sites of the matrix can dictate the use of strict extraction conditions.

Extraction from air samples, only of marginal interest in the context of oils and fats, usually involves identification and determination of organic compounds from emission fumes or the atmosphere itself. The classical approach to this involves two independent steps (66,84–90):

1. Pollutants are trapped on a solid-phase sorbent.
2. The trapped pollutants are then either thermally desorbed and analyzed by GC or eluted with a suitable liquid solvent before their separation by HPLC.

Liquid Samples

Direct extraction of liquid samples has been performed utilizing high-performance liquid chromatography (HPLC) columns with meshed frits in the end caps. For extractions of liquids Hedrick and Taylor (91–93) have designed a special extraction cell.

These studies used a vessel that is conceptually the same as that reported by Ong et al. (94) but with significant improvements. The supercritical fluid is introduced from the bottom by passing the tubing in from the top (see Fig. 19.5). Using this cell design, Hedrick and Taylor extracted phosphonates and various substituted phenols from water. A static extraction, where the fluid was recirculated through the solution, was employed for the phosphonates. Incorporation of sodium chloride (as a salting-out agent) into the aqueous matrix allowed complete removal of the analyte at the 200-ppb level in less than 5 min. On-line supercritical fluid chromatography with Fourier transform infrared spectroscopy and flame ionization detection (SFC–FTIR–FID) conclusively identified the phosphonate and afforded a quantitative method of analysis (91). Dynamic extraction of solutions containing triprolidine and pseudoephedrine, originally in the form of hydrochloride salts, in the presence of a molar excess of tetrabutylammonium hydroxide (to form free bases) was qualitatively done using CO_2, at 340 atm and 50°C. Extraction of phenol from water using CO_2 at different densities was also demonstrated. Using conditions of 100, 125, and 150 atm at 50°C, recoveries improved from 17% to 62% to 68% respectively when analytes were collected in a liquid solvent. Violent bubbling of the solvent resulted in sporadic losses, so a collection arrangement involving deposition onto a solid support was engineered in order to remove this problem. In this arrangement the analyte was deposited onto a solid-phase extraction cartridge, which was cooled by the CO_2 expansion from a 50 μm restrictor, providing a 2 mL/min flow rate that cooled the trap to −30 to −35°C. This arrangement provided a significant improvement in analyte collection compared to liquid solvent collection, as 80% recoveries were achieved for 150 atm and 50°C extraction conditions (93).

Also studied was the feasibility of using supercritical CO_2 for both static and dynamic extraction of organic bases from an aqueous matrix. This task was expected to be difficult, because the pH of water decreases to approximately 3.5 as CO_2 is introduced into the solution. The extraction of basic compounds is more difficult than that of acidic compounds because as bases protonate, they become more soluble in water, further reducing their solubility in supercritical CO_2. This was found to be the case for the nitrogen-containing analytes of succinonitrile, 2,5-lutidine, and picolinic acid. In similar experiments performed on caffeine and nicotine, the caffeine was dynamically extracted from water successfully, whereas nicotine was not. In contrast, larger organic bases such as triprolidine, pseudoephedrine, and sulfamethazine, were found to be directly extractable from water with supercritical CO_2. Sulfamethazine could be recovered from whole milk at 250 atm and 50°C for 15 min (20 mL CO_2) with a 95% recovery. The triprolidine and pseudoephedrine were extracted from water [to which tetramethylammonium hydroxide (TMAOH) had been added to neutralize the acid salts]. The presence of TMAOH was necessary for SFC to be performed subsequently on the free bases that had been extracted.

The mechanism by which bases are extracted from water with CO_2 is not well understood. Extractability of a base depends on the presence of a significant, strictly hydrocarbon moiety in the analyte, rather than on the pK_a of the base. At the

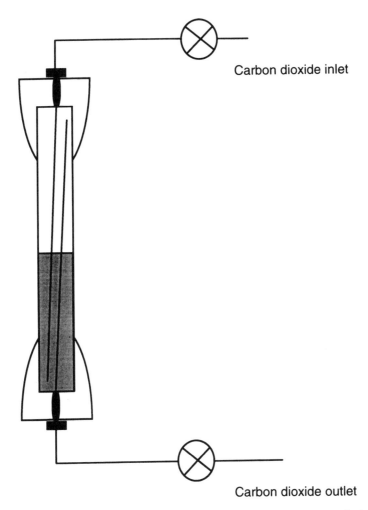

Carbon dioxide inlet

Carbon dioxide outlet

Fig. 19.5. Extraction vessel for aqueous samples. (*Source:* Adapted from Hedrick, J.L., and L.T. Taylor, *J. High Resol. Chrom. 13*:312 (1990).)

fluid–water interface, the organic base may become aligned with its polar side towards the aqueous phase and its lipophilic side (if any) towards the supercritical fluid. The greater the amount of lipophilic material, the more likely the analyte is to partition into the supercritical fluid (93).

A different arrangement for the extraction of aqueous matrices was employed by Thiebaut et al. (95). A phase segmenter and subsequent phase separator were used to extract spiked 4-chlorophenol from a urine sample with supercritical CO_2 at 118 bar and 40°C; phenol was extracted from water under the same conditions. Extraction efficiencies of the test compounds were greater than 85%.

Solid Samples

The properties of supercritical fluids make them very useful for the rapid and quantitative extraction of solid matrices, particularly for extracting organic pollutants from environmental solids, organic flavor and fragrance compounds from natural products, contaminants from foodstuffs, and additives in polymers (including compounds absorbed by a polymer or present in controlled levels). Examples from such matrices are numerous, and many, such as the extraction of PAHs from coal, particulates and sediments (96–100), are not very relevant to the field of lipids.

Relevant applications include the following:

- Tocopherol was extracted from soybean sludge using CO_2 (101);
- A hypolipidemic drug was recovered from animal feed using CO_2 (102);
- Using CO_2, phenolic lipids were extracted from a matrix of cashew nut (103); and
- With CO_2 and CO_2/dichloromethane (CH_2Cl_2), essential oils were extracted from the aromatic plants savory, dragonhead, and peppermint (104).

Extraction from polymers is of relevance in the food industry, because the SFE of additives and low oligomers can be considered to mimic their migration into food, particularly fats and oils. CO_2 was used to extract oligomers from polymeric materials such as poly(ethylene terephthalate) (105) and to extract polymer additives from polypropylene (106).

Influence of the Physical Nature of the Matrix

The physical morphology of the substrate undergoing SFE can have a pronounced influence on the efficiency of the extraction and the rate at which it is conducted. Grinding solid samples before using them enhances extraction by increasing the surface area exposed to the supercritical fluid, thereby shortening internal diffusional path lengths over which the extracted solutes must travel to reach the bulk fluid phase. Studies have shown that the geometric size of the matrix particles can influence the speed and efficiency with which an SFE can be conducted (107). Precautions must be taken however, in order to avoid large pressure drops in the extraction cell that results from sample compaction (the presence of fines is not recommended) and plugging. If small particles are used, the cell ends should contain inert materials such as glass wool, filter paper, or glass beads, in order to prevent particles from being swept out of the extraction cell. Blending the sample with such materials also reduces the likelihood of sample compaction. Inert extraction enhancers can include materials that absorb moisture as well as providing a homogenized extraction bed, as in the use of pelletized diatomaceous earth in the extraction of pesticides from foods with CO_2 (108).

Chemical Composition of the Sample Matrix

The chemical composition of the sample matrix can have either an enhancing or a retarding effect on the results obtained with SFE. The recovery rate is therefore a function of both the chemical nature of the solute and that of the matrix itself. A solid with numerous active sites will highly adsorb polar solutes, often resulting in a poor extraction rate. Such an example is the reduced recovery of explosive compounds from soil samples when the organic content of the soil is increased (109). One of the most difficult matrices to extract is fly ash (99). Organics inside the material can also affect the extraction by being more easily extracted than the target analyte or by plugging the restrictor. One example of restrictor plugging is noted for the extraction of PAHs from a reference sample [marine sediment, standard reference material (SRM) 1941], which contained 2% (w/w) of elemental sulfur whose deposition in the restrictor on fluid decompression caused blockage (110).

One major parameter that influences the composition of the extract is the moisture level in the sample matrix. Depending on the nature of the solute, the presence of water may affect the extraction efficiency. Onuska and Terry (67) observed that with 2% methanol-modified CO_2, the same recovery of 2,3,7,8-tetrachlorodibenzo-*p*-dioxin took twice as long from a wet, spiked sediment [19.8% (w/w) water content) as from a dry, spiked sediment (0.3% (w/w) water content). Similar results were obtained by Hawthorne et al. (111) for the CO_2 extraction of PAHs from petroleum waste sludge. Water usually hinders the extraction of nonpolar compounds by blocking the surface of the matrix and acting as a barrier to CO_2 penetration. This reasoning was confirmed by a comparison between extractions using R22 ($CHClF_2$) and CO_2 for the SFE of PAHs from petroleum waste sludge. It was found that the R22 removed much more water from the sample than did CO_2, thus increasing the amount of analyte exposed to the supercritical solvent and thereby leading to higher extraction efficiencies.

In contrast to the foregoing situation, water can also enhance extraction of polar compounds by competing for active adsorption sites. The differences between extractions of phenol with modified and unmodified CO_2 from dry and wetted soils are summarized in Table 19.3 (112). Methanol at 2 mole% in CO_2 improves the extraction of phenol from a dry soil sample *via* an increase in the polarity of the mixture (H-bonding occurs between phenol and methanol). However, this organic solvent has no effect when a wetted soil is being extracted, because methanol dissolves in the water contained in the soil and no longer increases and no longer increases the polarity of the SF. In contrast, benzene is virtually insoluble in water, so it favors the supercritical phase over the wetted soil phase. Hence, almost all the phenol can be extracted from the wetted soil via π–π interactions between benzene and phenol.

The effect of moisture on the SFE of analytes from biological tissues has been a topic of controversy for some time among researchers. However, it appears that partial dehydration of the sample matrix will allow SFE to be performed more rapidly, because highly hydrophilic matrices inhibit contact between the supercritical fluid and the target analytes (113).

Other approaches considered in overcoming the effects of the chemical nature of the matrix include competition with or alteration of the matrix structure in order to make extraction easier. The foregoing examples enhanced solubility by use of a cosolvent that could solubilize the target analyte. A specific cosolvent added to a super-

TABLE 19.3   Influence of Modifiers on the SFE of Phenol From Dry and Wetted Soil[a]
(113)

| | Distribution coefficient | |
SC fluid	Dry soil	Wetted soil
CO_2	0.35	0.35
CO_2/2 mole% benzene	0.8	>7
CO_2/2 mole% methanol	>7	0.35

[a]Extraction conditions: 25°C and 150 bar. The distribution coefficient is defined as the ratio of the weight fraction of phenol in the supercritical phase to the weight fraction in the soil.

critical fluid may aid in the desorption of analytes (from a highly adsorptive sample), displacing the analyte from the surface rather than increasing its solubility in the supercritical fluid. The molecular mechanism appears to be one of competitive adsorption among the cosolvent, the main fluid substance, and the adsorbed analyte on the surface of the sample (114). For a similar effect Hills and Hill (115) used a reactive solvent modifier, the derivatization agents hexamethyldisilane and trimethylchlorosilane, on two standard reference materials (NIST 1649 and NRCC HS-3) in their determinations of PAHs. In that study and others, improved yields were observed for the analytes that reacted with the derivatization reagent. Furthermore, the investigators noticed enhanced yields for compounds that were inert to the derivatization reagents used. These yields were attributed to the interaction of the derivatization reagents, not with the analyte molecules themselves, but rather with adsorption sites with which the analyte was associated (59). In certain situations the matrix can be pretreated with acid in order to reduce analyte interactions with sorptive sites on the matrix (116). It is not unusual to use pretreatment of the matrix in combination with modified fluids as was done for the extraction of PAHs from soil with a high carbon content (50%); in that study 0.35 g of soil was pretreated with 1 M hydrochloric acid before CO_2 was introduced at 400 atm and 100°C or 140°C, with toluene modifier added directly to the matrix at different levels (117). Formic or hydrochloric acid treatment of municipal fly ash was used to aid the release of chlorinated dibenzofurans from the sample surface prior to extraction with N_2O, which was modified with either 2% methanol or 5% toluene (118). This method was optimized for this sample and was compared with Soxhlet extraction of similar acid-treated samples. It was found that the methanol-modified N_2O was more efficient than the Soxhlet method. Thomson and Chesney (119) found that treatment of plant tissue (barley seed and straw) with 17% orthophosphoric acid at 100°C for 4 h (to hydrolyze the matrix and release the analyte of interest, 2,4-dichlorophenol) was beneficial when extracting with CO_2.

Another extraction strategy that is employed with difficult matrices containing non–thermally sensitive components is to use high extraction temperatures. Langenfeld et al. (37) showed that increasing the extraction temperature (to 200°C from 50°C) is far more effective in obtaining quantitative extraction efficiencies for PAHs and PCBs, from three different certified reference materials, than it was by increasing the fluid pressure. The analyte boiling points on SFE efficiencies were different for the three samples used in this study; therefore it appears that analyte volatility considerations are not adequate in themselves for explaining the increased extraction efficiencies achieved by raising the extraction temperature. It appears more likely that the trends in extraction efficiencies with extraction temperature are highly dependent on the sample matrix and possibly on the relative concentrations of the pollutant species. Since increasing the extraction temperature dramatically raises the recoveries of PCBs from river sediment and of PAHs from urban air particulates, it appears that desorption, not solubility, of the analyte from these sample matrices controls the extraction efficiencies. However. it should be noted that once the energy barrier of desorption is overcome (87), solubility of the analyte of interest in the fluid can become the limiting factor in SFE because solubility will determine the partitioning equilibria between the fluid and the sample matrix.

Küppers (120) used temperature variation in SFE of polymers for the selective extraction of low-molecular-weight components from poly(ethylene terephthalate). It was felt that temperature would have significant effects, because the diffusion coefficient is temperature-dependent, and reported SFC experiences showed that the temperature range of interest may be between 40°C and >150°C for different mobile phases and samples (123–125). For the particular sample studied, the optimum condition for the extraction of low-molecular-weight material occurred at a temperature slightly above the polymer's glass transition temperature, T_g, in combination with a high solvating power at 360 bar. The temperature optimum was that at which the target molecules' diffusion rate in the polymer matrix was at its highest. This meant that extraction temperatures should be kept below the crystallization temperature of the polymer and, in addition, the extraction should be below the polymer melting point—above which the extraction efficiency is impaired because of the reduction of polymer surface area.

Efficiency of the Solute-Trapping System in Off-Line SFE

It is important to distinguish the efficiency of the trapping step from the efficiency of the extraction step. It is easy to recognize that a poor collection step could be mistaken for poor extraction recoveries, and this may have been the case for some reported data. An example is given in Fig. 19.6 for the SFE of n-alkanes from Tenax, collecting in 2 mL of dichloromethane (DCM) (7). The more volatile alkanes were not quantitatively trapped in the DCM during a 30-min extraction with a CO_2 flow rate of 1 mL/min (500 mL/min of gas) at 400 atm. By changing the solvent to hexane or by increasing the DCM volume, the n-alkanes could be efficiently trapped.

Apart from liquid trapping, several other techniques have been adapted for the trapping and concentration of volatile substances (124):

1. Cold trapping and cryogenic trapping.
2. Trapping on a solid sorbent.
3. Trapping in a liquid stationary phase coated on a solid sorbent.
4. Chromatographic evaporation of a solvent in a capillary tube.

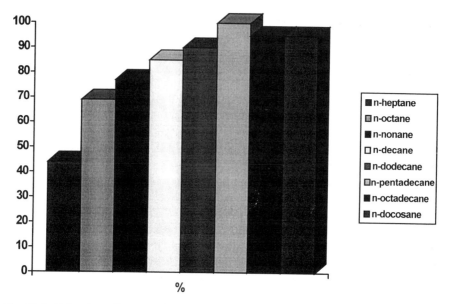

Fig. 19.6. Collection efficiency of *n*-alkanes in 2 mL dichloromethane during a 30-min SFE using carbon dioxide. (*Source:* Hawthorne, S.B., et al., in *Hyphenated Techniques in Supercritical Fluid Chromatography and Extraction, J. Chrom.* Library Series, Vol. 53, edited by K. Jinno, Elsevier, Amsterdam, 1992, p. 233.

Two different methods of extract collection are commonly used with off-line SFE: On depressurization of the fluid, extracts are collected either by a liquid trap or by a solid surface.

With a liquid trap the restrictor is simply placed in a vial containing a suitable liquid. The analyte is gradually dissolved in the solvent while the supercritical fluid substance is discharged to the atmosphere. In the solid surface method the extracted analytes are trapped on a solid surface (often glass vials or stainless steel or glass beads); cryogenically cooled, either by the expanding fluid or by another source (CO_2 or liquid nitrogen); and then rinsed from the surface for further analysis. In some instances, the trapping may involve a solid-phase sorbent (usually chromatographic packing material); after being cryogenically and chemically trapped, the solutes are eluted from the sorbent with a small volume of solvent.

Liquid Trapping. It is essential in quantitative SFE that all of the extractables are trapped on depressurization (although ≥90% recovery is often reported as quantitative). Liquid solvent accumulation has probably been the most popular method used to date. Hawthorne and coworkers (99) have shown high recoveries, in comparison with Soxhlet results, for PAHs from urban dust, river sediment, and fly ash when collecting into 2 mL of DCM spiked with 0.5 µg of 4,4′-dichlorophenol as internal standard.

Flow rates were controlled by a 10-cm length of 20–30 μm fused silica capillary tubing, resulting in extraction efficiencies with <10% standard deviation.

Lopez-Avila et al. (69) reported the extraction of spiked PAHs and organochlorine pesticides from sand, using hexane as a collection solvent. Recoveries ranged from 22 to 107% for the PAHs and were excellent for 38 of the 41 organochlorine pesticides with relative standard deviations (RSDs) from 2.4 to 29.2%. Again, flow was controlled with a fused silica restrictor, whose dimensions were 60 cm × 50 μm internal diameter; an unaccounted volume of hexane was used. In both examples the internal standard was added to the collection solvent before the extraction was performed in order to account for gross loss of analyte due to bubbling. Any losses of analyte due to bubbling presumably affect the internal standard in the same way, and no net effect on the analysis was predicted to result.

A 1-mL hexane trap was used for the extraction of polychlorinated-*p*-dioxins and dibenzofurans from municipal incinerator fly ash samples by Alexandrou and Pawlisyn (68). Recoveries of these compounds ranged from 70 to 117% with RSDs of approximately 10% when 10% benzene was added to the CO_2. "Ice-cold" hexane was the collection solvent used for the extraction of polychlorinated organics from biological tissue samples (125). At the 250-ppb level, 100% recoveries were reported for a variety of polychlorinated organics. Hexane has also been used as a collection solvent for the extraction of

1. PCBs from whole blood and milk; triazine herbicides, thiophosphate, and carbamate pesticides from dairy biomass.

2. *n*–Alkanes and PAHs from tissue samples.

3. Dioxin from liver tissue (126).

Langenfeld et al. (81) studied the collection efficiencies of DCM, chloroform, acetone, methanol and hexane for a wide number of semivolatile pollutants using CO_2 at 400 atm and 50°C; the time for each extraction was 40 min. The results show that DCM and chloroform, in spite of their differing boiling points, perform equivalently (losses 10 to 25%), acetone also performed similarly. Losses using methanol were 35 to 50%, depending on the analyte, but hexane gave the poorest results for several of the most volatile species.

The effect of possible analyte purging through extraction was examined using a trap of DCM, which contained flavor and fragrance components standard spike. Flow rates of 0.3, 0.6, and 1.2 mL/min were tested over 30 min. The analyte most affected was α-pinene and losses were 6% at 0.3 mL/min, increasing to 24% at 1.2 mL/min; only 4 to 7% loss was recorded for the remainder of the components. The results of this purge study demonstrated that excessively high flow rates, coupled with long extraction times, may result in lower overall recoveries.

The effect of the collection volume and height was found to be neglible for most of the compounds, but α-pinene once again was affected the most. The rationale for conducting this study was based on the presumption that at a constant rise rate for bubbles, a greater height of solvent should permit longer solvent-analyte contact time, thus increasing trapping in the collection solvent.

Porter et al. (71) quantitatively extracted PCBs from sewage sludge and PAHs from marine sediment (SRM 1941) and mussel tissue (SRM 1974) using CO_2 modified with 8% isopropanol. The collection solvent was isopropanol, and it was found that collection solvent temperature, as well as volume, influences trapping of the extracted analyte. Cooler temperatures resulted in increased recoveries because of increased viscosity of the collection solvent, which produced smaller bubbles (higher surface area), which had a longer contact time in the collection solvent. Cooler temperatures also lowered the vapor pressure for extracted analytes, resulting in their improved trapping. These results are similar to those observed by Langenfeld et al. (81).

Less conventional liquid collection devices include an argon-pressurized collection vessel containing a few mL of hexane for the collection of extracts from brown algae (127). Meyer et al. (117) employed an analyte collection arrangement that combined liquid and solid trapping. Essentially this is the commercially available "dual chamber" vial (Dionex, Salt Lake City, UT), with the transfer tube modified to contain silica gel supported between silanized glass fiber wads, as shown in Fig. 19.7.

The extractions from soil high in carbon content used pure or toluene-modified CO_2 at various temperatures. The analytes of interest were PAHs, and this method was shown to be more efficient than conventional analyte collection in pure organic solvent. The CO_2 extractions were found to be strongly influenced by temperature, but they reached the efficiency of conventional Soxhlet extraction with *n*-hexane/acetone only when using toluene-modified CO_2.

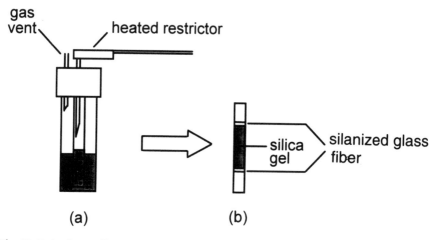

Fig. 19.7. Analyte collection method employed in the study by Meyer et al. *Source:* Meyer, A., et al., *J. High Resol. Chrom. 16*:491 (1993). (*a*) Dual-chamber vial (*b*) recently developed liquid-solid trap. *Source:* Vannoort, R.W., et al., *J. Chrom 505*:45 (1990).

Trapping on Cryogenically Cooled Surfaces. This type of trapping uses a cryogenically cooled, inert solid surface and is amenable to both on-line and off-line extraction. Trapping efficiencies for PAHs on the cooled surfaces of a volumetric and a round-bottomed flask were reported by Wright et al. (129). They found that the geometry of the vessel and sealing of the vessel affected extraction recoveries. The efficiencies for the open vessel, maintained at 0°C, ranged from 0 to 8.2%. The mechanism for solute loss was attributed to solute aerosol formation. When a sealed, liquid nitrogen–cooled vessel was used, collection ranged from 25 to 95%. Wheeler and McNally (130) made use of a glass vial as a collection device for the extraction of linuron and diuron from Sassafras soil. Extractions with pure CO_2 yielded less than 1% recovery, while addition of modifier directly to the matrix resulted in 90 to 100% recoveries for both compounds.

A novel, previously untried solid surface trap has encompassed the essentials of on-line SFE–GC. Vejrosta et al. (131) used a system in which the components were manufactured "in house." The trap was a 30-cm length of 500-μm i.d. fused silica tubing into which the fused silica capillary restrictor/postextraction cell is inserted *via* a connection union. Extracted analytes were cryofocused within the collection tubing, but the flash heating used in on-line SFE–GC was replaced by microvolume solvent rinsing, with final collection of solvent in microvials. Quantitative trapping (95%) of the test substance, fluoranthene, was achieved when extracted from glass beads, with RSDs of 3.4% obtained from ten repeated extractions. The advantages of using a low volume of solvent are high reproducibility of the off-line procedure and improved analyte sensitivity.

Another variation of the more common approach to surface trapping was demonstrated successfully and quantitatively by Miller et al. (132). Quantitative (>95%) collection efficiencies for spiked samples of *n*-alkanes, PAHs, PCBs, gasoline, and diesel fuels on sea sand were obtained. This was achieved by solventless collection of analytes by rapid depressurization after static SFE. A 0.07-in (1 in. = 0.0254 m) i.d. stainless steel tube was extended 3 mm into a vial on penetration of a Teflon-faced silicone septum. After a period of static extraction, the fluid inlet valve was closed to the extraction cell and the postextraction cell valve was opened, spraying the cell contents into the vial. Typically, this took approximately 30 sec, and the vial was immediately rinsed with 5 mL of DCM. Solventless collection of PCBs, PAHs, gasoline, and diesel fuel by SFE generally showed high (>90%) collection efficiencies and recoveries that normally exceeded those obtained by using dynamic SFE with collection in a liquid solvent. It should be noted that no effort to cool the vial was made. Even wet samples showed no evidence of blocking the outlet, in contrast to SFE using small internal-diameter flow restrictors. King et al. (8) have found the use of a receiving vessel to be successful in collection when using a micrometering valve to depressurize the supercritical fluid, in cases where large quantities of bulk matrix are extracted, as in the case of SFE of fats from meats.

Trapping onto a Solid Sorbent. Another type of collection system involves a solid-phase sorbent as a trap, usually some kind of chromatographic packing material. After being cryogenically and chemically trapped, the solutes are eluted from the sorbent with a small volume of solvent. This system provides two trapping mechanisms: cryogenic trapping and adsorption. Consideration of a suitable desorption solvent and the possibility of slow desorption rates for some polar solutes from adsorptive materials must be made (133). Schantz et al. (134) packed a short section (5 cm) of 0.25-in outer diameter stainless steel tubing with U-Bondapak C_{18} to trap PCBs extracted from air particulates. A C_{18} column was used in tandem to detect any breakthrough of extracted PCBs from the first column. Rinsings of the second column showed no evidence of breakthrough when analyzed. The trap column itself was rinsed with 40 mL of solvent using an HPLC pump, and the concentrated rinsings were analyzed by GC. Taylor and Hedrick (92) used solid-phase extraction (SPE) tubes as traps for the extraction of phenol from water. A silica trap was used and was compared with a liquid trap (5 mL 50/50 methanol and water). Collection of phenol on the solid trap resulted in 80% recovery (9% RSD) and 60% in the liquid trap (15% RSD). The poor recovery and RSDs of the latter were caused by high compressed flow rates (>1 mL/min).

A variety of normal and reverse SPE tubes used as traps were evaluated by Mulcahey et al. (133) for the off-line SFE of hydrocarbons, phenols, and a polarity "test mix." The C_{18} and C_8 reverse-phase tubes gave excellent recoveries at –20°C for the n-alkanes (C_{10} through C_{26}). Both the cryotrapping and sorption mechanisms are applied because the most volatile alkanes would not have been trapped efficiently at –20°C without adsorption. Normal-phase traps such as diol, cyano, and amino gave lower recoveries with the lower-molecular-weight hydrocarbons. A polar "test mix" incorporating acetophenone, N,N-dimethylaniline, n-decanoic acid, 2-naphthol, and n-tetracosane was also studied; the extractables trapped on both reverse and normal phase adsorbent. Of the reverse phase traps, the phenyl trap performed slightly better than the octyl trap. Of the normal phase traps, diol, silica, and amino, the diol and silica performed best, although all three traps were comparable to the reverse phase trap. Decanoic acid was the only component poorly trapped; it was not trapped at all by the silica and amino sorbent traps.

Mulcahey and Taylor (135) studied the effect of modifier on solid-phase trapping efficiency. The supercritical fluid was methanol-modified CO_2 (1, 2, 4, and 8%) the trap contained octadecyl silica (ODS) sorbent, and the analytes were from the same polar "test mix" as mentioned above in the foregoing paragraph. Trap temperatures studied ranged from 5 to 80°C, but it was found that temperatures of <20°C were necessary for volatile analytes (boiling point <200°C); the less volatile components were trapped with almost 100% recoveries, regardless of the trap temperature, at the 1 to 2% modifier level. However, at greater methanol concentrations, trap temperatures of at least 40 to 50°C were required to obtain efficient trapping. For methanol concentrations of 2% or less on the ODS trap, it was not necessary to maintain the trap temperature above the boiling point of methanol. At concentrations of methanol >2%, high temperatures were required for efficient trapping; low temperatures caused the methanol to condense on the trap surface, thereby reducing its efficiency. Recovery maxima were obtained in the 30 to 50°C range, depending on the methanol concentration.

Ashraf-Khorassani et al. (136) studied the trapping performance of a cryogenically cooled adsorbent trap and compared it with empty impinger and liquid-filled impinger trap arrangements. The test compounds were aliphatic hydrocarbons (C_{10} to C_{40}), PAHs, and pesticides which were prepared as standard solutions spiked into the extraction cell. For the compounds studied, poor collection efficiencies were reported when the sample was collected into an empty vial. High collection efficiencies (>90%) were achieved when using the impinger containing solvent, provided lower liquid CO_2 flow rates (<1 mL/min) were used. Collection of the same analytes in a cryogenically cooled vessel filled with adsorbent showed an efficiency of >90%, and this was not affected in the same way by alteration of the flow rate. The recoveries were maintained at this level with a 5 mL/min liquid CO_2 flow rate. For collection of volatile analytes (aliphatic hydrocarbons down to *n*-decane), the cryogenically cooled adsorbent traps required temperatures as low as –40°C to trap *n*-decane at the 75% level. This study also demonstrated the importance of the suitability of the rinse solvent used to wash the adsorbent trap. In this case dichloromethane was found to be more efficient in washing the sorbent than was methanol.

When using sorbent trapping with off-line SFE, each solute will have an optimal trapping sorbent/rinse solvent combination. Unfortunately, difficulties arise for "real" samples that contain multiple extractable components, because it is not practical to change rinse solvents and traps for each component.

Coupled (On-Line) SFE

SFE can be coupled to a variety of chromatographic techniques, such as gas chromatography (SFE–GC) and supercritical fluid chromatography (SFE–SFC) (1,137). The success of such on-line couplings depends on the collection efficiency of the extract and its quantitative transfer into the chromatographic column in narrow bands to achieve good chromatography.

When using on-line SFE–capillary gas chromatography (CGC), the simplest method involves an on-column injector, where the restrictor is inserted directly into the analytical column *via* the injector (9,88–90,137,138). For more concentrated samples or larger sample sizes, the supercritical fluid can be depressurized into a conventional split-splitless injector port *via* a heated transfer line (139–143). Other interfaces that have been developed include:

- A T-piece between the restrictor and the column (with a short retention gap) (144,145);
- A thermodesorption cold-trap injection system (146);
- A programmed-temperature vaporizer (PTV) injector (147,148); and
- A six port valve (149).

SFE has also been coupled with both capillary and packed-column SFC (SFE–CSFC and SFE–PSFC). With capillary columns the simplest method of injection is an injector loop; with this arrangement only a portion of the extract is injected directly into the column (150,151). Other interfaces have usually required preconcentration of the analytes after depressurization of the fluid. This can be carried out in an injector loop (150,151), via a cryogenically cooled capillary (109,152) or actually in the column (153). With packed columns, the injector loop has been used successfully because it allows part or all of the extract to be injected into the column (95,154–156). Other methods require a preconcentration step, which is carried out in the chromatographic column itself (157), in a pre-column (158,159), or in a cryogenic trap (160,161).

SFE has been coupled to SFC followed by GC (SFE–SFC–GC) for the trace analysis of pesticide residue from fatty food samples (162). The inclusion of a packed column for the SFC stage was essentially to provide a cleanup step, thereby allowing fractionation of the organochlorine and organophosphorus pesticides from the coextracted fatty materials. High-resolution separation of the pesticides was then possible on the GC column after transfer of the pesticide fraction using an electron capture or nitrogen-phosphorus detector. The fats were removed at the end of the analysis by passing CO_2 through the SFC column at 415 atm routing them to work through a FID, therefore bypassing the GC column.

Other on-line combinations of SFE with analytical methods have been demonstrated. Stalling et al. (163) utilized on-line SFE size exclusion chromatography (SEC) for the cleanup of environmental and biogenic samples. They also determined the efficiency of the SFE–SEC method and studied the effect of extraction fluid flow rate and sample size on resultant peak resolution in the SEC stage. Sub-part per million (ppm) levels of chlorinated phenols in various solid matrices were measured by Liu et al. (164) using SFE coupled on-line with liquid chromatography (LC). Speed of analysis and selectivity of this approach compared favorably with traditional methods.

SFE has been combined with flow injection analysis (FIA) using a membrane phase separator to remove CO_2 from the extracted sample (165), for the analysis of chloramphenicol and penicillin G. Kirschner and Taylor (166) directly coupled SFE with FTIR using a high-pressure IR flow cell, requiring no intermediate chromatographic separation steps. Quantitative analysis was performed by integrating the IR signal area of the extracted analyte against time. The detection limit for the solute, n-tetracosane, using this method was reported to be 74 ng. SFE has also been combined with various bioassay procedures; for example, France and King (167,168) combined SFE with an enzyme assay to screen meat products for pesticides.

Limiting Factors in SFE

Experimental factors

As an evolving technique, analytical SFE has some unique capabilities and limitations. Compared to liquid extractions, SFE has many more experimental factors that have an impact on the resultant extraction. These must be controlled and their effect must be understood to achieve reproducible results.

Collection technique and sample size are determined by the physical nature of the expected extract and sample homogeneity, respectively. Extracts containing volatiles are not always efficiently collected by employing a simple phase separation, so deposition on a solid phase sorbent may be a more useful approach. Small sample sizes ought to be avoided with nonhomogeneous samples, because the final extract may not accurately reflect the content of the sample. Using larger sample sizes favors off-line methodology rather than on-line, because the latter is limited by the analyte concentration that can be successfully analyzed. As the number of parameters complicates the optimization of a method, more experiments are required; it is often useful to employ a statistical experimental design method (169) to reduce the need for experimentation.

Coextraction of unwanted solutes, along with those that have been targeted, frequently occurs in analytical SFE. These interferences can be removed by conventional sample cleanup methods or by the use of a sorbent column downstream from the extraction cell. Target analytes can be selectively desorbed from coextracted background matrix components, provided that there are sufficient differences in their respective breakthrough volumes for the sorbent in the presence of a supercritical fluid. Alternatively, an analytical step can be used in which detection is analyte-specific and thus "blind" to the interfering species. King and France (113) gave an example of this in the on-line SFE–SFC of DDT from a fat matrix, comparing detection by electron capture and flame-ionization detectors. The former shows high specificity for the chlorinated pesticide, whereas the latter shows only a large, unresolved triglyceride profile.

Physical and Kinetic Factors

Extractions have quite often been discussed in terms of maximum solubilities. This parameter is important when the target analyte represents the bulk of the matrix. However, more often the analyte of interest is present at levels well below the solubility limit of the supercritical fluid; hence solubility in the extracting fluid is not the limiting factor. In many cases, the extraction of an analyte depends upon its distribution between the supercritical fluid phase and sorptive sites on the sample matrix. Hence, the ability of the supercritical fluid to compete with the analyte for these sites may be more important than solubility considerations (7). Thus, overall rate of removal of a solute from a matrix is a function of its solubility in the fluid media and its rate of mass transport out of the sample matrix.

King and France (113) devised a simple extraction model for the SFE of an analyte from a single particle as a way to visualize the rate-inhibiting mechanisms that impact on the extraction. Figure 19.8 shows four major mass transport mechanisms that must be considered:

1. Analyte diffusion through the internal volume of the sample.
2. Surface desorption of the analyte.
3. Diffusion of the analyte through a surface boundary layer.
4. Transport in the bulk supercritical fluid phase.

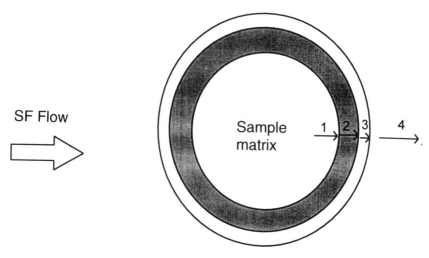

SF Flow

Fig. 19.8. Steps in the mass transport of an analyte from a porous matrix. (*Source:* King, J.W., and J:E. France, in *Analysis with Supercritical Fluids: Extraction and Chromatography,* edited by B. Wenclawiak, Springer-Verlag, Berlin, 1992, p. 52.)

If the rate-determining step is controlled by intraparticle diffusion, the rate of extraction will be a function of matrix sample particle size. It should be recognized that some sample matrices swell when exposed to supercritical fluids thereby facilitating mass transport of the analyte from within the matrix (170). An excellent example of this was demonstrated by Chiou et al. (171) in their observation of polymeric films that were plasticized by supercritical fluids. This has been used to good effect by Cotton et al. (172) in their successful analysis of additives present in plastics.

Surface desorption of an analyte by a supercritical fluid is an important step in SFE for many sample types. For certain analyte-matrix combinations, the solvent strength of the supercritical fluid alone will not be sufficient to ensure a rapid and complete extraction. It is recognized from regeneration studies of adsorbents that many compounds are not completely recovered with pure supercritical fluids and that desorption times will be prohibitively long. This difficulty has led to the use of co-solvents, such as water or methanol, which accelerate the process of analyte desorption from the surface of the sample matrix. As discussed in an earlier section, the work of Wheeler and McNally (130) has shown that extraction efficiencies for herbicides from soils can be increased by addition of µL quantities of methanol or ethanol to the sample prior to extraction.

The third step, diffusion through a surface film, may also influence the rate of extraction of an analyte. King (173) and Parcher and Strubinger (174) observed that many solid samples promote condensation of the supercritical fluid as a dense surface layer at the fluid-solid interface. The density of the adsorbed surface film depends partly on the pressure applied to the fluid and partly on the affinity of the sample matrix for the fluid. The condensed fluid film at the surface of the matrix can aid in the recovery of certain analytes through competitive adsorption at the sample interface (173) as well as inhibit the transport of the analyte into the fluid phase. The kinetics of this transport are dependent on the film thickness and total surface area of the sample matrix.

Transport away from the matrix particle in the bulk fluid is governed primarily by the diffusional coefficient of the analyte in the fluid medium. As identified earlier, the diffusion coefficients of solutes in supercritical fluids are intermediate between those they exhibit in liquid and gaseous media. This factor is independent of the sample matrix and is one of the main criteria for selecting SFE over conventional solvent extraction.

References

1. Hawthorne, S.B., *Anal. Chem. 62:*633A (1990).
2. Lee, M.L., and K.E. Markides, *Analytical Supercritical Fluid Chromatography and Extraction,* Chromatography Conferences, Provo, Utah, 1990.
3. Bartle, K.D., and A.A. Clifford in *Advances in Applied Lipid Research, Vol. 1,* edited by F.B. Padley, JAI Press, London, 1992, p. 217.
4. Jinno, K., *Hyphenated Techniques in Supercritical Fluid Chromatography and Extraction, J. Chrom.* Library Series, Vol. 53, Elsevier, Amsterdam, 1992.
5. Dean, J.R., *Applications of Supercritical Fluids in Industrial Analysis,* Blackie, Glasgow, 1993.
6. Westwood, S.A., *Supercritical Fluid Extraction and its Use in Chromatographic Sample Preparation,* Blackie, Glasgow, 1994.
7. Hawthorne, S.B., D.J. Miller, and J.J. Langenfeld, in *Hyphenated Techniques in Supercritical Fluid Chromatography and Extraction, J. Chrom.* Library Series, Vol. 53, edited by K. Jinno, Elsevier, Amsterdam, 1992, p. 233.
8. King, J.W., J.H. Johnson and J.P. Friedrich, *J. Agric. Food Chem. 37:*951 (1989).
9. Hawthorne, S.B., and D.J. Miller, *J. Chrom. 403:*63 (1987).
10. Raynie, D.E., and T.E. Delaney, *J. Chrom. Sci. 32:*298 (1994).
11. Bartle, K.D., A.A. Clifford, S.A. Jafar, and G.F. Shilstone, *J. Phys. Chem. Ref. Data 20:*713 (1991).
12. Chrastil, J., *J. Phys. Chem. 86:*3016 (1982).
13. Bamberger, T., J.C. Erickson, and C.L. Cooney, *J. Chem. Eng. Data 33:*327 (1988).
14. Brunetti, L., A. Daghetta, E. Fedeli, I. Kikic, and L. Zanderighi, *J. Am. Oil Chem. Soc. 66:*209 (1989).
15. King, M.B., D.A. Alderson, F.H. Fallaha, D.M. Kassim, K.M. Kassim, J.R. Sheldon, and R.S. Mahmud, in *Chemical Engineering at Supercritical Fluid Conditions,* edited by M.E. Paulaitis, J.M.L. Penninger, R.D. Gray and P. Davidson, Ann Arbor Science, Ann Arbor, Michigan, 1983, p. 31.
16. Kramer, A., and G. Thodos, *J. Chem. Eng. Data 33:*230 (1988).
17. Ohgaki, K., I. Tsukahara, K. Semba, and T. Katayama, *Kasaku Kogaku Ronbunshu 13:*298 (1987).
18. Schmitt, W.J., and R.C. Reid, *Chem. Eng. Comm. 64:*155 (1988).

19. Czubryt, J.J., M.N. Myers, and J.C. Giddings, *J. Phys. Chem. 74:*4202 (1970).
20. King, M.B., T.R. Bott, M.J. Barr, R.S. Mahmud, and N. Sanders, *Sep. Sci. Tech. 22:*1103 (1987).
21. Fattori, M., N.R. Bulley, and A. Meisen, *J. Am. Oil Chem. Soc. 65:*968 (1988).
22. Friedrich, J.P., U.S. Patent 4,466,923 (1984).
23. Stahl, E., K.W. Quirin, A. Glatz, D. Gerard, and G. Rau, *Ber. Bunsenges. Phys. Chem. 88:*900 (1984).
24. Klein, T., and A. Schutz, *Ind. Eng. Chem. Res. 28:*1073 (1989).
25. Stahl, E., and K.W. Quirin, *Fluid Phase Equil. 10:*269 (1983).
26. Stahl, E., E. Schutz , and H.K. Mangold, *J. Agric. Food Chem. 28:*
27. Friedrich, J.P., and E.H. Pryde, *J. Am. Oil Chem. Soc. 61:*223 (1984).
28. de Filippi, R.P., *Chem. Ind.* 390 (1982).
29. Stahl, E., and A. Glatz, *Fette Seifen Anstrichm. 86:*346 (1984).
30. Schafer, K., and W. Baumann, *Fresenius Z. Anal. Chem. 332:*122 (1988).
31. Bartle, K.D., A.A. Clifford, S.B. Hawthorne, J.J. Langenfeld, D.J. Miller, and R. Robinson, *J. Supercrit. Fluids 3:*143 (1990).
32. King, J.W., *J. Chrom. Sci. 27:*355 (1989).
33. Giddings, J.C., M.N. Myers, L. McLaren, and R.A. Keller, *Science 162:*67 (1968).
34. Giddings, J.C., M.N. Myers, and J. W. King, *J. Chrom. Sci. 7:*976 (1969).
35. Ashraf, S., K.D. Bartle, A.A. Clifford, R. Moulder, M.W. Raynor, and G.F. Shilstone, *Analyst 117:*1697 (1992).
36. McNally, M.E.P., and J.R. Wheeler, *J. Chrom. 447:*53 (1988).
37. Langenfeld, J.J., S.B. Hawthorne, D.J. Miller, and J. Pawliszyn, *Anal. Chem. 65:*338 (1993).
38. Andrews, A.T., R.C. Ahlert, and D.S. Kosson, *Environ. Progress 9:*204 (1990).
39. Johnston, K.P., D.H. Ziger, and C.A. Ekert, *Ind. Eng. Chem. Fundam. 21:*191 (1982).
40. Peter, S. and G. Brunner, *Angew Chem. Int. Ed. Engl. 17:*746 (1978).
41. Joshi, D.K., and J.M. Prausnitz, *AIChE J. 30:*522 (1984).
42. Kurnik, R.T., and R.C. Reid, *Fluid Phase Equil. 88:*93 (1982).
43. Johnston, K.P., and S. Kim, in *Supercritical Fluid Technology,* edited by J. Penninger, Elsevier, Amsterdam, 1985.
44. Dobbs, J.M., J.M. Wong, and K.P. Johnston. *J. Chem. Eng. Data 31:*303 (1986).
45. Dobbs, J.M., J.M. Wong, R.J. Lahiere, and K.P. Johnston, *Ind. Eng. Chem. Res. 26:*56 (1987).
46. Foster, N.R., H. Singh, S.L.J. Yun, D.L. Tomasko, and S.J. Macnaughton, *Ind. Eng. Chem. Res. 32:*2849 (1993).
47. McNally, M.E.P., and J.R. Wheeler, *J. Chrom. 435:*63 (1988).
48. Ho, J.S., and P.H. Tang, *J. Chrom. Sci. 30:*344 (1992).
49. Li, S., and S. Hartland, *J. Supercrit. Fluids 5:*7 (1992).
50. Engelhardt, H., and P. Haas, *J. Chrom. Sci. 31:*13 (1993).
51. Sanagi, M.M., U.K. Ahmed, and R.M. Smith, *J. Chrom. Sci. 31:*20 (1993).
52. Brunner, E., W. Hultenschmidt, and G. Schlickthrale, *J. Chem. Therm. 19:*273 (1989).
53. Deye, J.F., T.A. Berger, and A.G. Anderson, *Anal. Chem. 62:*615 (1990).
54. Hawthorne, S.B., D.J. Miller, D.E. Nivens, and D.C. White, *Anal. Chem. 64:*405 (1992).
55. Gholson, A.R., R.H. St. Louis, and H.H. Hill, Jr., *J. Assoc. Off. Anal. Chem. 70:*897 (1987).
56. Hills, J.W., and H.H. Hill, Jr., in *Proceedings of the International Symposium on*

Supercritical Fluid Chromatography and Extraction, Park City, UT, Jan. 14–17, 1991, p. 113.

57. Levy, J.M., A.C. Rosselli, D.S. Boyer, and K. Cross, in *Proceedings of the International Symposium on Supercritical Fluid Chromatography and Extraction,* Park City, UT, Jan. 14–17, 1991, p. 21.

58. White, D.C., D.E. Nivens, D. Ringelberg, D. Hedrick, and S.B. Hawthorne, in *Proceedings of the Third International Symposium on Supercritical Fluid Chromatography and Extraction,* Park City, UT, Jan. 14–17, 1991, p. 43.

59. Hills, J.W., H.H. Hill, Jr., and T. Maeda, *Anal. Chem. 63:*2152 (1991).

60. Venema, A., and T.J. Jelink, *J. High Resol. Chrom. 16:*166 (1993).

61. Lee, H.-B., T.E. Peart, and R.L. Hong-You, *J. Chrom. 605:*109 (1992).

62. Lee, H.-B., T.E. Peart, and R.L. Hong-You, *J. Chrom. 636:*263 (1993).

63. Field, J.A., D.J. Miller, T.M. Field, S.B. Hawthorne, and W. Giger, *Anal. Chem. 64:*3161 (1992).

64. King, J.W., J.E. France, and J.M. Snyder, *Fresenius J. Anal. Chem. 344:*474 (1992).

65. Cai, Y., R. Alzaga, and J.M. Bayona, *Anal. Chem. 66:*1161 (1994).

66. Hawthorne, S.B., M.S. Krieger, and D.J. Miller, *Anal. Chem. 61:*736 (1989).

67. Onuska, F.I., and K.A. Terry, *J. High Resol. Chrom. 12:*357 (1989).

68. Alexandrou, N., and J. Pawliszyn, *Anal. Chem. 61:*2770 (1989).

69. Lopez-Avila, V., N.S. Dodhiwala, and W.F. Beckert, *J. Chrom. Sci. 28:*468 (1990).

70. Burford, M.D., S.B. Hawthorne, D.J. Miller, and T. Braggins, *J. Chrom. 609:*321 (1992).

71. Porter, N.L., A.F. Rynaski, E.R. Campbell, M. Saunders, B.E. Richter, J.T. Swanson, N.B. Nielsen, and B.J. Murphy, *J. Chrom. Sci. 30:*367 (1992).

72. Hawthorne, S.B., D.J. Miller, M.D. Burford, J.J. Langenfeld, S. Eckert-Tilotta, and P.K. Louie, *J. Chrom. 642:*301 (1993).

73. Camel, V., A. Tambute, and M. Caude, *J. Chrom. 642:*263 (1993).

74. Hawthorne, S.B., D.J. Miller, D.D. Walker, D.E. Whittington, and B.L. Moore, *J. Chrom. 541:*185 (1991).

75. Furton, K.G., and Q. Lin, *J. Chrom. Sci. 31:*201 (1993).

76. Kane, M., J.R. Dean, S.M. Hitchen, C.J. Dowle, and R.L. Tranter, *Anal. Chim. Acta 271:*83 (1993).

77. Thomson, C.A., and D.J. Chesney, *Anal. Chem. 64:*848 (1992).

78. Furton, K.G., and J. Rein, *Anal. Chim. Acta 248:*263 (1991).

79. Furton, K.G., and J. Rein, *Chromatographia 31:*297 (1991).

80. Furton, K.G., and Q. Lin, *Chromatographia 34:*185 (1992).

81. Langenfeld, J.J., M.D. Burford, S.B. Hawthorne, and D.J. Miller, *J. Chrom. 594:*297 (1992).

82. Anderson, M.R., J.W. King, and S.B. Hawthorne in *Analytical Supercritical Fluid Chromatography and Extraction,* edited by N.L. Lee and K.E. Markides, Chromatography Conferences, Provo, UT, 1990, p. 328.

83. King, J.W., J.E. France, and S.L. Taylor, in *Proceedings of the 11th International Symposium on Capillary Chromatography,* edited by P. Sandra and G. Redant, Hüthig, Heidelberg, 1990, p. 595.

84. Raymer, J.H., and E.D. Pellizzari, *Anal. Chem. 59:*1043 (1987).

85. Raymer, J.H., E.D. Pellizzari, and S.D. Cooper, *Anal. Chem. 59:*2069 (1987).

86. Wong, J.M., N.Y. Kado, P.A. Kuzmicky, H-S. Ning, J.E. Woodrow, D.P.H. Hsieh, and J.N. Seiber, *Anal. Chem. 63:*1644 (1991).

87. Alexandrou, N., M.J. Lawerence, and J.W. Pawliszyn, *Anal. Chem. 64:*301 (1992).
88. Hawthorne, S.B., D.J. Miller, and M.S. Krieger, *J. Chrom. Sci. 27:*347 (1989).
89. Hawthorne, S.B., D.J. Miller, and M.S. Krieger, *J. High Resol. Chrom. 12:*714 (1989).
90. Hawthorne, S.B., and D.J. Miller, *J. Chrom. Sci. 24:*258 (1986).
91. Hedrick, J.L., and L.T. Taylor, *Anal. Chem. 61:*1986 (1989).
92. Hedrick, J.L., and L.T. Taylor, *J. High Resol. Chrom. 13:*312 (1990).
93. Hedrick, J.L., and L.T. Taylor, *J. High Resol. Chrom. 15:*151 (1992).
94. Ong, C.P., H.M. Ong, S.F.Y. Li, and H.K. Lee, *J. Microcol. Sep. 2:*69 (1990).
95. Thiebaut, D., J-P. Chervet, R.W. Vannoort, G.J. De Jong, U.A. Th. Brinkman, and R.W. Frei, *J. Chrom. 477:*151 (1989).
96. Lanças, F.M., M.H.R. Natta, L.J. Hayasida, and E. Castillio, *J. High Resol. Chrom. 14:*633 (1990).
97. Smith, R.M., and H.R. Udseth, *Fuel 62:*466 (1983).
98. Schulten, H.R., and M. Schnitzer, *Soil Sci. Soc. Am. J. 55:*1603 (1991).
99. Hawthorne, S.B., and D.J. Miller, *Anal. Chem. 59:*1705 (1987).
100. Paschke, T., S.B. Hawthorne, and D.J. Miller, *J. Chrom. 609:*333 (1992).
101. Lee, H., B-H. Chung, and Y-H. Park, *J. Am. Oil Chem. Soc. 68:*571 (1991).
102. Messer, D.C., and L.T. Taylor, *Anal. Chem. 15:*238 (1992).
103. Shobha, S.V., and B. Ravindranath, *J. Agric. Food Chem. 39:*2214 (1991).
104. Hawthorne, S.B., M-L. Riekkola, K. Serenius, Y. Holme, R. Hiltunen, and K. Hartonen, *J. Chrom. 634:*297 (1993).
105. Bartle, K.D., T. Boddington, A.A. Clifford, and N.J. Cotton, *Anal. Chem. 63:*2371 (1991).
106. Daimon, H., and Y. Hirata, *Chromatographia 32:*549 (1991).
107. Snyder, J.M., J.P. Friedrich, and D.D. Christianson, *J. Am. Oil Chem. Soc. 61:*5475 (1984).
108. Hopper, M.L., and J.W. King, *J. Assoc. Off. Anal. Chem. 74:*661 (1991).
109. Engelhardt, H., J. Zapp, and P. Kolla, *Chromatographia 32:*527 (1991).
110. Hawthorne, S.B., D.J. Miller, and J.J. Langenfeld, *J. Chrom. Sci 28:*2 (1990).
111. Hawthorne, S.B., J.J. Langenfeld, D.J. Miller, and M.D. Burford, *Anal. Chem. 64:*1614 (1992).
112. Roop, R.K., R.K. Hess, and A. Akgerman, in *Supercritical Fluid Science and Technology,* edited by K.P. Johnston and J.M.L. Penninger, ACS Symposium Series, 406, American Chemical Society, Washington, DC, 1989, p. 468.
113. King, J.W., and J.E. France, in *Analysis with Supercritical Fluids: Extraction and Chromatography,* edited by B. Wenclawiak, Springer-Verlag, Berlin-Heidelberg, Germany, 1992, p. 52.
114. Dooley, K.M., C-P. Kao, K.P. Gambrell, and F.C. Knopf, *Ind. Eng. Chem. Res. 26:*2058 (1987).
115. Hills, J.W., and H.H. Hill, *J. Chrom. Sci. 31:*6 (1993).
116. Onuska, F.I., and K.A. Terry, *J. High Resol. Chrom. 14:*829 (1991).
117. Meyer, A., W. Kleiböhmer, and K. Cammann, *J. High Resol. Chrom. 16:*491 (1993).
118. Onuska, F.I., K.A. Terry, and R.J. Wilkinson, *J. High Resol. Chrom. 16:*407(1993).
119. Thomson, C.A., and D.J. Chesney, *Anal. Chem. 64:*848 (1992).
120. Küppers, S., *Chromatographia 33:*434 (1992).
121. Berger, T.A., and J.F. Deye, *J. Chrom. Sci. 29:*54 (1991).
122. Berger, T.A., *J. High Resol. Chrom. 14:*312 (1991).

123. Küppers, S., B. Lorenschat, F.P. Schrnitz, and E. Klesper, *J. Chrom. 475:*85 (1989).
124. Jursik, T., K. Stránský, and K. Ubik, *J. Chrom. 586:*315 (1991).
125. Nam, S.K., S. Kapilla, D.S. Viswanath, T.E. Clevenger and A.F. Yanders, *Chemosphere 19:*33 (1989).
126. Nam, S.K., S. Kapilla, A.F. Yanders, and R.K. Puri, *Chemosphere 20:*871 (1990).
127. Subra, P., and P. Boissinot, *J. Chrom. 543:*413 (1991).
128. Vannoort, R.W., J.-P. Chervet, H. Lingeman, G.J. De Jong, and U.A.T. Brinkman, *J. Chrom. 505:*45 (1990).
129. Wright, B.W., C.W. Wright, R.W. Gale, and R.D. Smith, *Anal. Chem. 59:*38 (1987).
130. Wheeler, J.R., and M.E.P. McNally, *J. Chrom. Sci. 27:*534 (1989).
131. Vejrosta, J., A. Ansorgová, M. Mikeová, and K.D. Bartle, *J. Chrom. 659:*209 1994).
132. Miller, D.J., S.B. Hawthorne, and M.E.P. McNally, *Anal. Chem. 65:*1038 (1993).
133. Mulcahey, L.J., J.L. Hedrick, and L.T. Taylor, *Anal. Chem. 63:*2225 (1991).
134. Schantz, M.M., and S.N. Chesler, *J. Chrom. 363:*397 (1986).
135. Mulcahey, L.J., and L.T. Taylor, *Anal. Chem. 64:*2352 (1992).
136. Ashraf-Khorassani, M., R.K. Houck, and J.M. Levy, *J. Chrom. Sci. 30:*361 (1992).
137. Hawthorne, S.B., M.S. Krieger, and D.J. Miller, *Anal. Chem. 60:*472 (1988).
138. Hawthorne, S.B., D.J. Miller, and M.S. Krieger. *Fresenius Z. Anal. Chem. 330:*211 (1988).
139. Ashraf-Khorassani, M., L.T. Taylor, and P. Zimmerman, *Anal. Chem. 62:*1177 (1990).
140. Hawthorne, S.B., D.J. Miller, and J.J. Langenfeld, *J. Chrom. Sci. 28:*2 (1990).
141. Levy, J.M., R.A. Cavalier, T.N. Bosch, A.F. Rynaski, and W.E. Hulak, *J. Chrom. Sci. 27:*341 (1989).
142. Levy, J.M., E. Storozynsky, and R.M. Ravey, *J. High Resolution Chrom. 14:*661 (1991).
143. Levy, J.M., and A.C. Rosselli, *Chromatographia 28:*613 (1989).
144. Wright, B.W., S.R. Frye, D.G. McMinn, and R.D. Smith, *Anal. Chem. 59:*640 (1987).
145. Lohleit, M., and K. Bachmann, *J. Chrom. 505:*227 (1990).
146. Nielen, M.W.F., J.T. Sanderson, R.W. Frei, and U.A.T. Brinkman, *J. Chrom. 474:*388 (1989).
147. Houben, R.J., H-G.M. Janssen, P.A. Leclercq, J.A. Rijks, and C.A. Cramers, *J. High Resol. Chrom. 13:*669 (1990).
148. Blatt, C.R., and R. Ciola, *J. High Resol. Chrom. 14:*775 (1991).
149. Onuska, F.I., and K.A. Terry, *J. High Resol. Chrom. 12:*527 (1989).
150. Anton, K., R. Menes, and H.M. Widmer, *Chromatographia 26:*221 (1988).
151. Raynor, M.W., I.L. Davies, K.D. Bartle, A.A. Clifford, A. Williams, J.M. Chalmers, and B.W. Cook, *J. High Resol. Chrom. & Chrom. Commun. 11:*766 (1988).
152. Xie, Q.L., K.E. Markides, and M.L. Lee, *J. Chrom. Sci. 27:*365 (1989).
153. Mitra, S., and N.K. Wilson, *J. Chrom. Sci. 28:*182 (1990).
154. Engelhardt, H., and A. Cross, *J. High Resol. Chrom. & Chrom. Commun. 11:*38 (1988).
155. Engelhardt, H., and A. Cross, *J. High Resol. Chrom. & Chrom. Commun. 11:*726 (1988).
156. Jahn, K.R., and B. Wenclawiak, *Chromatographia 26:*345 (1988).
157. Ramsey, E.D., J.R. Perkins, D.E. Games, and J.R. Startin, *J. Chrom. 464:*353 (1989).
158. Liebman, S.A., E.J. Levy, S. Lurcott, S. O'Neill, J. Guthrie, T. Ryan, and S. Yocklovich, *J. Chrom. Sci. 27:*118 (1989).
159. Ryan, T.W., S.G. Yocklovich, J.C. Watkins, and E.J. Levy, *J. Chrom. 505:*273 (1990).

160. Ashraf-Khorassani, M., M.L. Kumar, D.J. Koebler, and G.P. Williams, *J. Chrom. Sci.* *28:*599 (1990).
161. Ashraf-Khorassani, M., D.S. Boyer, and J.M. Levy, *J. Chrom. Sci. 29:*517 (1991).
162. Nam, K.S., and J.W. King, *J. High Resol. Chrom. 17:*577 (1994).
163. Stalling, D.L., S. Saim, K.C. Kuo, and J.J. Stunkel, *J. Chrom. Sci. 30:*486 (1992).
164. Liu, M.H., S. Kapilla, K.S. Nam, and A.A. Elseewi, *J. Chrom. 639:*151 (1993).
165. Brewster, J.D., R.J. Maxwell, and J.W. Hampson, *Anal. Chem. 65:*2137 (1993).
166. Kirschner, C.H., and L.T. Taylor, *Anal. Chem. 65:*78 (1993).
167. France, J.E., and J.W. King, *J. Assoc. Off. Anal. Chem. 74:*1013 (1991).
168. Nam, K.S., and J.W. King, *J. Agric. Food Chem. 42:*1469 (1994).
169. Fisher, R.J., *Food Technol. 43:*90 (1989).
170. Fahmy, T.M., M.E. Paulatis, D.M. Johnson, and M.E.P. McNally, *Anal. Chem. 65:*1462 (1993).
171. Chiou, J.S., J.W. Barlow, and D.R. Paul, *J. Appl. Polym. Sci. 30:*3911 (1985).
172. Cotton, N.J., K.D. Bartle, A.A. Clifford, S. Ashraf, R. Moulder, and C.J. Dowle, *J. High Resol. Chrom. 14:*164 (1991).
173. King, J.W., in *Supercritical Fluids: Chemical and Engineering Principles and Applications,* edited by T.G. Squires and M.E. Paulaitis, American Chemical Society, Washington, DC, 1987, p. 150.
174. Parcher, J., and J.R. Strubinger, *J. Chrom. 479:*251 (1989).

Index